Contemporary Developments in Agricultural Cyber-Physical Systems

G.S. Karthick
Department of Software Systems, PSG College of Arts and Science, India

IGI Global
PUBLISHER of TIMELY KNOWLEDGE

A volume in the Advances in Environmental Engineering and Green Technologies (AEEGT) Book Series

Published in the United States of America by
IGI Global
Engineering Science Reference (an imprint of IGI Global)
701 E. Chocolate Avenue
Hershey PA, USA 17033
Tel: 717-533-8845
Fax: 717-533-8661
E-mail: cust@igi-global.com
Web site: http://www.igi-global.com

Library of Congress Cataloging-in-Publication Data

Names: Karthick, G. S., 1990- editor.
Title: Contemporary developments in agricultural cyber-physical systems / edited by G. S. Karthick.
Description: Hershey, PA : Engineering Science Reference, [2023] | Includes bibliographical references and index. | Summary: "Contemporary Developments in Agricultural Cyber-Physical Systems provides a forum for researchers and practitioners to exchange ideas and achieve progress in cyber-physical systems by highlighting agricultural applications, advances, and research challenges. The book features chapters on all aspects pertaining to this multidisciplinary paradigm, in particular in its application to sustainable agriculture developments. Covering topics such as automation, monitoring systems, and smart agriculture, this premier reference source is an excellent resource for scientists, healthcare professionals, data analysts, computer scientists, students and educators of higher education, researchers, and academicians"-- Provided by publisher.
Identifiers: LCCN 2023029577 (print) | LCCN 2023029578 (ebook) | ISBN 9781668478790 (hardcover) | ISBN 9781668478806 (paperback) | ISBN 9781668478813 (ebook)
Subjects: LCSH: Alternative agriculture. | Agricultural innovations. | Internet of things.
Classification: LCC S494.5.A65 C665 2023 (print) | LCC S494.5.A65 (ebook) | DDC 338.1/62--dc23/eng/20230721
LC record available at https://lccn.loc.gov/2023029577
LC ebook record available at https://lccn.loc.gov/2023029578

This book is published in the IGI Global book series Advances in Environmental Engineering and Green Technologies (AEEGT) (ISSN: 2326-9162; eISSN: 2326-9170)

British Cataloguing in Publication Data
A Cataloguing in Publication record for this book is available from the British Library.

Advances in Environmental Engineering and Green Technologies (AEEGT) Book Series

Sang-Bing Tsai
Zhongshan Institute, University of Electronic Science and Technology of China, China & Wuyi University, China
Ming-Lang Tseng
Lunghwa University of Science and Technology, Taiwan
Yuchi Wang
University of Electronic Science and Technology of China Zhongshan Institute, China

ISSN:2326-9162
EISSN:2326-9170

Mission

Growing awareness and an increased focus on environmental issues such as climate change, energy use, and loss of non-renewable resources have brought about a greater need for research that provides potential solutions to these problems. Research in environmental science and engineering continues to play a vital role in uncovering new opportunities for a "green" future.

The **Advances in Environmental Engineering and Green Technologies (AEEGT)** book series is a mouthpiece for research in all aspects of environmental science, earth science, and green initiatives. This series supports the ongoing research in this field through publishing books that discuss topics within environmental engineering or that deal with the interdisciplinary field of green technologies.

Coverage

- Policies Involving Green Technologies and Environmental Engineering
- Cleantech
- Alternative Power Sources
- Waste Management
- Air Quality
- Water Supply and Treatment
- Biofilters and Biofiltration
- Contaminated Site Remediation
- Radioactive Waste Treatment
- Electric Vehicles

IGI Global is currently accepting manuscripts for publication within this series. To submit a proposal for a volume in this series, please contact our Acquisition Editors at Acquisitions@igi-global.com or visit: http://www.igi-global.com/publish/.

The Advances in Environmental Engineering and Green Technologies (AEEGT) Book Series (ISSN 2326-9162) is published by IGI Global, 701 E. Chocolate Avenue, Hershey, PA 17033-1240, USA, www.igi-global.com. This series is composed of titles available for purchase individually; each title is edited to be contextually exclusive from any other title within the series. For pricing and ordering information please visit http://www.igi-global.com/book-series/advances-environmental-engineering-green-technologies/73679. Postmaster: Send all address changes to above address.

Titles in this Series

For a list of additional titles in this series, please visit: http://www.igi-global.com/book-series/advances-environmental-engineering-green-technologies/73679

Perspectives on Ecological Degradation and Technological Progress
Veli Yilanci (Faculty of Political Sciences, Canakkale Onsekiz Mart University, Turkey)
Engineering Science Reference • © 2023 • 300pp • H/C (ISBN: 9781668467275) • US $225.00

Artificial Intelligence Applications in Water Treatment and Water Resource Management
Victor Shikuku (Kaimosi Friends University College, Kenya)
Engineering Science Reference • © 2023 • 300pp • H/C (ISBN: 9781668467916) • US $225.00

Handbook of Research on Improving the Natural and Ecological Conditions of the Polesie Zone
Anatoliy Rokochinskiy (National University of Water and Environmental Engineering, Ukraine) Lyudmyla Kuzmych (Institute of Water Problems and Land Reclamation of the National Academy of Agrarian Sciences, Ukraine) and Pavlo Volk (National University of Water and Environmental Engineering, Ukraine)
Engineering Science Reference • © 2023 • 479pp • H/C (ISBN: 9781668482483) • US $270.00

Environmental and Technological Aspects of Redox Processes
Gheorghe Duca (Institute of Chemistry, Moldova State University, Moldova) and Ashok Vaseashta (International Clean Water Institute, USA)
Engineering Science Reference • © 2023 • 410pp • H/C (ISBN: 9798369305126) • US $295.00

Multidisciplinary Approaches in AI, Creativity, Innovation, and Green Collaboration
Ziska Fields (University of Johannesburg, South Africa)
Engineering Science Reference • © 2023 • 405pp • H/C (ISBN: 9781668463666) • US $270.00

Opportunities and Challenges in Climate-Friendly Clean Water and Energy Technologies
J. Sadhik Basha (Department of Process Engineering, National University of Science & Technology, Oman) Jimoh Kayode Adewole (Department of Process Engineering, National University of Science & Technology, Oman) Anteneh Mesfin Yeneneh (Department of Process Engineering, National University of Science & Technology, Oman) and Luqman Kolawole Abidoye (Department of Process Engineering, National University of Science & Technology, Oman)
Engineering Science Reference • © 2023 • 344pp • H/C (ISBN: 9781668473030) • US $250.00

IGI Global
PUBLISHER of TIMELY KNOWLEDGE

701 East Chocolate Avenue, Hershey, PA 17033, USA
Tel: 717-533-8845 x100 • Fax: 717-533-8661
E-Mail: cust@igi-global.com • www.igi-global.com

Editorial Advisory Board

Table of Contents

Detailed Table of Contents

In this chapter, smart agriculture irrigation monitoring systems using IoT principles are described. The general information for precision agriculture systems, the integration of the internet of things with irrigation systems, and measuring parameters have been illustrated. Two framework models, such as Agriculture Factor-Based Relevance Vector Analysis (ARV) and advanced irrigation systems for soil moisture prediction systems for smart agriculture irrigation, have been illustrated in this chapter. The environmental factors for the irrigation system have also been monitored and controlled to improve the production rate. The framework model for an advanced irrigation system and the ARV model for prediction and control of water supply to the land based on soil moisture has also been illustrated.

Agriculture is the backbone of human activity, and it is playing a vital role in human survival. Smart agriculture has a favorable impact on society as a result of technological advancements. The major goal is to increase production quality and quantity. Many industries, including healthcare, law, real estate, banking, and other applications are being transformed by using blockchain technology. On the other hand, agriculture is a known application that uses blockchain technology to totally transform the lives of human being in the world. With such a growing population it is obvious that the food demand will increase significantly. To fulfill the food demand of the world population, the agriculture industry is making use of Artificial Intelligence (AI), blockchain, and other food automation techniques and technologies to meet the demands through innovative production. This chapter presents an overview of various blockchain technology use cases, benefits and obstacles of food industry, and deployment of blockchain technology in agriculture.

The ideals of precision farming, agronomy, and automation in the food supply chain have also been established because of robotization and agricultural advancements. The rigidity, networking, standardization, robustness, skill requirements, absence of real-time data, and actuation are some of the areas where present systems still have serious flaws. The development of robotics and independent systems was prompted using artificial intelligence, and intelligent ICT robotic systems in cyber-physical systems will be necessary. These systems can communicate, compute, track, and control physical elements and processes in a variety of operations. This system uses ICT robots equipped with the necessary sensors to monitor the entire agricultural area, transferring data to edge nodes. While the procedure was still in progress, it began updating the field and storing the information in the cloud, and it transferred updates to the farmers. This chapter explores the various developments of robotics and autonomous systems in the agricultural field.

Agriculture cyber-physical systems are becoming increasingly significant in improving crop quality and output while employing the least amount of farmland possible. Many agricultural components and methods have been automated to produce faster and superior-quality items. As a result, several methods and techniques to help prevent or reduce plant diseases have been created. Imaging analysis tools and gas sensors are increasingly being included in innovative cyber agribusiness for plant disease monitoring. This chapter develops intelligent cyber-physical system-based plant disease detection for successfully preventing and managing plant diseases and decision-making. To extract the important features linked to the sick region, a novel multi-statistical feature extraction technique has been developed. The proposed methodology is implemented using a Raspberry Pi board running Stretch OS. Metrics like detection and classification accuracy are used to assess the efficacy of the multi-statistical feature extraction methodology on the plant disease detection system.

The usage of cyber-physical systems is well known to play a vital role in the development of precision agriculture, which in turn raises overall output. When a network of Unmanned Aerial Vehicle (UAV) is deployed for these applications, they are considered CPS. One of the most crucial things that needs to be handled is covering the area that needs to be travelled with the least number of UAV traversals possible. UAV traversal is a crucial part of precision agriculture because it enables farmers to gather crucial information and make informed decisions that may lead to higher crop yields and cheaper costs. Thanks to improved algorithms and techniques, UAVs can traverse agricultural land more effectively and precisely. As a result, farming activities become more productive and environmentally friendly. This chapter explores the various approaches that might be used to determine the UAV's best route when it is deployed over farmland.

Unmanned Aerial Vehicles (UAV) or drones are a growing technological boon to major fields which possess high complexity in logistics such as defense systems, cargo transportation, mapping geographic locations and terrain, border surveillance, medical emergency shipments, and many other areas where humans find it inaccessible or inefficient to work. Drones can cope with tasks that demand precisive services and tasks that are inefficient under human intervention. Precision agriculture deals with efficiency as a key factor and it uses information technology to keep track of optimum health of the crops and productivity of a particular crop field using precisive approach which can be done with the help of purposely modified drones for precision agriculture. This chapter analyses and evaluates the significance of drones in precision agriculture, challenges possessed in scaling them, and solutions to implement them in large-scale environments. In the future, cognitive analysis and sensory evaluation can be used to further reduce the complexity of accessing drones in various environments.

Foreword

It is with great pleasure and anticipation that I introduce the book *Contemporary Developments in Agricultural Cyber-Physical Systems*. In an era defined by remarkable advancements in technology and an urgent need for sustainable agricultural practices, this book offers a comprehensive exploration of the intersection between agriculture and cyber-physical systems (CPS).

The integration of CPS into agriculture represents a significant shift in the way we approach farming and addresses the pressing challenges faced by the industry. This book serves as a valuable resource for researchers, practitioners, and policymakers, providing a deep understanding of the latest developments in agricultural CPS and their potential to transform the future of agriculture.

As our world continues to experience exponential population growth and rapid urbanization, the need for sustainable food production has never been more critical. The adoption of advanced technologies, such as IoT, robotics, AI, data analytics, and automation, offers tremendous opportunities to enhance productivity, reduce resource consumption, and mitigate environmental impacts.

Through the chapters of this book, readers will gain insights into the fundamental principles and concepts that underpin agricultural CPS. The architectural frameworks, communication protocols, and interoperability standards discussed in these pages lay the foundation for the seamless integration of various components in CPS, enabling efficient data exchange and collaboration between physical and digital systems.

The exploration of IoT devices and sensors demonstrates their vital role in collecting and monitoring data, enabling real-time decision-making and precise control of agricultural processes. The integration of robotics and automation technologies showcases their potential in optimizing farming operations, from planting and harvesting to crop monitoring and livestock management.

One cannot overlook the transformative power of AI and machine learning algorithms in agriculture. Through data-driven insights and predictive analytics, these technologies unlock opportunities for improved crop management, disease detection, yield prediction, and more. Furthermore, the challenges associated with managing and analyzing big agricultural data are addressed, shedding light on the utilization of cloud computing and edge computing technologies for efficient data processing and storage.

The book also delves into the significance of sustainable farming practices, as agricultural CPS provides innovative solutions for resource management, water-efficient irrigation systems, and resilient crop cultivation techniques. By harnessing the power of CPS, we can promote environmentally conscious farming practices, optimize resource utilization, and reduce the ecological footprint of agriculture.

Looking towards the future, the emerging trends and applications discussed in this book provide a glimpse into the exciting possibilities of agricultural CPS. The integration of blockchain technology for supply chain transparency, the utilization of drones and satellite imagery for precision agriculture, and the potential implications of CPS for rural development and social equity are areas ripe for exploration and innovation.

In conclusion, *Contemporary Developments in Agricultural Cyber-Physical Systems* serves as a comprehensive guide that invites readers to embark on a journey through the cutting-edge advancements in agricultural CPS. The book encompasses a multidisciplinary approach, drawing upon the expertise of researchers, engineers, and professionals from various fields to offer a holistic view of this dynamic and transformative field.

I commend the authors and contributors for their dedication in bringing forth this invaluable resource. I am confident that this book will inspire and inform readers, fostering innovation, collaboration, and the adoption of sustainable practices in agriculture. Let us embrace the potential of agricultural CPS and work together towards a future where technology and nature intertwine harmoniously, ensuring a prosperous and sustainable global agricultural landscape.

E. Chandra
Department of Computer Science, Bharathiar University, India

Preface

BACKGROUND

Cyber Physical Systems (CPS) are the new generation of collaborative computational entities, with a prime focus on integration of the physical world and cyber space. Through a feedback mechanism, the system adapts itself to new conditions in real time. The scope of this book is to share the research experiences of experts in CPS infrastructure systems, incorporating sustainability by embedding computing and communication in agricultural applications. The integration of CPS with Blockchain, Artificial Intelligence, Internet of Things, Big Data, Cloud Computing and Communication, lay a foundation for the fourth agricultural revolution, Agriculture 4.0.

There are several contexts in which CPS can exist, one of which is the agricultural industry. When agriculture integrates with CPS, this is termed as precision agriculture. It can be defined as a methodology of enhanced decision making, which associates effective management of farm practices, in concert with sensing devices, positioning systems from satellites and the integration with computers to provide information. The cultivation of crops plays a very important role in agriculture. However, proper maintenance and management are required. Lack of such management would lead to crop loss or reduced crop yields. Hence, the ability to detect and identify the diseases on infected crops is a problem of increasing concern. Real-time disease detection systems do not exist in the current scheme of things. It requires tremendous amounts of work, expertise in the plant diseases, and also requires excessive processing time. Using precision agriculture techniques, combined with artificial intelligence, a great deal of manual work is reduced. This book will be of immense use to practitioners with a focus on autonomous and adaptive configuration, and on optimization, leading to increased agility, elasticity and cost effectiveness.

ABOUT THIS BOOK

The contributors of this book include renowned academics, industry practitioners and researchers. It offers a rigorous introduction to the theoretical foundations, techniques and practical solutions, through case studies. Building Agricultural CPS with effective communication, control, intelligence and security is discussed in terms of societal and research perspectives.

The objective of this book is to provide a forum for researchers and practitioners to exchange ideas and to achieve progress in CPS by highlighting agricultural applications, advances and research challenges. It is highly recommended to be used as a reference book for graduate and post-graduate level programmes in universities, with a focus on research in computer science-related courses.

TARGET AUDIENCE

The book seeks to develop the core system science needed to enable development of complex Agricultural CPS upon which people can depend with high confidence, and also aims to foster a research community committed to advancing research and education in CPS and to transitioning applicable CPS science and technology into engineering practice. Researchers, academicians, students, faculty, scientists and healthcare professionals will find this book beneficial for research exposure and new ideas in the field of Agricultural Cyber-Physical Systems.

ORGANIZATION OF THE BOOK

This book organized into 13 chapters. Introduction section reveals how the connected IoT-enabled healthcare devices are stepped up even under the presence of stumbling blocks. All the contributors to this book are the upcoming researchers in the field of IoT. The summary of the chapter contents is depicted as follows:

1. Cyber-Physical Systems: Foundations, Design Principles, Challenges, and Applications – Comprehensive Insights

This chapter provides a comprehensive understanding of cyber-physical systems (CPS), exploring their foundations, design principles, challenges, and applications. It serves as a foundational piece for readers, introducing them to the seamless integration of computation, communication, and physical processes that define CPS, and highlighting the transformative potential of this interdisciplinary field.

2. Cyber-Physical Systems in Agriculture: Applications, Challenges, and Future Perspectives

Focusing on agriculture, this chapter explores the applications, challenges, and future perspectives of cyber-physical systems. It emphasizes how CPS technology can revolutionize farming practices, enabling precision agriculture and addressing critical challenges such as resource optimization, crop monitoring, and environmental sustainability. By highlighting the potential of CPS in agriculture, the chapter aims to inspire researchers and practitioners to harness its power for a smarter and more productive farming industry.

3. Energy-Efficient and High-Performance IoT-Based WSN Architecture for Precision Agriculture Monitoring Using Machine Learning Techniques

This chapter delves into the design and implementation of an energy-efficient and high-performance IoT-based wireless sensor network (WSN) architecture for precision agriculture monitoring. It explores the integration of machine learning techniques to optimize resource utilization, enhance data processing capabilities, and improve overall system efficiency. The chapter provides insights into how IoT and machine learning can synergistically contribute to the development of sustainable and intelligent solutions for precision agriculture.

4. Cyber-Physical System Framework for Efficient Management of Indoor Farming Production

Indoor farming presents unique challenges in environmental control, resource management, and production optimization. This chapter introduces a cyber-physical system framework tailored for efficient management of indoor farming production. It explores the integration of sensors, actuators, and control systems to create an intelligent and automated environment that maximizes crop yield, minimizes resource consumption, and ensures optimal growth conditions.

5. Long-Range and Low-Power Automated Soil Irrigation System Using Internet of Things: An Experimental Study

This chapter presents an experimental study on a long-range and low-power automated soil irrigation system that leverages the Internet of Things (IoT). It highlights the development and implementation of a system that optimizes water usage and maintains soil moisture levels using IoT technologies. The chapter provides valuable insights into the practical application of IoT in automating soil irrigation processes.

6. Smart Agriculture Irrigation Monitoring System Using Internet of Things

Focusing on smart agriculture, this chapter explores the implementation of an irrigation monitoring system that utilizes IoT technologies. It highlights the integration of sensors, wireless communication, and data analysis to optimize irrigation practices, conserve water, and improve crop yield. The chapter showcases the potential of IoT in revolutionizing irrigation management in agriculture.

7. Futuristic Adoption of Blockchain Technology in the Smart Agriculture and Food Industry: Basic Concepts, Structure, Characteristics, Applicabilities

This chapter explores the futuristic adoption of blockchain technology in smart agriculture and the food industry. It provides a comprehensive overview of blockchain's basic concepts, structure, characteristics, and applicabilities in the context of agriculture. By examining the potential benefits and challenges of blockchain implementation, the chapter offers insights into how this technology can enhance transparency, traceability, and trust in the agricultural sector.

8. Integration of Technology and Nanoscience in Precision Agriculture and Farming

Focusing on precision agriculture and farming, this chapter explores the integration of technology and nanoscience. It highlights the potential of nanotechnology to revolutionize farming practices, enhance crop productivity, and improve resource efficiency. The chapter provides an overview of the current applications and future prospects of nanotechnology in precision agriculture.

9. Internet of Things-Based Micro Climate Monitoring System for Irrigation Water Optimization

This chapter focuses on an Internet of Things (IoT)-based microclimate monitoring system for optimizing irrigation water usage. It explores the integration of IoT technologies, sensors, and data analytics to monitor and analyze microclimate conditions. By optimizing irrigation water based on real-time data, the chapter showcases the potential of IoT in enhancing water management practices in agriculture.

10. Precision Agriculture and Farming Using Cyber-Physical Systems: A Systematic Study

This chapter presents a systematic study of precision agriculture and farming using cyber-physical systems (CPS). It explores the integration of CPS technologies in agriculture to enable data-driven decision-making, precision farming techniques, and resource optimization. The chapter highlights the benefits and challenges of implementing CPS in precision agriculture and provides a comprehensive overview of its applications and potential impact.

11. Smart Cyber-Physical System-Based Plant Disease Detection for Agriculture

Focusing on plant disease detection, this chapter explores the implementation of a smart cyber-physical system (CPS) for agriculture. It highlights the integration of CPS technologies, sensors, and data analytics to detect and diagnose plant diseases in real-time. The chapter showcases the potential of CPS in improving crop health monitoring, disease prevention, and overall agricultural productivity.

12. Effective Path Planning of Cyber-Physical Systems for Precision Agriculture

This chapter addresses the effective path planning of cyber-physical systems (CPS) in precision agriculture. It explores the integration of CPS technologies, such as drones and autonomous vehicles, for optimized path planning and resource management in agricultural operations. By highlighting the challenges and potential solutions, the chapter provides insights into the efficient deployment of CPS in precision agriculture.

13. Analysis of Drones Applications in Precision Agriculture

This chapter focuses on the analysis of drone applications in precision agriculture. It explores the use of drones for various agricultural tasks, including crop monitoring, soil analysis, and pesticide spraying. The chapter highlights the benefits, challenges, and future prospects of using drones in precision agriculture, offering valuable insights for researchers and practitioners in the field.

The contributions compiled in this book presents the review reports, discussions, issues, implementation tactics, most recent research, as well as case studies focusing on the design and development of Agricultural Cyber Physical Systems.

G.S. Karthick
Department of Software Systems, PSG College of Arts and Science, India

Acknowledgment

Writing this part is probably the most difficult task. Although the list of people to thank heartily is long, making this list is not the hard part. The difficult part is to search the words that convey the sincerity and enormity of our gratitude and love.

First of all, I would like to thank the almighty.

Acknowledgment when sincerely given is a reward without price. I am thankful to the publishing team of IGI Global for accepting to publish this edited book. I would like to thank the entire development team, for helping us in bringing this project to a successful end.

I would like to express our deepest appreciation to Editorial Advisory Board members for extending their valuable suggestions and support to accomplish the task in a successful manner.

I would like to express appreciation and thank the reviewers for their careful reading of our manuscript and their many perceptive comments and suggestions. I want to acknowledge the authors for their valuable contributions and hereby I convey sincere thankfulness.

Editor Dr.G.S.Karthick owes his sincere thanks to his grandparents who are always blessing him from heaven. He expresses the deepest gratitude to his parents Mr.G.Selvaraj and Mrs.S.Saraswathi for their moral support, blessings, and encouragement throughout his life. Above all he expresses his thanks to his wife Mrs.K.Haripriya for her constant support.

He is extremely thankful to the Management, Principal, Secretary, and Vice-Principal of PSG College of Arts & Science, Coimbatore for providing encouragement and support to complete this project. Also, he expresses his deepest gratitutde to the Faculty Colleagues for their moral support during the hard times. He is greatly indebted to his sister Mrs.K.Geetha, brother Dr.M.Sridhar and other family members for their constant support in every aspect of life and also responsible for encouraging him to finish this book.

Introduction

In an era characterized by rapid technological advancements and the growing complexity of global challenges, the field of agriculture has not remained untouched. The advent of cyber-physical systems (CPS) has revolutionized the way we approach agriculture, ushering in an era of unprecedented innovation, efficiency, and sustainability. This book, titled "Contemporary Developments in Agricultural Cyber-Physical Systems," explores the intersection of agriculture and CPS, highlighting the transformative impact it has on the industry.

Agricultural CPS represents the convergence of the physical and digital worlds, where interconnected devices, sensors, machines, and algorithms collaborate to optimize agricultural processes (Dumitrache et al., 2017). This integration of advanced technologies, such as the Internet of Things (IoT), robotics, Artificial Intelligence (AI), data analytics, and automation, empowers farmers and agricultural stakeholders to make data-driven decisions, enhance productivity, reduce resource consumption, and mitigate environmental impacts.

The objective of this book is to provide a comprehensive overview of the contemporary developments in agricultural CPS, shedding light on the various components, applications, and challenges that arise in this dynamic field. It serves as a valuable resource for researchers, practitioners, and policymakers seeking to understand the latest advancements and leverage the potential of CPS in agriculture.

The chapters in this book cover a wide range of topics, each addressing a unique aspect of agricultural CPS. We delve into the fundamental concepts and principles underlying CPS in agriculture, exploring the architectural frameworks, communication protocols, and standards that enable seamless integration and interoperability among diverse components. Additionally, we examine the role of IoT devices and sensors in data collection, monitoring, and control of agricultural processes, as well as the integration of robotics and automation for precision farming.

Moreover, this book explores the transformative power of AI and machine learning algorithms in optimizing agricultural operations, including crop management, livestock monitoring, disease detection, and yield prediction. We discuss the challenges and opportunities associated with big data analytics and the utilization of cloud computing and edge computing technologies in handling and processing vast amounts of agricultural data.

Furthermore, as the world faces pressing challenges related to climate change, water scarcity, and food security, this book investigates how agricultural CPS can contribute to sustainable farming practices. We examine the potential of CPS in resource management, water-efficient irrigation systems, smart greenhouse technologies, and resilient crop cultivation techniques.

Lastly, this book delves into the emerging trends and future directions of agricultural CPS, including the integration of blockchain technology for traceability and transparency in supply chains, the adoption of drones and satellite imagery for precision agriculture, and the implications of CPS for rural development and social equity.

As we embark on this journey through the contemporary developments in agricultural CPS, we hope to foster a deeper understanding of the potential and challenges associated with this transformative field. By harnessing the power of CPS, we have the opportunity to revolutionize the agriculture industry, ensuring sustainable and efficient food production while preserving our precious resources and safeguarding our environment. Together, let us explore the cutting-edge innovations and envision a future where agricultural cyber-physical systems drive the next agricultural revolution.

G. S. Karthick
Department of Software Systems, PSG College of Arts and Science, India

REFERENCE

Dumitrache, I., Caramihai, S. I., Sacala, I. S., & Moisescu, M. A. (2017, May). A cyber physical systems approach for agricultural enterprise and sustainable agriculture. In *2017 21st International Conference on Control Systems and Computer Science (CSCS)* (pp. 477-484). IEEE. 10.1109/CSCS.2017.74

Chapter 1
Cyber-Physical Systems: Foundations, Design Principles, Challenges, and Applications

G. S. Karthick
https://orcid.org/0000-0002-9340-6928
PSG College of Arts and Science, India

V. Sumathi
Sri Ramakrishna College of Arts and Science, India

ABSTRACT

Cyber-physical systems (CPS) are the combination of physical, networking, and computational processes. With feedback loops where physical processes influence calculations and vice versa, embedded systems and networks monitor and manage the physical processes. With the use of abstractions and modelling, design, and analytic approaches, CPS blends the dynamics of physical processes with those of software and networking. Standards, protocols, and test techniques that facilitate the discovery, interoperability, and composition of components required to create these CPS will foster innovation while also improving economic viability and enable systems to become more efficient and resource efficient. This chapter presents the basic mechanisms of the CPS such as workflow, characteristics, simple structure, and major applications of CPS. Also, the chapter analyzes the design considerations of CPS based on the security, trust management, energy-efficiency, reliability, service composition, and resource provisioning aspects.

OVERVIEW OF CYBER PHYSICAL SYSTEMS

Cyber-physical systems (CPS) is a term first introduced by Helen Gill in 2006 at National Science Foundation (NSF). It stems from the word Cybernetics which was introduced by Norbert Wiener. CPS acts as a bridge between cyber world and physical world. The principal goal of CPS is to monitor the physical process and according to the need they will be actuated to change its behaviour (Edward Lee., 2015). There is a tight integration between cyber and physical world. This tight integration differentiates

DOI: 10.4018/978-1-6684-7879-0.ch001

it from desktop computing, traditional embedded systems, real-time systems and wireless sensor network. Many researchers have proposed different definitions of CPS. The definition of CPS is as follows:

"Cyber Physical Systems is defined as the interaction between tightly coupled physical processes and cyber processes in a controlled and coordinated manner to extract deeper knowledge, which enables more accurate actions and tasks and thus provide a set of proactive as well as intelligence-rich services which forms the basis for building smart applications."

It can also be described as the integration of computing with physical protocols, with both physical and cyber components equally defining the behaviour (Volkan Gunes et al., 2014). A mechanism called a cyber-physical system is one that is meant to be anticipated and managed by computer-based algorithms. The entire CPS system is tangibly connected to the Internet; as a result, the word "cyber" is included in it. Software and physical components work side by side via various temporal and spatial scales in a CPS. A CPS is designed to draw attention to crucial problems like contiguous engineering ethnicities in the physical and cyberspace.

Figure 1. Definition of cyber-physical systems

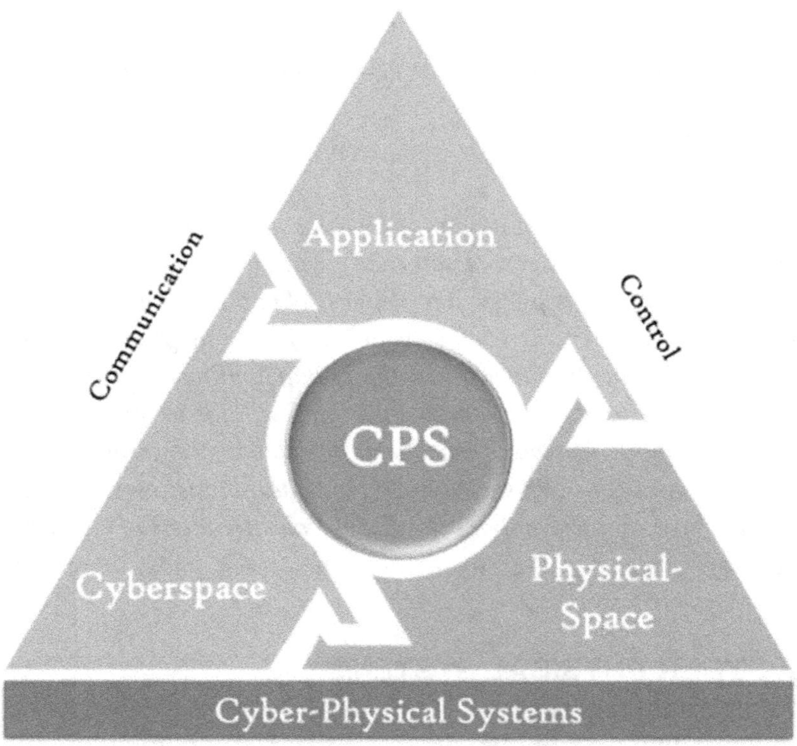

WORKFLOW OF CYBER-PHYSICAL SYSTEMS

CPS consistently integrates sensor networks with embedded systems to monitor and regulate the physical environment, including feedback loops that allow this external stimulus to self-activate communication, control, or computing (Eric Ke Wang et al., 2010). An overall workflow of CPS is written off as four key phases and illustrated in Figure 2.

Figure 2. Workflow of cyber physical system

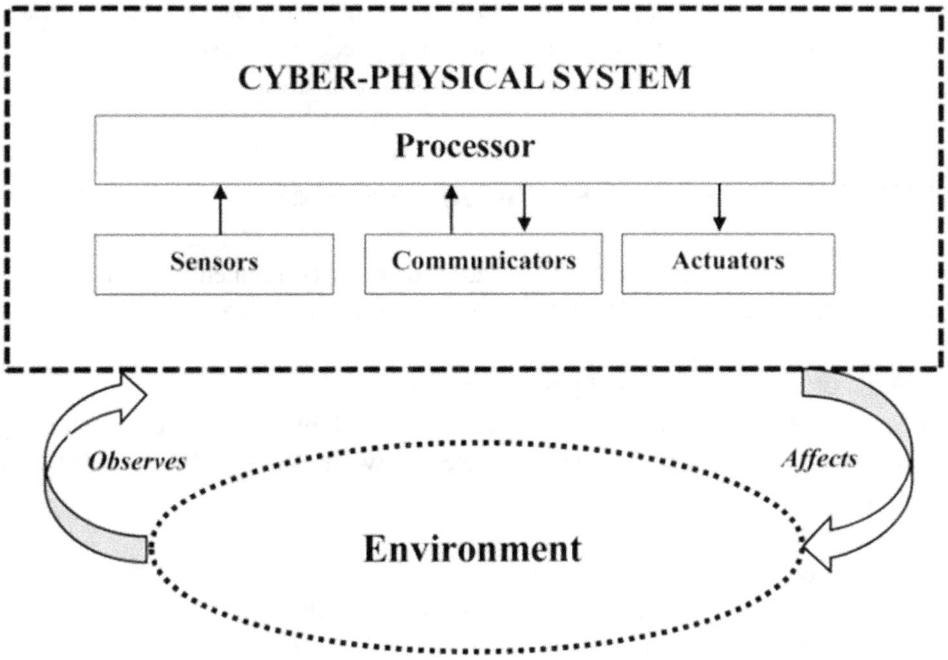

Monitoring

It's the preliminary stage of the CPS process, when historical data is analysed to monitor the physical system from beginning to end. The CPS will then utilise this information to ensure that every method involved is secure for use in the future. Any flaw that CPS notices in the physical system must be fundamentally fixed.

Networking

Data collection is to be done at this stage. Numerous RF modules, sensors, and/or wireless devices may be implicated in a cyber-physical system. Real-time data are produced by sensors and/or RF modules. Later, all of the generated data must be gathered together for further processing. Numerous applications are also interacting and communicating across the network in the meantime.

Computing

During this phase, the obtained data is synopsized and examined during the overseeing review and testing of the physical system's circumstances. Actions must be carried out while running through the physical system's circumstances. For example, utilization of a data prediction model to anticipate valve controller in storage tank operations based on the algorithm intended in accordance with historical data.

Actuation

Actions planned in the computing stage will be carried out in this phase. Actuation will actuate multiple actions such as physical process control, cyber behaviour correction, and so on. As an example, action is to regulate and govern the valve in the storage tank based on historical data.

CHARACTERISTICS OF CYBER PHYSICAL SYSTEMS

A general workflow of CPS can be decomposed into four main components which are sensing, communication, computation and control. The inherent characteristics of CPS are as follows:

- Close integration of computational and physical processes
- Heterogeneous in terms of devices, computing platforms and inter-connection technology
- Unreliable networking as it consists of wired/wireless network, WLAN, Blue- tooth, GSM
- Exhibits mobility
- A tight environmental coupling
- Cyber capability in every physical component
- Resource constraint in terms of computation, network bandwidth and so on
- Complexity at multiple temporal and spatial scales
- Large scale network centric
- Dynamic operating conditions

CPS is now recognised by the world as technology enabler as they enable numerous innovative applications and for this many countries have listed CPS research on high priority (Imre Horvath and Bart H. M. Gerritsen, 2012). Practically its applications span across all domains including smart infrastructure, assisted living, precision agriculture, avionics, communication and information systems, automotive systems, energy conservation, distributed robotics, medical, defense systems, critical infrastructure health monitoring of power grid, water resource, environment and heritage, and has its own research challenges. The desired requirements of any CPS application are high avail- ability, high reliability, real-time, autonomics and safety.

There are various background theories present behind CPS which are shown in Figure 3, but CPS is different from all of them in various aspects. The embedded systems emphasize more on computational elements than the intense link between computational and physical elements whereas Machine to Machine (M2M) communication is all about wired and wireless communication with other devices of same capability without or with limited human intervention(Hong Chen, 2017). Wireless Sensor Network (WSN) is mostly field specific and the focus is on collecting and managing sensed data whereas CPS crosses several fields, it does intra as well as inter WSN communication focusing on the discovery of new knowledge across multiple sensing domains and building the intelligent applications. Many times, CPS is referred as a technology similar to Internet of Things (IoT).

Figure 3. Contextual theories of cyber-physical systems

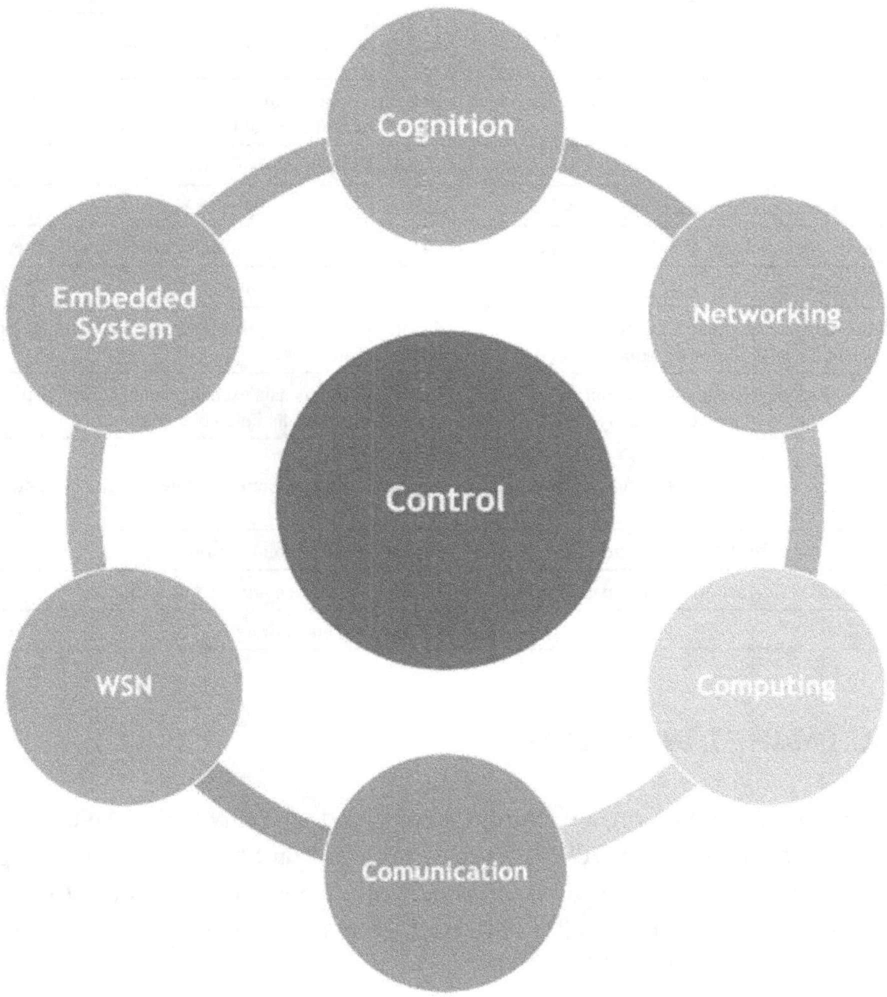

Amongst the fraternity of researchers there are various opinions about comparison be- tween IoT and CPS, but it lacks in clear differentiation. The core of IoT is all about how objects interact over Internet, whereas CPS is more about how to extract the intelligence in the close interaction between physical world and cyber world and use it for decision making which can control physical world. The differences between CPS and IoT are enlisted as shown in Table 1.

Table 1. Differences between CPS and IoT

CPS	IoT
Emphasis is on physical systems, information exchange and feedback	Emphasis is on interconnection of objects.
Main components are sensing, communication, computation and control.	Main components are sensing, heterogeneous access, information processing and applications as well as services.
Usually a closed loop system	Usually an open loop system
They are controllable and partially predictable by simulation	Difficult to control or predict system behavior
Sensors and communication tools are embedded in products	Uses special sensors to identify products and materials.
Human in the loop is present	No human intervention
Combines cognition, networking, control, communication, embedded systems and wireless sensor networks	Combines wireless sensor networks, internet and mobile communication
Core challenge is abstraction, modeling, simulation, programming and verification of the com- plex systems	Core challenge is data exchange among large scale heterogeneous network elements and effective integration
Service is environment dependent because cyber world takes into consideration feedback from physical world and accordingly modifies the physical process.	Service is often dynamic, mobile, less reliable and object dependent.
Timed communication and information exchange	Time delayed communication
Well-known and controlled interaction between participants	Interaction between unknown participants
Changes are controlled	Changes are uncontrolled

CPS VERSUS MANET, WSN, M2M

In (Fang Jing Wu et al., 2011) a review learning is prepared on WSN and MANET and also the way they have driven in the direction of CPS's progress. In (Wai Wai Shein et al., 2013) explored features of M2M and CPS, and also indicated correlations between them. Characteristics of M2M architecture, design challenges and application of M2M have been evaluated. Table 2 portrays a qualitative assessment between CPS, WSN, M2M and MANET.

Table 2. Comparison between MANET, M2M, WSN, and CPS

Networking Criteria	Features	MANET	M2M	WSN	CPS
Network Formation	Random Deployment	Yes	Yes	Yes	Yes
	Dynamic Topology	Yes	Yes	No	Yes
	Time-Vary Deployment	No	Yes	No	Yes
	Interconnection among Multiple Network	No	Yes	No	Yes
	Internet Supporting Network	No	No	Yes	Yes
Communication Pattern	Query Response Flows	Yes	Yes	Yes	Yes
	Arbitrary Communication Flows	Yes	Yes	No	Yes
	Cross network Communication Flows	No	No	No	Yes
	Cross domain Communication Flows	No	No	No	Yes
	Deterministic Delay communication Flows	No	No	No	Yes
Power Management Network	Opportunistic sleep	Yes	Yes	Yes	Yes
	Multiple Sleep Modes of Nodes	No	Yes	No	Yes
	Power Management Techniques	No	Yes	No	Yes
Network Connectivity and Coverage	Connectivity	Yes	Yes	Yes	Yes
	Coverage	No	Yes	Yes	Yes
	Heterogeneous Coverage	No	No	No	Yes
Knowledge Mining	Data Mining and Data Base Management	No	No	Yes	Yes
	Multi-Domain Data Sources	No	No	Yes	Yes
	Data Privacy and Security	No	No	No	Yes
Quality of Services	Networking QOS	Yes	No	Yes	Yes
	Multiple Data Resolution	No	No	Yes	Yes

SIMPLE STRUCTURE OF CYBER-PHYSICAL SYSTEMS

A simple assembly of CPS is demonstrated in Figure 4, which comprises of three prominent sections i.e., Physical Plant, Network Fabric and Computational Platforms.

- ❖ Human operators, chemical or biological protocols, or mechanical components are all part of the Physical Plant. This section excludes digital networks.
- ❖ Second, network fabric provides strategies and processes that are used in communication protocols. Finally, computational platforms are comprised of several computers and operating systems, actuators, sensors and so on.
- ❖ Computational platforms and network fabric compose the "cyber" component of the Cyber-Physical System.

Figure 4. Simple structure of cyber-physical systems

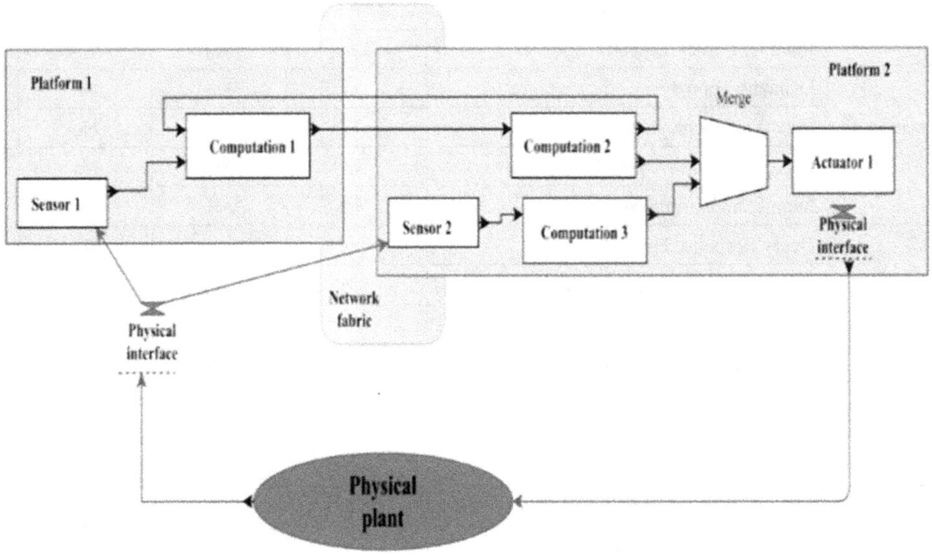

Dual networked platforms individually with its own sensor and/or actuator is also depicted. An actuator's activity habitually affects the sensors data delivered by means of a physical plant. As depicted, Platform 2 regulates the physical plant by way of Actuator 1. Physical plant progressions are measured via Sensor 2. The section Computation 2 implies a control law, which concludes and mainly cantered on the sensor data whatever commands to be imposed and delivered to the actuator (Lim Shinyoung et al., 2011). This loop is termed as 'feedback control loop'. Platform 1 knocks off added surplus measurements through Sensor 1, and directs messages to Platform 2 using network fabric. Computation 3 comprehends an additional control law, amalgamated with that of Computation 2, conceivably prompting to pre-empty it.

DESIGN AND DEVELOPMENT OF CYBER-PHYSICAL SYSTEMS

The widespread adoption of CPSs in several sectors such as energy, healthcare, manufacturing has motivated several concerns about their security, sustainability, efficiency and resilience in these domains. The integration of computation and communication technologies with the physical world demands a detailed investigation into the development of CPSs to obtain the desired performance metrics (Carlos Andres Macana et al., 2011). This section presents an overview of the requirements and challenges in the design and development of CPSs. In particular, a broad perspective of the four CPS design objectives considered in this dissertation: security, energy efficiency, trust, and reliable operation.

CPS Security

CPS uses a communication network to exchange information among its subsystems. This network is potentially vulnerable to adversities such as time delays, packet losses and various cyber-attacks launched by adversaries. Adversaries may manipulate data packets to initiate a cyber-attack on CPS that may cause economic losses or casualties. Besides, control systems are at the core of CPSs in critical infrastructures such as industrial plants, which are vulnerable to security attacks. Adversaries can launch attacks on CPSs by exploiting the dynamics of CPS control systems (Shah Ahsanul Haque et al.,2014). CPS security threats can be grouped into three categories, which are Disclosure attacks, Deception attacks, and Disruption attacks.

Disclosure Attacks

Disclosure attacks are initiated by adversaries to obtain confidential information from a system. The disclosed information may be directly used by the adversary for malicious activities, initiate attacks in the future, or exchange information for profit. Disclosure attacks are vital in CPS security as they take a longer time to detect. An example is the Inference attack, also called plaintext attack, which targets critical system components to gain confidential information such as the system transfer function. Further, in some cases, attacks that utilize confidential information can de-stabilize the CPS control system. Other examples of disclosure attacks include intrusion attempts such as eavesdropping, snooping, or packet sniffing. Some of the prominent CPS cyber-attacks are listed in Table 3.

Deception Attacks

Deception attacks consist of adversaries injecting false data into the network such that the signal values differ from their actual values. Adversaries can initiate these false data injection attacks at the sensor level, actuator level, or control blocks (Yongqi Ge et al.,2012). In sensor attacks, sensors are sabotaged by subverting the fidelity of sensor measurements. Whereas in actuation attacks, the control signals are corrupted by injecting false data. In controller attacks, the control law of the CPS control system is targeted. Deception attacks are among the most potent attacks regarding the magnitude of the damages inflicted on the CPS control system.

Disruption Attacks

In Disruption attacks, the adversary intentionally tampers the system such that it leads to the non-availability of data and control signals at CPS subsystems. These attacks are also referred to as Denial-of-Service (DoS) attacks or jamming attacks. Adversaries can attack a CPS using DoS attacks by targeting its sensor data or actuation signals by sabotaging the underlying CPS communication network. The increasing likelihood of cyber-attacks that sabotage CPS operations heightens the concern regarding the security of CPS used in mission-critical applications such as aerospace, military, medicine and manufacturing.

The CPS security mechanisms can be classified as cryptographic techniques and control-theoretic techniques. Cryptographic techniques are based on information security and form a major branch of CPS security, and they are based on widely popular encryption algorithms and privacy standards (Wai Wai Shein et al., 2013). In comparison, control-theoretic techniques use insights from control theory to secure CPSs. These schemes focus on resolving control issues in the CPS environment and safeguarding the control system from attacks on the control algorithms.

Table 3. Comparison of cyber-attacks in CPS

Attack Type	CPS Target	Impact of Attack
DoS	Medical Devices	Patients will not receive the expected therapy
False Data Injection	Insulin Pumps	Wrong Medical Decisions
Packet Injection	Engine Control Unit	Loss of Control, Safety-Critical Consequences
Replay Attack	Engine Control Unit	Safe-Critical Situation
Relay Attack	Remote Keyless Entry	Theft, Unauthorized Access
Malware Injection	Programmable Logic Controller	Physical Damage, Access Denial

CPS Energy Efficiency

CPSs in critical infrastructures are often required to operate for extended periods with minimal maintenance and recharging facilities. Hence, energy harvesting from the physical environment can supplement traditional energy sources and potentially enable CPSs to extend their operational lifetime (Syed Hassan Ahmed et al.,2013). However, uncertainties in the physical environment often lead to erratic energy supply, necessitating the development of energy- efficient computing techniques. As CPSs find applications in many energy-constrained devices, there is a pressing need to develop energy-efficient and sustainable designs. The following principal aspects characterize a sustainable CPS.

- Energy-Efficient Operation
- Low Operating Costs
- Restrained Energy Consumption

Traditional CPSs do not factor in energy efficiency as a primary objective as most de- vices are powered continuously during their operation. Nowadays, CPSs have found applications in many battery-powered devices with restrained energy consumption requirements. Moreover, CPSs are vulnerable to different types of cyber-attacks, necessitating secure mechanisms for data security (Park et al.,2012). As security and energy efficiency are contradicting requirements, improving security has an adverse impact on energy efficiency. Consequently, it is essential to balance the trade-offs between energy-efficient operation and CPS security. Several notable works in the area of CPS energy efficiency are listed in Table 4.

Table 4. Comparison of energy-efficient techniques in CPS

Type of CPS	Objective	Techniques
IoT Network	Energy-Efficient Routing	Multi-hop Zone-Based Routing Algorithm
Smart Grid	Energy-Efficient Load Scheduling	Bio-Inspired Load Scheduling Algorithm
Unmanned Aerial Vehicle (UAV)	Engine-Efficient Relaying	Optimizing Communication Time and UAV Trajectory
WSN	Energy-Efficient Routing	Optimized Cluster Head Selection
Wireless Body Area Network	Extending Lifetime	Energy-Efficient MAC Protocol

CPS Trust Management

The resilience of a system is defined as its ability to withstand and recover from the impact of cyber-attacks. In many systems, resilience is not an intrinsic property and thus needs to be imparted through suitable design methods. Among the numerous methods to improve resilience, trust-based approaches are of particular interest. Trust-based methods have been applied not only to problems pertaining to security but also where some subsystems cannot be trusted due to the activities of adversaries. Recently, emerging attacks such as Advanced Persistent Threats (APTs) have surfaced in the realm of cyber warfare. They use covert and advanced attack techniques to access the network and remain undetected for a lengthy time period, with potentially harmful consequences (Kyoung-Dae Kim and P.R. Kumar, 2013). When specific CPS components are jeopardized by novel attack techniques such as APTs, conventional defense mechanisms, such as cryptographic techniques, are no longer adequate to provide resilience to cyber-physical infrastructures.

Existing works that use encryption and digital signatures assume implicit trust in the interactions between cyber-physical components. Since adversaries may forge false data and unauthorized commands, data and control signals may not be explicitly trusted. Mitigating and defending against novel and sophisticated cyber-attacks entail developing novel approaches for improving the resilience of CPSs. Some of the significant works in the area of CPS trust management are given in Table 5.

Table 5. Comparison of trust-based techniques in CPS

Type of CPS	Objective	Techniques
IoT Network	Security and Trust Assessment	Multi-Layered Approach using SDN/NFV
Vehicular CPS	Trust-Based Recommendation Scheme	Cooperative Computing-Based Filtering Algorithm
WSN	Trust and Reputation Model	Fuzzy Sets
UAV	Improve Security of Data Routing	Trust-Based Active Detection Scheme Based on Information Theory
IoT	Improve Security of Data Routing	Trust and Reputation Based Secured Routing Protocol.

Reliable CPS Operation

As CPSs incorporate various heterogeneous components to establish distributed control, they are often subject to numerous design challenges. CPSs are composed of different hardware (sensors, actuators and embedded devices) and software components (both proprietary and open-source). Integrating multiple components to facilitate real-time operation becomes increasingly challenging with the scale and the growing complexity of CPSs. Since CPS operations are controlled and coordinated by computing and communication elements, there is a close coupling between cyber and physical domains (Siddhartha Kumar Khaitan and James,2014). This coupling leads to undesired interactions among the different cyber and physical elements of CPS. This may result in the physical properties of CPS components, causing unexpected system behaviour in the cyber domain. These include limitations in network bandwidth, network delays, signal interference and data quantization issues. These factors degrade the system performance and adversely impact system reliability. Besides, the dynamic and complex nature of CPSs lead to several non-idealities such as uncertainties and randomness in network traffic and noise from the physical environment. These factors contribute to asynchronous interaction between cyber and physical components. Furthermore, CPS components may be susceptible to unexpected faults and failures; consequently, CPSs must be robust against system uncertainties for stable and reliable real-time operation. Table 6 compares some of the existing solutions that address uncertainties in CPS.

Table 6. Comparison of solutions for uncertainties in CPS

CPS Uncertainty	Objective	Techniques
Network Delay	Wide Area Control in the presence of Network Delays	Delay-Aware Control Design
Network Congestion	Congestion Control	Channel-Aware Congestion Control Algorithm
Signal Interference	Interference Mitigation	Power Control-Based on Game Theory
Bandwidth Allocation	Dynamic Bandwidth Allocation	Optimization Theory
Data Quantization	Improve Control Performance	Fuzzy Logic

Service Composition in CPS

Service Oriented Computing (SOC) and Service Oriented Architecture (SOA) are very popular because of their features like reusability, flexibility, platform independence as well as transparency, interoperability and adoptability (Lichen Zhang et al.,2013). Service composition is defined as follows:

"Service composition means combining more than one individual service to complete a desired goal when a single service alone is not sufficient to satisfy the specific goal." Software service and service provided by CPS are different from each other in various aspects. The differences between software service and CPS service are enlisted as shown in Table 7 and it indicates that traditional SOA cannot be directly applied to CPS.

Table 7. Comparison of software service and CPS service

Aspect	Software Service	CPS Service
Dependency on Physical Environment	Are not dependent on physical environment so does not impact on availability of service and in turn QoS	Highly dependent on physical environment so impacts on availability and QoS
Multiple Instances	Can have multiple instances	Can have multiple instances but only one will be served at a time
Contention Problem	Contention problem may not occur	Contention problem occurs because only one physical service can be served at a time
Context Consideration	Need not take into consideration the context while describing service	Context needs to be considered while describing the service.
Predictability of Service Execution	More predictable in terms of successful service execution	Less predictable in terms of successful service execution because of joint dynamics involved in cyber and physical space
Real Time Factor	Need not require real time consideration	Requires real time consideration
Side Effects	The side effect of a software service generally, can be specified indepen dent of other software services	The side effect of CPS service can not specified independently
Resource Consideration	No physical resources are considered as a part of definition	Physical resources need to be considered

Service composition is very well studied in various domains like Web Service, Pervasive Computing, Opportunistic Network, Cloud Computing, Distributed Systems, Wireless Sensor Network, Internet of Things and many more, but in the CPS domain, service composition still lacks in maturity. In the beginning years of CPS, few researchers have applied SOA in CPS. In (Borzoo Bonakdarpour, 2008) presented three tier service-oriented CPS architecture which has environmental tier, control tier and service tier. They have also discussed design methods of each layer. The authors conclude that with service-oriented CPS, the heterogeneity and dynamic composition issues can be solved. They have discussed safety guard case study with theoretical discussion. In (Jiafu Wan et al., 2013) applied SOA to enhance sustainability. They have proposed real time SOA middleware which focus on service level and workflow level predictability and global resource management. In (Fang Jing Wu et al.,2013) applied SOA for adoptability, dependability, trustworthiness and context awareness investigation while proposing service-oriented view.

Resource Provisioning in CPS

Resource management is defined in terms of operations like resource provisioning, re- source scheduling and resource monitoring. Resource provisioning has been studied previously in contexts like Cloud Computing and Grid Computing. Various steps involved in resource management consists of resource discovery, resource selection, resource allocation, resource mapping and monitoring for the resource usage. Resource provisioning is defined as: "Resource Provisioning means the selection, deployment, and run-time management of resources for ensuring guaranteed performance for applications."

The main objective of resource provisioning is making the resources available, as and when required as per the demand of application. Few researchers have worked on different aspects of resource management which can be summarised as follows:

In (Vivek Sehgal and Anubhav Patrick.,2014) have developed performance metric to characterize formally the system behaviour under different overload condition and proposes algorithm for distributed system. They have formulated the problem as criticality aware bin packing problem and ductility matrix is used to capture the mixed criticality property.

In (Chengyuan Yu et al.,2012) have reviewed some methods of system resource allocation in CPS domains and also proposed performance optimization model with resource constraints. In the successive work they have solved this optimization problem using particle swarm optimization, by considering only periodic tasks. A case study of unmanned vehicle with WSN navigation is presented.

In (Simmon Eric et al.,2013) have proposed a resource-centric CPS model. Also, the notation for specifying resources along with its dependencies, and constraints is shown. They have discussed the case study of ROBOT.

In (Schmidt Douglas et al.,2014) have proposed environment adaptive scheduler for CPS which can schedule the task on multiprocessors. It gives detailed study of used structures, models and reasoning methods. It is carried out in two phases. In first phase it calculates the quality of scheduling tasks and in second phase it allocates these tasks on multiprocessors. This is applied on an obstacle-avoiding robot case study. In (Niels Reijers et al., 2011) have given extension of their previous work presented in along with details of methods of task scheduling along with the case study of scheduling criticality aware tasks in radar system. In (Minyoung Kim et al., 2010) have discussed P-RSM that is a probabilistic resource space model to deal with uncertainty in managing resources in CPS. Operations and integrity constraints are also developed which helps in effective management of resource space.

CHALLENGES OF CYBER PHYSICAL SYSTEMS

CPS expectations are huge which includes robustness, self-organization, efficiency, and interoperability. Mostly different challenges are driven by expectations and requirements. A subset of challenges is in this section.

Context-Aware Systems

This challenge represents extensive context-conscious system demands (Jessica Diaz et al., 2016). The context-aware characteristic is essential to recognize, analyze and interpret the intentions of various components of the CPS. In other words, CPS purpose will be successfully satisfied only if it is applied in a right context.

Cooperative Production Systems

CPS is a complex interconnected unit (Quan Shen et al., 2014). The challenge in CPS is that the way the components are connected together such that the outcome is good. This challenge is closely linked to context-aware systems.

Prediction of Dynamic Systems

Extending the present techniques or developing new ones (particularly in the case of interconnected CPSs) should be given.

Elective and Well-Operating Systems

Multiple parts (sensors, actuators, control units) generated by many manufacturers can be used to assemble CPSs. Proper assembly of these parts, as well as appropriate knowledge and exploitation of the respective information models is one of the most significant problems that are needed to make CPSs operate faultless and effective.

ADVANTAGES OF CYBER PHYSICAL SYSTEMS

Integration

CPS addresses integration of network features such as medium access control mechanisms, middleware and software. The integration of WSNs and Distributed Computing aspects is also a significant aspect of CPS.

Machine and Human Communication

Modelling and predicting situational information and its changes are important for decision making processes.

Handling Uncertainty

The validity of the design is proved by certainty mechanisms. Cyber physical systems are developed to function correctly in new and unreliable situations.

Better System Performance

CPS can deliver better performance with the feedback provided by the close interaction between the sensor and cyber infrastructure (Kalasapur, Swaroop et al., 2007). It may assist organisations in optimising their operations and increasing the efficiency and dependability of their systems. An industrial control system, for example, may monitor and regulate manufacturing operations in real time, decreasing waste and increasing productivity.

Scalability

CPS is capable of handling the growth in the amount of computation it can perform by extending its resources.

Flexibility

CPS is capable of adapting to the modifications that might take place in the system. It can provide for greater flexibility in the design and operation of systems. A smart grid, for example, may be built to be more robust and responsive to changing energy demands and conditions.

Quick Response

CPS will provide rapid response and promote fast detection of failures. Also, it can give real-time monitoring and management of physical processes, allowing organisations to discover and handle problems faster. This can increase system dependability and performance.

APPLICATIONS OF CYBER-PHYSICAL SYSTEMS

Cyber Physical Systems has its applications in various arenas such as agriculture, medical monitoring, sovereign automobile systems, process control systems, robotics systems, instinctive pilot avionics, smart grids, traffic logistics systems and so on (which is shown in Figure 5).

Robotics

MIT showed the application of CPS in distributed robotics using a distributed robot gardening system. A team of robots administers to a tomato garden using decentralised sensing, manipulation, navigation, and wireless communication.

Manufacturing

CPS are mostly utilised in factories to self-monitor manufacturing processes and operations (Tamhane Sagar A et al., 2012). The exchange of information across machines, business systems, supply chains, suppliers, and customers significantly improve the manufacturing process. This is referred to as intelligent or smart manufacturing. This enhances supply chain transparency and control, resulting in increased safety and integrity.

Smart Home

CPS have an important role in raising agricultural yield and decreasing hunger. The system is designed to be flexible in terms of light intensity, temperature, watering, and humidity. These options are designed to react to specific computer programmes in order to stimulate growth. The system also gives the user ongoing feedback, enabling them to know the condition of the house at all times. Feedback can be managed remotely using network services. Additionally, temperature, soil moisture, light sensor, and humidity sensors are fitted to the designs as station sensors. Furthermore, both temperature and humidity sensors are placed, enabling sprinkler systems and fans to be regulated to either raise or lower the temperature.

Figure 5. Applications of cyber-physical systems

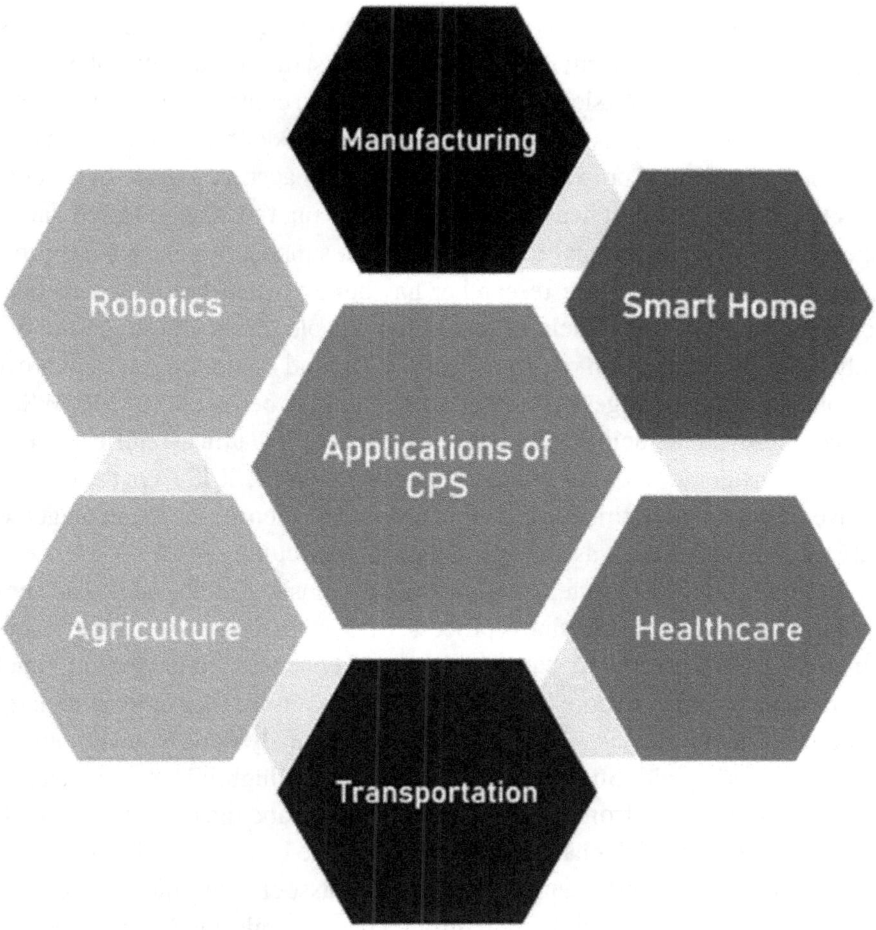

Healthcare

In the medical field, CPS are widely used for continuous monitoring of patients and sensor control. This leads to improved treatment for the elderly and disabled, as well as a reduction in patient hospitalisation. To improve workflow and system security, a network closed loop is combined with a human loop in this system.

Transportation

Cars may communicate with one another using CPS in order to share information such as traffic congestion, location, and route concerns in order to avoid accidents and endorse safety. A further instance of transportation is the use of an ABS or anti-skid braking system to avoid an accident and bring the automobile to a complete stop.

Agriculture

In the field of agricultural scenario, CPSs is interlinked with the Agricultural Cyber- Physical Systems (ACPSs). In order to build integrated farm management systems, the applications of electronic technologies support toward interact with a physical environment that also controls the ideal environment values. Also, they have suggested that there is a need of an effective model that should be capable to collect data about soil, climate change and crops information with high accuracy ratio. In order to manage the external resources like humidity, plant health, watering and so on, thus the collected data has been supported to manage and control the external resources. But, to support or enhance the profitable energy systems with autonomous behaviour, the researcher has believed that needs to enhance the SPVWS into ACPS conceptual model. With the help of earlier sensing platforms, the suggested model is highly complex especially with the integration of remote services, cloud computing with photovoltaic facility services. In this context, a study suggested a model toward the cyberization of SPVWS. Furthermore, they have integrated photovoltaic facilities into ACPS architecture via this significant design, cyberization and optimization considerations were assumed (Wu Taotao et al., 2017). On the other hand, some of the researchers have designed an optimized model using ACPS. Hence, for the management of various services, a variety of ACPSs has been assumed in precision agriculture.

In the field of Smart Farming (SF) system, it's specifically considering the huge data analytics process across the agricultural value chain and business process. SF system reviews the relation among variable, concepts, and functions in the eco-system that support toward enhancing their working functionality. The expansiveness of the degree makes it more useful for agribusinesses that are more associated with supply sources of info and post-gather exercises. In this methodology, analysis of huge amount of data will be done to increase the farmer's yield. Studies on this methodology highlight the prescient estimation of enormous information examination from a business point of view and the manners in which it empowers proactive activities to moderate quick changes that influence the business cycle. Accentuation is set on the ideas of the information chain and organization the board association. Information chain is clarified as the action arrangement from information catch through the dynamic and information advertising, it is a portrayal of the information cycle and remembers all necessary exercises for information. The portrayal of the information cycle reveals insight into the idea of information in agribusiness through information catch, stockpiling, move, change, examination and advertising. farms and agribusinesses can utilize it as a rule in the detailing and improvement of enhancement systems in their exchange cycle. The cycle perceives two different ways business measure targets are finished, which are administration and plan of action. Administration alludes to the organization of authoritative cycles that comprise a piece of the partner organization, and they can be either formal or casual. Such course of action as it identifies with enormous information remembers arrangements for the accessibility of information, the responsibility for, who is obligated for the treatment of information and the security of different informational indexes and issues identified with privacy.

Recently, the technological enhancement like data transmission, electronic system, has acquainted extremist changes with the rural workplace. These progressions request refreshed data from creation frameworks and from business sectors and specialists associated with creation to give dynamic data to create just as for the vital and administrative issues included (Iria Estevez-Ayres et al., 2009). Shrewd cultivating (SF), in light of the merging of information and correspondence innovations into hardware, gear, and sensors in agrarian frameworks, permits a huge amount of evidence and data to be produced with the reformist addition of computerization into the cycle. Keen planting depends on data transmission and the convergence of information in far off capacity frameworks to empower the mix and review of different homestead information for dynamic. Segment patterns, including maturing populaces and is preceded with the movement of individuals from rural to metropolitan regions, have pulled in consideration of specialists, since work issues may turn into a shortage factor in agribusiness. Notwithstanding these patterns, the strengthening of environmental change will keep on adjusting developing conditions, for example, precipitation, dampness of soil and temperature in less unsurprising ways. SF apparatuses can help lessen these effects, keep them consistent or diminish creation costs in horticultural exercises, and they can help with limiting ecological requirements.

SF is an idea that started with programming and software engineering that showed up with the expansion of figuring advances and the transmission of information from farming, inside a general climate of essentially universal registering. These figuring components are implanted in articles and interconnected with one another and the web. The SF field includes different terms with comparative implications, for example, brilliant farming. As needs are, covering interfaces and advancements exist and incorporate thoughts, for example, exactness agribusiness and the executive's data frameworks in horticulture, which have been gotten from the possibility of the homestead the board data framework (FMIS). FMIS is characterized as a framework that is intended for gathering, preparing, putting away, and scattering information in a necessary configuration to perform activities and capacities on country properties. The utilization of SF devices is conceivable because of the utilization of sensors in farming. A sensor is an electrotechnical gadget that estimates actual amounts from the climate and converts these estimations into a sign that can be perused by an instrument. Among the estimations read by sensors are the accompanying: humidity, pressure, absence or presence of different types of objects, temperature, light, noise levels, mechanical stress levels, direction, speed and object size. The summary of the IoT technologies for smart agriculture is tabulated in Table 8.

Table 8. IoT technologies in smart agriculture

IoT Technology	Application in Agriculture	Benefits in Agriculture
Radio communication capabilities and sensor nodes	They have been used different physical parameters that integrated with a sensor network.	Sufficient amount of data is collected related to agricultural field and managed from sensor network
Internet-based cloud computing platform	They provide the data and share processing resources to devices on-demand and computer networks	Easily manage the collected data from cloud environment such as cloud storage, agriculture field's maps, and so on.
Analyze and process the enormous amount of data using big data analytics framework.	Able to access the different type of data.	Data correlations, customer preferences, uncover patterns, market trends and some of other significant data.
A computer system with a combination of both hardware and software (also known as embedded systems)	It performs some particular type of task like data controlling, activity management and data monitoring activities.	Minimized the production cost to a remarkable level that will increase data sustainability and profitability.
The communication protocols used to enable data connectivity which is the backbone of IoT systems.	This model is adaptable to facilitate data interchange with different formats.	Will manage huge collection of data that collected from cloud computing services, sensors, cloud storage and so on.

CONCLUSION

Overall, the near future of CPS is promising with an increasing emphasis on IoT connectivity, the adoption of AI and ML capabilities, and more emphasis on security and privacy. Cloud-based CPS are becoming more popular because they offer better flexibility and scalability over on-premises solutions. It is progressively focused on these areas as the relevance of data security and privacy grows. This involves employing secure communication methods and putting in place security measures to guard against cyber-attacks. This chapter provided the deep insights on the basic mechanisms of the CPS which includes workflow, characteristics, simple structure and major applications of CPS. Also, the design considerations of CPS based on the security, trust management, energy-efficiency, reliability, service composition and resource provisioning aspects has been analyzed.

REFERENCES

Ahmed, S. H., Kim, G., & Kim, D. (2013). Cyber Physical System: Architecture, Applications and Research Challenges. IEEE IFIP Wire- less Days (WD), (pp. 1-5). IEEE.

Bonakdarpour, B. (2008). Challenges in Transformation of Existing Real-Time Embedded Systems to Cyber-Physical Systems. *ACM SIGBED Review*, *5*(1), 1–2. doi:10.1145/1366283.1366294

Chen, H. (2017). Applications of Cyber Physical Systems: A Literature Review. *Journal of Industrial Integration and Management*, *12*(3), 1–28. doi:10.1142/S2424862217500129

Diaz, J., Perez, J., Perez, J., & Garbajosa, J. (2016). Conceptualizing a Framework for Cyber-Physical Systems of Systems Development and Deployment. *Proccedings of the 10th European Conference on Software Architecture Workshops (ECSAW),* (pp. 1-7). ACM. 10.1145/2993412.3004852

Douglas, S., White, J., & Gill, C. (2014). Elastic Infrastructure to Sup- port Computing Clouds for Large-Scale Cyber Physical Systems. *7th IEEE International Symposium on Object/Component/Service-Oriented Real-Time Dis- tributed Computing,* (pp. 56-63). IEEE.

Eric, S., Kyoung-sook, K., Ryong L., & de Vaulx , F. (2013). A Vision of Cyber-Physical Cloud Computing for Smart Networked Systems. *Technical Resport NISTR -7951*. NISTR.

Estevez-Ayres, I., Basanta-Val, P., Valls, M. G., Fisteus, J. A., & Almeida, L. (2009). QoS-Aware Real-Time Composition Algorithms for Service-Based Applications. IEEE Transactions on Industrial Informatics, 5(3), 278-288. IEEE. doi:10.1109/TII.2009.2026422

Jing Wu, F., Fen Kao, Y., & Tseng, Y. (2011). From Wireless Sensor Network to wards Cyber Physical Systems. *Pervasive and Mobile Computing, 7,* 397413 2011 doi:10.1016/j.pmcj.2011.03.003

Ge, Y., Dong, Y., & Zhao, H. (2012). A Cyber-Physical Energy Sys- tem Architecture for Electric Vehicles Charging Application. 12th International Conference on Quality Software, (pp. 246-250). ACM.

Gunes, V., Pete, S., Givargi, T., & Vahid, F. (2014). A Survey on Con- cepts, Applications and Challenges in Cyber Physical Systems. *KSII Transactions on Internet and Information Systems*, *8*(2), 4242–4268.

Haque, S. A., Aziz, S. M., & Rahman, M. (2014). Review of Cyber-Physical System in Healthcare. *International Journal of Distributed Sensor Networks*, *10*(4), 217415. doi:10.1155/2014/217415

Horvath, I., & Gerritsen, B. H. M. (2012). Cyber Physical Systems: Concepts, Technologies and Implementation Principles. *Proceedings of TMCE, Voume*, *1*, 7–11.

Kalasapur, S., Kumar, M., & Shirazi, B. A. (2007). Dynamic Service Composition in Pervasive Computing. *IEEE Transactions on Parallel and Distributed Systems*, *18*(7), 907–918. doi:10.1109/TPDS.2007.1039

Khaitan, S. K., & McCalley, J. D. (2014). Design Techniques and Ap- plications of Cyber Physical Systems: A Survey. *IEEE Systems Journal*, 1–15.

Kim, K.-D., & Kumar, P. R. (2013). An Overview and Some Challenges in Cyber- Physical Systems [A Multidisciplinary Reviews Journal]. *Journal of the Indian Institute of Science*, *93*(3), 342–352.

Kim, M., Stehr, M.-O., Kim, J., & Ha, S. (2010). An Applica- tion Framework for Loosely Coupled Networked Cyber Physical Systems. *Proceedings IEEE/IFIP International Conference on Embedded and Ubiquitous Com- puting (EUC),* (pp. 144-153). IEEE.

Lee, E. (2015). The Past, Present and Future of Cyber-Physical Systems: A Focus on Models. *Sensors (Basel)*, *15*(3), 4837–4869. doi:10.3390150304837 PMID:25730486

Macana, C. A., Quijano, N., & Nava, E. M. (2011). A Survey on Cyber Physical Energy Systems and their Applications on Smart Grids. IEEE PES Conference on Innovative Smart Grid Technologies (ISGT), (pp. 1-7). IEEE. 10.1109/ISGT-LA.2011.6083194

Park, K.-J., Zheng, R., & Liu, X. (2012). Cyber-Physical Systems: Mile- stones and Research Challenges. *Computer Communications*, *36*(1), 1–7. doi:10.1016/j.comcom.2012.09.006

Reijers, N., Wang, Y.-C., Shih, C.-S., Hsu, J. Y., & Lin, K.-J. (2011). Building Intelligent Middleware for Large Scale CPS Systems. *Proceedings of IEEE International Conference on Service-Oriented Computing and Applications (SOCA),* (pp. 1-4). IEEE. 10.1109/SOCA.2011.6166238

Sehgal, V., & Patrick, A. (2014). A Comparative Study of Cyber Physical Cloud, Cloud of Sensors and Internet of Things: Their Ideology, Similarities and Differences. IEEE International Advance Computing Conference (IACC), (pp. 708-716). IEEE. 10.1109/IAdCC.2014.6779411

Shein, W. W., Cheng, Z., Tan, Y., & Lim, A. O. (2013). Study of Temperature Control Using Cyber Physical System Approach in Home Environment. *IEEE 1st International Conference on Cyber-Physical Systems, Networks and Ap- plications (CPSNA), (vol. 3,* pp. 78-83). IEEE. 10.1109/CPSNA.2013.6614250

Shen, Q., Qiao, X. V., & Athanasios, V. (2014). Web Service Composition- A Decades Overview. *Informatics Sciences*, *250*, 218–238. doi:10.1016/j.ins.2014.04.054

Shinyoung, L., Chung, L., Han, O., & Kim, J. H. (2011). An Inter- active Cyber Physical System for People with Disability and Frail Elderly People. *ACM Proceedings of the 5th International Conference on Ubiquitous Information Management and Communication*, (pp. 113). ACM.

Tamhane, S. A., Kumar, M., Passarella, A., & Conti, M. (2012). Service Composition in Opportunistic Networks. IEEE International Conference on Computing and Communications (GreenCom), (pp. 285-292). IEEE.

Wan, J., Chen, M., Xia, F., Di, L., & Zhou, K. (2013). From Machine-to-Machine Communications towards Cyber Physical Systems. *Computer Science and Information Systems*, *10*(3), 1105–1128. doi:10.2298/CSIS120326018W

Wang, E. K., Ye, Y., Xu, X., Yiu, H., & Chow, K. P. (2010). Security Issues and Challenges for CPS. IEEE/ACM International Conference on Green Computing and Communications and on Cyber, Physical and Social Computing, (pp. 733-738). IEEE.

Wu, T., Dou, W., Hu, C., & Chen, J. (2017). Service Mining for Trusted Service Composition in Cross-Cloud Environment. *IEEE Systems Journal*, *11*(1), 283–294. doi:10.1109/JSYST.2014.2361841

Yu, C., Jing, S., & Li, X. (2012). An Architecture of Cyber Physical Sys- tem Based on Service. *Proceedings International Conference on Computer Science and Service System (CSSS),* (pp. 1409-1412). IEEE.

Zhang, L., He, J., & Yu, W. (2013). Challenges, Promising Solutions and Open Problems of Cyber-Physical Systems. *International Journal of Hybrid Information Technology*, *6*(2), 65–73.

Chapter 2
Cyber-Physical Systems in Agriculture:
Applications, Challenges, and Future Perspectives

Sasikala Chinnasamy
KPR Institute of Engineering and Technology, Coimbatore, India

L. Dharani
KPR Institute of Engineering and Technology, Coimbatore, India

S. Thanuja
KPR Institute of Engineering and Technology, Coimbatore, India

P. S. Kavishree
KPR Institute of Engineering and Technology, Coimbatore, India

ABSTRACT

Advanced mechatronic systems have undergone a significant transition to Cyber-Physical Systems (CPS) over the past 20 years. CPS are designed systems that rely on the seamless fusion of computational algorithms and physical elements to function. Agriculture is a basic need and a crucial part of our life. In order to feed the world, CPS will definitely play a significant role in the agricultural sector. CPS is a combination of multiple embedded systems assigned with significant works. CPS is a heterogeneous components that includes electronic chips, software systems, sensors and actuators. The capabilities, adaptability, scalability, resilience, safety, security, and usability of CPS will greatly outpace those of today's uncomplicated embedded systems. The architecture of the agricultural cyber-physical system is covered in this chapter. A detailed discussion of the uses of CPS in many agricultural domains, including plant data collecting, soil water monitoring, field monitoring, field information gathering, and other applications is presented in this chapter.

DOI: 10.4018/978-1-6684-7879-0.ch002

INTRODUCTION

Helen Gill of the US National Science Foundation first used the phrase "Cyber-Physical System" (CPS) in 2006. (Henshaw, 2016). CPS, as its name implies, combines physical (mechanism) and cyber (software control) components. A network of physical and computational components is interconnected by a Cyber Physical System (CPS), which is built for seamless interaction. These technologies will serve as the cornerstone of our vital infrastructure and enhance many aspects of our quality of life. In a CPS, computing components work together and interact with sensors, which keep an eye on actuators and indicators. Through the use of robotics, embedded systems, sensors, and actuators, they connect the physical world to the digital realm of processing information.

Embedded systems, the Internet of Things (IoT), Artificial Intelligence (AI), Machine Learning (ML), big data, and numerous cloud services are instances of the future generation of computational components that are coupled with physical systems. Other examples include machine controls for farming practices, real-time data collection, sensing devices, and Fifth-Generation (5G) communications networks. The world is undergoing a significant digital transition on a worldwide scale, which has greatly improved our daily ease, convenience, and security. (Teodora Sanislav, 2012). As the cyber-physical systems view things from physical and virtual perspective, they can autonomously evaluate operational conditions and support subsequent decision making and operational adaptations, these will be done under human supervision. Fig 1 shows the cycle that describe the communication between physical system and cyber system and vice versa.

Figure 1. Relationship between physical layer and cyber layer

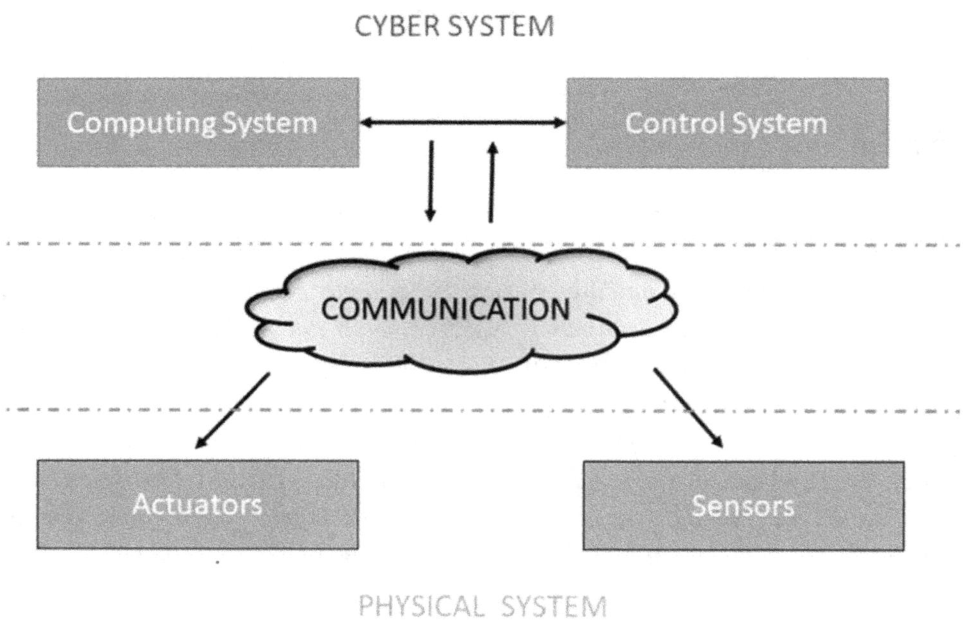

The demand for agricultural production is increasing due to a growing global population and a long average life expectancy in order to achieve the objective of eradicating hunger. The adoption of enhanced, optimised, and sustainable agricultural practises that expand agricultural production, real-time disease detection and pest management, etc., are so necessary. As a result, using contemporary technologies, farmers have streamlined agricultural operations together in sustainable manner.

Due to its enormous potential for helping the farmers, this integration has increased the quantity of agricultural goods. The drive for technical improvement has brought back old agricultural practises and produced farming that is efficient, sustainable, and friendly to the environment. For the enhancement of agriculture, can also employ CPS.(An.,AT D. Wu, S. Ci, H. Luo, V. Adamchuk, Z. Xu,2017] In order to create more precise systems of agricultural management, agriculture cyber-physical systems (ACPSs), the CPSs built and deployed in agriculture, may gather essential and timely information on the weather, the land, and the vegetation at great precision.

Crop losses or lower crop yields would result from a lack of such management. As a result, a problem of growing concern is the ability to detect and identify the illnesses on afflicted crops. There are currently no systems in place for real-time disease detection. As a result, a topic of growing concern is the ability to recognize and diagnose the infections in afflicted crops. It needs a lot of laborious effort, knowledge of plant diseases, and lengthy processing times.

There is a significant reduction in manual labour when precision agriculture techniques are used. In order to support already-existing small-scale ecosystems (Tauber et al.,) precision farming, an emerging process, uses cyber-physical systems to enhance product quality and safety, field tracking, crop scouting, in addition to declining agricultural expenditures for the steadily rising food demand resulting from population growth.

CYBER PHYSICAL SYSTEM IN AGRICULTURE

(Selmani et al.,2018)The CPS is a very promising technology that might be integrated with certain other techniques to offer a variety of useful farming applications. In order to create integrated farm management systems that can communicate with the physical environment and preserve ideal quality of the environment, CPS applications are built on cutting-edge electronic technology and agricultural infrastructure. They must be able to accurately and comprehensively collect relevant and necessary information on the soil, the crops, and the climate. To control external resources like irrigation, temperature, crop growth, etc., the data collected are employed. So in order to accurately evaluate farming needs the systems are highly dependent on timely analysis of the collected soil and environmental data.

Structure of ACPS

Agriculture cyber-physical system can be broken down into three tiers of services, each of which includes a set of smaller services.

1. Sensing
2. Communication
3. Analysis and Action

An innovative energy and environmental framework for greenhouse ecosystems that might be applied to improve energy effectiveness and ecological sustainability. Additionally, the optimization of sensor deployment is a common feature of ACPS architecture. The architecture is illustrated in the Figure 2.

Figure 2. Architecture of cyber physical system

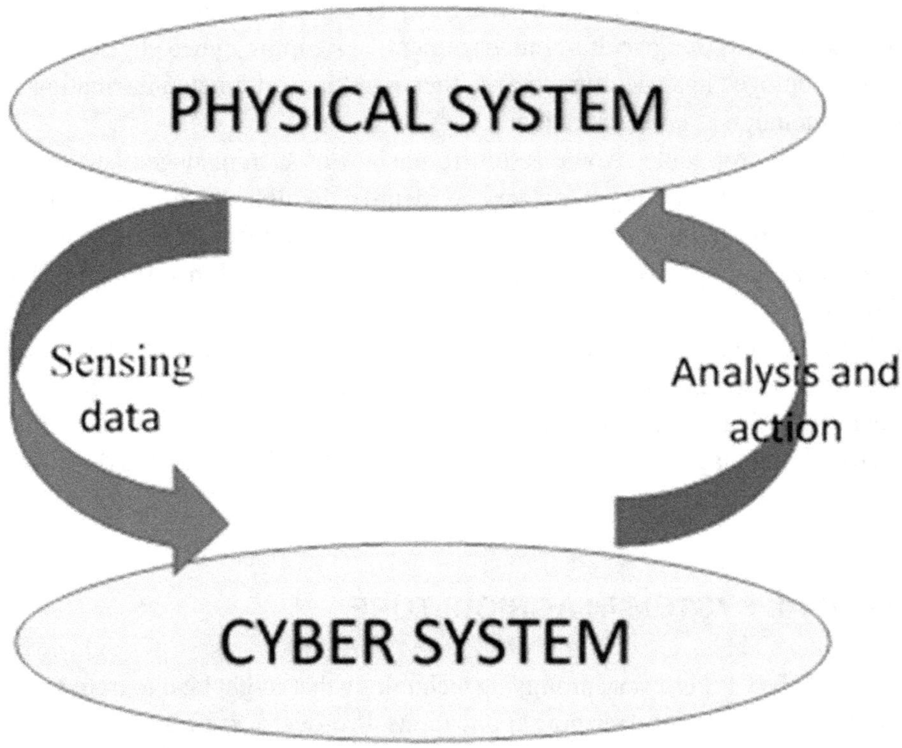

1. Sensing

The author (Tushar et al., 2023) first developing a digital realm of a physical system for monitoring, analysis, and decision-making is to gather precise and trustworthy data from various physical system components. For the process of gathering information, sensors are used. At-ground level sensors will be used to gather data. Additionally, it continuously monitors fields with nearly single-plant resolution, forecasts productivity, and assists farmers in controlling their usage of fertiliser and water. Uses (R Fresco et.al, 2018) include whole-field monitoring with sensors mounted on unmanned aerial vehicles, wearable plant sensors, soil water potential sensors, plant and soil nutrient sensors, low-cost/high-performance field sensors, process control, and analytic engines.

IoT is now a common data collecting technology for precise measurement and collection. The internet of things concept is formed from networked smart devices, as well as an integrated IoT enables accurate real-time monitoring and sensing of a CPS's numerous environmental and mechanical characteristics. This can be executed as the appropriate response by identifying plant performance and identifying the cause at the same time. Agronomists can benefit from sensor-based engineering solutions. These sensors for the soil and plants were created by humans.

Modern technology combines many sensors, models, and controls to understand and forecast plant-soil dynamics with great and unparalleled detail. Useful data is being produced for decisions on the management, planning, and administration of water and fertiliser at different rates along the pivot point of a water supply.

2. Communication

For precision agriculture, wireless connectivity is a crucial technology (Tusharet al., 2023). Employing the right communication protocols, such 5G, to send the data to the server. For the CPS's smart communication level, two crucial variables must be taken into account. First, a smooth and tether-free technique for data collecting must be created, taking into account the various forms of data that result from various CPS components. The verification of safe communication protocols is the second step in transferring data from IoT devices to the server. Additionally, it is crucial to choose the right sensors (specifications) to gather data from various pieces of equipment and components. Although it is a high-tech system, it also provides farmers with a practical means of increasing yields and enhancing sustainability.

In agriculture, there are numerous activities involved in the harvesting of the crops that include analysing the soil, adding manure and fertilizers, irrigation and storage. So for these processes CPS system is used to connect the crops and the farmers. So the communication between the farmers and the crops may include the following steps. The first step involved in the communication is that farming facilities including fertilization, irrigation and insecticide spraying, pesticides supplying. All these techniques are connected with the control system. These control system with all these information is connected with control centre. The control centre may include user, console (a computer terminal where a user may input the commands and view the output for that inputted commands), server and a database. These database is capable of storing numerous data, so the server can easily analyse the data and give the solutions for the problem. These are all connected with the internet that is cloud. From the cloud the analysed solution can be sent as the output using the connecting devices like Wi-Fi.

3. Analysis and Activation

In order to create the digital twin of the physical structure, data is being pushed from every linked component over the network. A digital twin, as defined by IBM, is a representation of a system that is constructed and updated using real-time data and that employs simulation, machine learning, and reasoning to support decision-making. This level of the CPS manages, analyses, and benchmarks all information, which helps to optimise the decision.

In reference[Wayes Tushar, Chau Yuen, 2023] making informed decisions based on thorough understanding of the system being monitored and the information being analysed in order to take the right actions at the physical level. In accordance with the choices made, the system generated control signals to carry out the various actions in the CPS's physical structure.

The system's configuration module, which manages the CPS as a whole, will oversee the corrective and preventive actions made. Therefore, the monitoring and recording systems will acquire the data, assess it at the cyber level, and plan the necessary measures if there is any change in the physical level situations. The control operations will be executed in the physical space with quality and efficacy adjusted by the configuration system.

APPLICATION OF CYBER PHYSICAL SYSTEM

The proposed farming activities of smart agriculture, such as gathering plant information, using unmanned aerial vehicles to monitor fields, deploying sensors to collect field information, monitoring soil water, planning fertilisation by soil mineral monitoring, etc., can be carried out because the Agriculture CPS provides the essential technology.

Plant Information Gathering

It is being known that, a cyber-physical system (CPS) combines cyber and physical systems. The physical system learns about the plant and its surroundings. In agriculture,(Rad et al.,2018) keeping an eye on the plants is more essential to keep them healthy and safe out of any harmful diseases. It begins by sensing the agricultural field, which is a physical system, and communicating it to the cyber system. The cyber system then analyses it and sends some instructions to the physical system, which then takes the appropriate necessary actions.

In reference (S. Maksuti et al.,2021)Sensors are employed in the field to gather data on temperature, humidity, wind, and soil nutrition. This data is then sent via wi-fi and the internet to the cloud, where it is used for weather forecasting by web crawlers and post-processing. Finally, algorithms such as AI, ML, and proposal expert systems are used to generate suggestions based on various databases, and farmers can engage in various interactions. A conclusion was drawn from this. As a result, the farmer's plants may be better protected.

CPS is utilised right from the beginning of the plantation process, when it advises the farmer to grow which kinds of plants by analysing database data and advising the farmer to grow plants that are suitable for the climate in question in order to maximise profits. to check for pest/disease infection, water shortage, nutrient insufficiency, and profit/loss estimation. Cameras deployed in fields, cameras mounted on UAVs (drones), and satellite imagery were used to monitor the physical environment.

Figure 3. Working model of cyber physical system

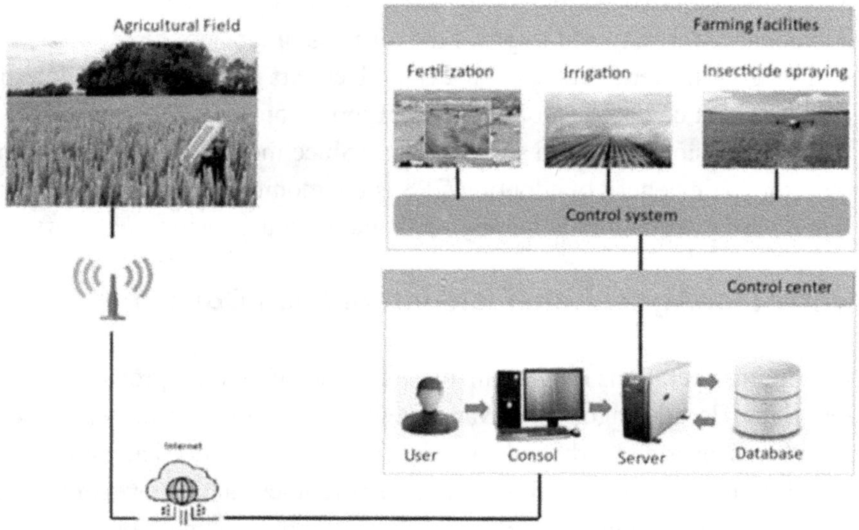

The methods include picture segmentation, image analysis, and growth modelling using time series data. Variety of photographs are being captured, including those of the root, stem, leaves, fruit, and flower, to see if the plant is being attacked by pests.As shown in Figure 3, Near-infrared images are very helpful in this regard. It also helps to determine the stage of the plant and to offer the right fertiliser to enhance plant growth. For this, a variety of machine learning (ML) algorithms, including SVM, KNN, ELM, and deep learning techniques, including Resnet, Googlenet, VGG, Inceptionnet, and Unet, are utilised.

Unmanned Aerial Vehicle in Agriculture

In traditional agriculture, the farmer must keep an eye on the entire field to determine the plants' current stage, whether they are afflicted by disease, and whether they require fertiliser for growth. In this case, the farmer's ears and vision serve as cameras to monitor the farm. However, CPS is currently utilised to assist farmers and lighten their workload. To do this, the farm must be monitored by using cameras and sensors. A drone is a type of unmanned aerial vehicle (UAV), which is an aircraft that lacks a human pilot, operator, or travellers.

This is the primary tool used to detect the farm. Additionally, satellite imaging, cameras mounted on UAVs (drones), and cameras positioned in fields. UAVs are a part of an unmanned aircraft system (UAS), that also includes the addition of a controller on the ground and a communications network. On looking at the conventional side, a farmer with a vast farm would require and over two servants to maintain a close eye on the field because they would not be able to check every crop each day. Therefore, if the farmer is not aware of the disease, it may impact the plant. Therefore, the CPS is required in this case to prevent such. Now that there is the ability to monitor every plant and to guard against any hazardous diseases affecting the plant.

Instead of routinely checking on the plants, this can free up the farmer to perform other tasks and lower the cost of the slaves' wages. Farmers may profit more as a result and be able to use the money for other beneficial projects. For farming, it contains the following details, such as environmental sensing, water management, threat attacks on plants, their stage of growth, and nutrition. Given that agriculture is a field where assistance is required and that by having advanced technology on our side, this technology might be used to advance farming and improve the fortunes of farmers.

The foundation of CPS is the aerial vehicles, which collect data in the form of numerous photos of plant components or films that can be seen on the control room computer. The farmers are assisted by cyber systems in making decisions based on server advice. Since there is no need for human labour in agricultural activities, the main benefit of adopting CPS is the monitoring system. Burning stubble can be managed by using a machine for stubble management and remote sensing to identify it.

Sensor Deployment for Agriculture Field Information Collection

The productivity of farming operations must be increased as the population grows. This idea was practically inconceivable a few decades ago, but advancements in wireless sensor networks have made it possible. Using a wireless sensor network, real-time data can be gathered. That could be helpful while making different decisions for farming preparation. In order to achieve a high level of productivity while using minimal resources and to prevent resource waste. Land, groundwater, crops, fans, fertiliser, drainage problems, and environmental concerns peculiar to a location are examples of agricultural factors.

To perceive the agricultural characteristics, the author (Y. Zhang, et al.,2018) employ a variety of sensors.Additionally, it gathers and transmits data through a wireless receiver that collects it and creates a module that is helpful for development of backward areas. Through the use of Wi-Fi, the server receives the data that the sensors have collected, stores it so that it can be processed, and then sends it back to the client. It is used to detect variations in the weather, including temperature and humidity, as well as any microorganisms that may have an impact on plant health and changes in plant development, etc. This has a very low operating cost and wirelessly controls all the settings from a central control room. Plants can be impacted by hazardous insects that farmers may not always see. Utilizing the sensors will eliminate this difficulty. Table 1 shows list of sensors and its usages.

Table 1. Sensors and its usages

S.No	Sensors Name	Description
1	Bio sensors	These sensors detect and measure biological molecules, such as DNA or proteins
2	Soil sensors	Measures the temperature, humidity of the soil
3	Accelerometers	Measures acceleration and tilt
4	Optical sensors	Detects light intensity or colour
5	LIDAR(Light Detection And Ranging	This sensor uses laser pulses to measure distances and create 3D maps of the environment
6	Gas sensor	These sensors detect the presence and concentration of gases in the environment

The sensors can identify pests and report this information to the server-side control room. so that the farm can be protected against such types of insect attack. When the farmer is not physically available in the field, animals and birds may also destroy plants. However, sensors in the field detect animal or bird invasions in the farm land and notify the farmer of them in a digital manner or another alarming that in-turn helps to safeguard the farmland. Watering the plants is a crucial aspect of farming. It is possible to measure the water distribution level so that the farm will receive the proper amount of water. This promotes the growth of healthy plants. The sensors also determine how much fertiliser the plant actually needs, allowing the right amount of pesticide or other fertiliser to be applied at the appropriate way.

Soil Water Monitoring

Focusing on the integration of agricultural facilities and sensing technologies, agricultural CPS's enable the implementation of reliable systems to monitor environmental factors that are important for the decision-making processes connected to intelligent and environmentally friendly water management systems. When a crop is watered it remains green, however, the amount of irrigation has a limit.

The farmer is in charge of all aspects of traditional agriculture, from planting through harvesting. Irrigation is crucial to this process, and watering intensity should also be taken into account. The plant won't develop in a healthy way if the water supply is low, and if the water level is excessive, the plant's roots may be affected, which might harm the entire field. A farmer cannot always deliver water accurately. CPS is the sophisticated system that is required in this situation. It has a number of drones and sensors that can regulate the water supply level. (G. Pandey et al.,) Sensors in the field that detect and provide results like when it requires water supply and the volume of water needed for the plant in order to prevent the roots from being harmed by the high water supply.

A plant typically requires as much water as is necessary to keep the soil wet for 3 to 4 inches around the plant stem. But depending on the type of plant, this can be different. Sensors may be able to detect this, in which case the field can be watered appropriately. The technique for watering also depends on the kind of soil. Because each soil has a particular set of characteristics, irrigation may be necessary. As a result, CPS is a well-developed system that can assist in all aspects of farming and enhances plant health by accurately sensing the water level required by the plant and the soil for growth.

Fertilization and Pesticides Scheduling

Agriculture is often acknowledged to be the most beneficial, practical, and honourable occupation for people. Agricultural terminology involves cultivation when it is just spoken. However, there are several minute steps in that cultivation, from planting to harvesting. In this cultivation method, a plant's health suggests a person's health.

The plant will provide all of its proteins to people if it is healthier and is not afflicted by any diseases. However, it may be stopped from happening by utilising certain nutrients if the plant is already afflicted by any microorganisms or diseases. Pests could also have an impact on the plants. In order to avoid this, insecticides that kill dangerous pests and shield plants from illness might be used. Plants need to be given a certain amount of insecticides or fertiliser.

The plant growth may be hampered if fertilisers or insecticides are applied in excess of or insufficiently compared to that amount. If more amount of pesticides are used, the pests will be killed quicker, but because there is more pesticide, it may linger within the plant. When the plant is being eaten, the pesticide may then enter our bodies and harm our health.

The formula and the primary nutrient need for nitrogen determine the rate of fertiliser application. In terms of nutrients, nitrogen is crucial. It is recommended to apply nitrogen fertiliser at a rate of 0.1 to

0.2 per 100 square feet. Furthermore, fertiliser is applied according to a timetable during specific plant stages. The time and quantity of fertiliser are already known by experience if a farmer has previous expertise with any conventional crop. However, they won't be aware of these information if they develop any new crops. CPS may assist the farmers in this region by letting them know the quantities of fertiliser. The CPS specifies which fertiliser to be used and at what stage of the plant, as well as how much .As known that CPS uses cameras, sensors, drones, and other devices, the protein content of the plant and its soil can be determined using this. Regarding this, the fertiliser amount is subtracted and given to the plants. This helps farmers avoid wasting money by purchasing big amounts of fertiliser. To determine the proper dosage of fertiliser and pesticides and to schedule their use for optimal administration to the plants, this approach is more beneficial in agriculture.

CHALLENGES IN AGRICULTURAL CYBER PHYSICAL SYSTEM

There are a number of difficulties that must be overcome before deploying any CPS application in order to guarantee ongoing functioning. In India, agriculture is important and is the nation's fastest-growing economic sector. It has a wide range of effects on society, including providing food, housing, and employment opportunities. 18% of the Gross Domestic Product of India is produced by the agricultural sector (GDP). Over 70% of India's people depended on agriculture at the time of independence.

In this sense, variables affecting the global economy like shifting commodity prices and trade disputes hamper farmer's business decisions. Field tests are carried out to assess how well the suggested sensor deployment approach performs in terms of crop yield. The applications of certain current agricultural cyber-physical systems, such as soil moisture monitoring for irrigation scheduling, soil mineral content monitoring for fertiliser scheduling, weather monitoring for frost avoidance, and harvest management, are examined. Farmers benefit from crop growth monitoring for disease protection since it is very challenging for farmers to effectively follow.

In digital aquaculture, the power supply requirement of sensors and edge systems is a persistently serious problem, particularly for offshore cages. Farmers may monitor soil temperatures using the smart agriculture system and irrigate the crops appropriately. Cyber-physical systems are used to monitor the soil environment, determine the appropriate water and fertiliser applications, and time the activation of watering systems. In addition, it is essential to take weather forecasts into account while caring for the soil in order to avoid excessive rains that might wash away the benefits of pesticides and fertiliser. The power supply demand of sensors and edge systems is always a severe issue in digital aquaculture, especially for offshore cages. With the intelligent agriculture system, farmers could detect soil temperatures and water the crops accordingly. Cyber-physical systems are employed to detect soil environment, adjust the amount of water and fertilizer accordingly, and judge the timing of watering system activation. Moreover, to take care of the soil, apart from the factors of watering timing, fertilizer, and farming, it is vital to consider weather forecasts to avoid heavy rain and washing down the effect of pesticide and fertilizer.

The accuracy and level of detail in data collecting are crucial for agricultural cyber-physical systems. Although there are many CPS ideas and definitions, they continue to go through a rigorous development process. The development of smart networks is complicated by the constantly changing characteristics of both the physical and digital components. Sensors, data storage systems, and linked wired and wireless systems are now more widely available and more reasonably priced because to the physical hardware systems' fast growth over the past 20 years. These advancements led to continued generation as a result.

The monitoring and actuator infrastructure, the communication network, the wiring, the power and energy, mobility, the bit rate, signal fading, and the computations and control operations, including processing speed, understanding data, precision, collection and regulation, as well as the deployment, including manageability, accessibility, stability, flexibility, and self-sufficiency, all present challenges.

Developing new structures and algorithms to cope with intricate cyber-physical systems is a difficult task. Three major issues, including synchrony, frequency, and granularity, might arise in a CPS with diverse systems. Current wireless networks use authentication and data encryption mechanisms on the application level, just like smart offices in major corporations use IP network level security solutions to protect private networks and data. Five different threats—criminal, financial, political, privacy, and physical—affect five different aspects of the system: the origin, the destination, the purpose, the vector, and the consequence. To consider both the cyber and physical components of security, CPS designers must alter the way they approach security. Making decisions in real time is necessary, and if they are not made immediately, it will negatively impact the network's overall state of defence. In order to survive an assault, networks need to act quickly in CPS.

Even in tightly regulated industrial contexts, the Internet of Things (IoT) devices and sensors that are part of the control system must be updated, removed, or replaced if they get damaged or are accidentally harmed. Farms, open spaces, and climatic disturbances like wind, rain, and heat can cause physical systems to break down or get damaged. When replacing something at the appropriate time or coming up with a backup plan to handle situations like these, it is important to conduct a more thorough practical analysis of the causes and find solutions to them in order to create a system that is fool-proof and works all year long in all kinds of weather.

Despite CPS's many benefits, putting it into practise presents significant difficulties. There are two layers in it. They are physical and cyber levels, respectively. It is primarily felt that surroundings is at the physical layer. Variety of tools are required to monitor the environment for this aim. It includes a sensor to detect plant growth, such as sprouting, an irrigation system for watering the crops at the appropriate time by examining the soil moisture content, and fertilising the plants to encourage plant growth and development so that crop plants can grow quickly and produce a high yield.

It keeps an eye on the plants and checks to see if any insects or pests are attacking. The system notifies the server of the situation if an attack is discovered. Therefore, every corner and crevice of the fields must have sensors, cameras, and drones for these objectives. They are gathered and put on a server for storage. The essential requirement is a wi-fi connection in order to transmit this collected data to the server. There would be a problem with the transmission and data loss if the wi-fi connection were to be interrupted in any way. There may be fundamental problems, such as coordination, in working as an integrated system for improving operations.

To address issues like the overuse of fertilisers and herbicides, as well as the mismatch between pesticides that might harm plants and creatures like earthworms that are essential to plant nutrition, real-time evaluation is necessary. For each component to endure environmental factors and other factors, resilience is required. There is a requirement for resilience design in this CPS due to the occurrence of the growing potential risks and assaults. Applications for data collecting, servers, and users all have complicated physical infrastructure requirements. Figure 4 shows the important issues of cyber physical system.

Figure 4. Issues in cyber physical system

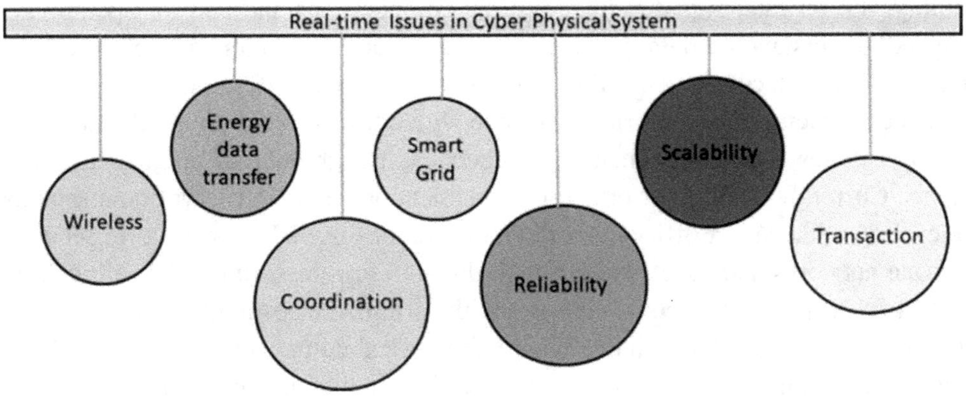

Coordination

When power electronic systems are integrated with cyber systems an efficient system can be created. As a result, the systems' energy management and control might become intelligent and trustworthy. To improve operations, these systems ought to operate as a single, integrated system as shown in Figure 5 major structural problems exist with CPS because of this.

Figure 5. Structural problems of cyber physical system

Energy Data Co-Transfer

The exchange of energy-related data between the various system components is essential in the field of agriculture. Because it necessitates the installation of numerous sensors, communications infrastructure, and processing devices, implementing an energy data co-transfer system in agriculture can be costly. It must be protected against unauthorized because this concerns sensitive information. Multiple components from various manufacturers may not work together, which will have a detrimental effect on performance.

Real Time Evaluation

(G. Pandey et al., 2021) challenging to handle and interpret in real time due to the volume of data that it contains. Evaluation may take longer as a result of this. In assessment, data quality is incredibly important. Complexity makes it difficult to conduct real-time evaluations since it necessitates a thorough comprehension of the system's interconnections and behaviour. Real-time processing requires a lot of computational power and resources, which makes it challenging to support in environments with little resources. Real-time review is needed to find solutions to issues like using excessive amounts of fertilisers and herbicides, as well as mismatched pesticides that might harm plants and beneficial species like earthworms that feed plants.

Reliability

The integration of several technologies, including wireless communication networks, sensors, actuators, and control systems, which increases the system's complexity and failure risk. Communication issues brought on by network and interference issues may affect it. Software and algorithms are essential. Any malfunctions might affect the system's overall performance and dependability. As it includes human contact and decision-making, the technologies used must be dependable. Continuous monitoring and maintenance are also required to prevent system failure.

Resilience Challenges

It involves combining a number of technologies, such as wireless communication networks, sensors, actuators, and control systems, which makes the system more complicated and prone to failure. It may be affected by network-related communication problems and interference problems. It is important to have software and algorithms. The reliability and general operation of the system may be impacted by any problems. It requires human interaction and decision-making, thus the technology employed must be reliable. System failure must also be prevented by ongoing maintenance and monitoring.

Scalability

To analyse data and make decisions, computing resources are needed, and these resources may have an influence on scalability. Wireless communication networks are employed in this case, and the scalability is impacted by network resources like bandwidth and capacity. Scalability is also impacted by the integration of new technology.

Transaction

The successful transaction is impacted by the vast volume of data exchanged and the assurance of data accuracy. For transactions, it is essential to guarantee the security and privacy of sensitive information including financial and personal data. The interoperability and integrity of various systems both have an impact on it. Transactions depend on effective coordination and communication between human and technical elements.

Additionally, it has concerns with applications for smart or micro grids, energy at the network edge, wireless EV chargers, IOT, drones, solid-state lighting, electric trains, renewable energy systems, smart transformers, and electric ships. During the implementation of cyber physical system, one should consider the list of points as follows.

1. Flexible CPS in all season with different fields.
2. This system should respond quickly for dynamic changes in agriculture.
3. Be resilient to emergent hazards.
4. Categorized for expected/unforeseen situations that a CPS could encounter through historical data
5. Visual indications are used to gather the CPS's current state (such as human and object identification).
6. Improves the decision-making of a CPS facing a complex scenario, ensures physical safety, and effectively enhances the human-machine team's productivity.
7. The confidentiality that is necessary to maintain the data securely.
8. Try to attempt to change the state of the physical system by communication channels between the sensors and the controller, and between the controller and the actuator.
9. Integrity, when data or resources can be changed without permission.
10. Availability, when there are failures in network and software.

FUTURE PROSPECTS OF CPS IN AGRICULTURE

Smart agriculture calls for CPS and the IoT to boost present production and provide significant assistance in enhancing product quality. Technology advancements are causing a transition to Agriculture 5.0, where conventional farming techniques are altered and enhanced through automation and the use of cutting-edge technological solutions. There are several development methods for enhancing life in various disciplines. By adopting a cyber-physical system, agricultural fields may be grown in this fashion. Despite the expense of deploying cameras and other equipment, there are brighter prospects for the future. Artificial Intelligence, Machine Learning, Big Data, Internet of Things, Wireless connection, and other related technologies are the main technologies for it.

To help students learn and explore the world for the better, Cyber Physical System has been made a topic in schools today. Such human-robot collaboration is made possible by CPS in the fields of dynamic task planning, proactive collision avoidance, and adaptable robots. Cyber-physical systems (CPSs) intimately connect the cyber elements of computers and communications with the physics and dynamics of physical systems working in the environment around us.

It competes with IOT in areas like smart grid optimization. New mathematical bases and formalisms will be defined by the new CPS science in order to design, analyse, verify, and validate systems that monitor and control physical objects and entities. CPS can help farmers to optimize their use of resources including water, pesticides and fertilizers by providing real time data on soil moisture, nutrients and crop growth.. It can help farmers to adapt the effects of climate change such as drought, flood. The data generated from CPS is very large in its volume with high frequency and multiple varieties. This data called, Big Data, has distinct 3V characteristics; Volume, Velocity, and Variety. Volume refers to the large amount of data. Velocity corresponds to the data rate or frequency at which data is generated and collected in real time. Variety of the Big Data is the different formats in which the data is generated. Big data techniques can provide large scale data analytics for data monitoring and visualization in near real time. Cloud computing offers storage of large volume of data and processing it with complex computational algorithms in cloud with lower latency. The benefits of cyber-physical systems are not limited to the fields of engineering and technology, rather, the science of CPS has the capability to impact technology in a wide variety of industries and organizations.

The proliferation of lower cost, smaller, more efficient, and more powerful computing technologies coupled with data transfer, storage, and management technologies, and supported by emerging techniques to make sense of voluminous and federated datasets, including AI, represents profound new capabilities and efficiencies for physical systems .It competes with IOT such as in optimization of smart Grids. New CPS science will define new mathematical foundations with formalisms to specify, analyse, verify and validate systems that monitor and control physical objects and entities. Digital representations of the physical world will be enabled by billions of embedded sensors and actuators that provide the necessary data and make it possible to control physical actions. All are moving to a future where it will also be possible to create digital representations Cyber-physical sensing systems and green communications put a new emphasis on the task of energy management for wireless communications and favour the use of energy-harvesting technologies. Reliable models and understanding of any uncertainties will be of increasing importance as more machines rely on digital representations and automation for their actions. Without any questions, CPS can be considered as an important step in the development of manufacturing systems. The proposed system allows farmers to follow the evolution of certain parameters of interest and take appropriate decisions in order to increase agricultural productivity Whether this step would be regarded as the fourth industrial revolution will be decided by the coming generations, but certainly, this will happen with no zero probability CPS may without a doubt be viewed as a crucial stage in the creation of manufacturing systems. The future generations will choose if this development qualifies as the fourth industrial revolution, but there is little doubt that it will occur.

Advantages

CPS are the growing technique that is the future of agriculture. The farmers are more efficient and productive while using fewer resources. Like conventional farming, smart farming can help farmers increase their yields, reduce their costs, and improve their environmental stewardship. So the following are the advantages to be noted:

Increasing the Yields

Traditional farming practices often yield lower crop yields due to pests and poor soil fertility. These new technologies helps farmers overcome these challenges using precision agriculture techniques. Precision agriculture is a farming management system that uses modern technologies to optimize activities such as planting, irrigation, and crop scouting. It is among the most revolutionary benefits.

Reducing the Costs

Farming is a capital-intensive industry, and farmers often incur high costs for inputs such as seeds, fertilizer, and fuel. The systems like CPS can help farmers reduce costs by improving input management and decision-making. Here, sensors are used that provides farmers real-time information about their crops, soils, weather, and market prices. This information can help farmers decide when to plant, how to irrigate, and what inputs to use.

Improved Environmental Stewardship

Farming is often associated with environmental degradation, but CPS farming promotes more sustainable farming. Some technology are used that prevents farm machinery from straying into protected areas. It can help farmers avoid accidental deforestation and soil erosion.

Data Collection and Analysis

Farmers can easily collect data about their crops, soils, and weather conditions using Internet of Things devices. This data can be used to improve farm management decisions. Data collected by CPS technologies can be used to improve crop breeding programs. By understanding the conditions under which a particular variety of crops performs well, breeders can develop new types that are better adapted to specific situations.

CONCLUSION

This chapter offered an integrated system architecture for precision agriculture management that uses CPS architecture and design technologies to monitor crop, soil, and water status. To maximise food production quality, quantity, and schedule while maintaining a balance with operational costs and resource usages, smart cyber-physical systems are used for controlled-environment agriculture. Farmers that use CPS may see their stress and workloads decrease. Here, CPS is utilised in agriculture, including the correct seeds are sown at the proper time, the plants are watered at the proper level and at the proper time, the plants are fertilised at the proper rate, and the soil protein and moisture levels are maintained. Numerous physical technologies, including cameras, drones, sensors, and others, may do this. The plants, in particular their stems, leaves, and roots, as well as their vegetables, fruits, seeds, and flowers, are photographed with great detail.

Numerous physical technologies, including cameras, drones, sensors, and others, may do this. The plants, in particular their stems, leaves, and roots, as well as their vegetables, fruits, seeds, and flowers, are photographed with great detail. The sensors are essential to CPS farming since they monitor the water, plants, soil, and everything else. Since CPS incorporates devices including cameras, sensors, and other systems. The implementation of such a system is a difficult undertaking that needs expertise in crop, administration, processing, and display of information technology, among other things, in order to be practical in daily use.

Real-time data and information on the soil/crop health enable farmers to take early corrective actions and provide opportunities to enhance productivity and reduce potential losses. Identifying crop stress in early stages also reduces the use of resources, e.g. it eliminates the need for massive widespread applications of chemicals to treat pests and diseases, which, consequently, reduces environmental impacts.

REFERENCES

Selmani., A., Oubehar., H., Outanoute, M., & Ed-Dahhak, A. (2018). Agricultural cyber-physical system enabled for remote management of solar-powered precision irrigation. *Biosystems Engineering.*

Ahmed, G. & Kim, D. (2013),*Cyber Physical System: Architecture, applications, and research challenges*, 11-15. IFIP Wireless Days.

Alguliyev, R., Imamverdiyev, Y., & Sukhostat, L. (2018). *Cyber-physical systems and their security issues -. Computers in Industry, 100*, 212–223. doi:10.1016/j.compind.2018.04.017

An, W., Wu, D., Ci, S., Luo, H., Adamchuk, V., & Z. Xu, Z. (2017). Agriculture Cyber-Physical Systems. *Cyber-Physical Systems Foundations, Principles and Applications Intelligent Data-Centric Systems*, 399-417. Elsevier.

Babu, J. S., Mohan, G. K., & Praveena, N. A. (2022). Flow of Cyber-Physical systems—A Comprehensive Survey. Smart Technologies in Data Science and Communication. *Proceedings of SMART-DSC.* Taylor and Francis.

Dineva, K., Parvanov, D., Atanasova, T., Mateeva, G., Petrov, P., & Kostadinov, G. (2021). Towards CPS/IoT System for Livestock Smart Farm Monitoring. *International Conference Automatics and Informatics.* IEEE. 10.1109/ICAI52893.2021.9639460

Fresco, R., & Ferrari, G. (2018). *Enhancing precision agriculture by internet of things and cyber physical systems*, 53-60. Atti Soc. Tosc. Sci. Nat. Mem. Supplemento.

Maksutil, S. (2021). Establishing a Chain of Trust in a Sporadically Connected Cyber-Physical System, *IFIP/IEEE International Symposium on Integrated Network Management*, (pp. 890-895). IEEE.

Mikhail,, V., Chester, B., & Allenby, R. (2020). Perspective: The Cyber Frontier and Infrastructure. *IEEE Access : Practical Innovations, Open Solutions.*

Osama Bin, S. (2023). A connective framework to minimize the anxiety of collaborative Cyber-Physical System. *International Journal of Computer Integrated Manufacturing* 1-19.

Pandey, G., Weber, R. J., & Kumar, R. (2018). Agricultural Cyber-Physical System: In-Situ Soil Moisture and Salinity Estimation by Dielectric Mixing. *IEEE Access : Practical Innovations, Open Solutions, 6*, 43179–43191. doi:10.1109/ACCESS.2018.2862634

Rad, C., Hancu, O., Takacs, I. A., & Teanu, G. (2015). Smart Monitoring of Potato Crop: A Cyber-Physical System Architecture Model in the Field of Precision Agriculture. *Agriculture and Agricultural Science Procedi.* Research Gate.

Rubala, J. I., & Anitha, D. (2017). Agriculture Field Monitoring using Wireless Sensor Networks to Improving Crop Production. *International Journal of Engineering Science.*

Sadiku, M. N. O., Wang, Y., Cui, S., & Musa, S. M. (2017). Cyber-Physical Systems: A Literature Review. *European Scientific Journal, 13*(36), 52. doi:10.19044/esj.2017.v13n36p52

Sanislav, T., Mois, G., Folea, S., Miclea, L., Gambardella, G., & Prinetto, P. (2014). *A cloud-based cyber-physical system for environmental monitoring*. IEEE. doi:10.1109/MECO.2014.6862654

Selmani, A., Oubehar, H., Outanoute, M., Ed-Dahhak, A., Guerbaoui, M., Lachhab, A., & Bouchikhi, B. (2019). Agricultural cyber-physical system enabled for remote management of solar-powered precision irrigation. *Biosystems Engineering*, *177*, 19–30. doi:10.1016/j.biosystemseng.2018.06.007

Sharma, V., Tripathi, A. K., & Mittal, H. (2022). Technological revolutions in smart farming: Current trends, challenges and future directions. *Computers and Electronics in Agriculture*, *201*, 107217. doi:10.1016/j.compag.2022.107217

Suma, D. N., Samson, S. R., Saranya, S., Shanmugapriya, G., & Subhashri, R. (2017). IOT Based Smart Agriculture Monitoring System. *International Journal on Recent and Innovation Trends in Computing and Communication*, *5*(2), 177–181.

Tauber, M., Gollan., B., Schmittner, C., & Knopf, P. (2023). Passive Precision Farming Reshapes the Agricultural Sector. *Computer*.

Tushar, W., Yuen, C., Saha, T. K., Nizami, S., Alam, M. R., Smith, D. B., & Poor, H. V. (2023). Poor. *A Survey of Cyber-Physical Systems From a Game-Theoretic Perspective. IEEE Access : Practical Innovations, Open Solutions*, *11*, 9799–9834. doi:10.1109/ACCESS.2023.3239834

Tyagi, A. & N. Sreenath, N. (2021). *Cyber Physical Systems: Analyses, challenges and possible solutions*, 22-23. Internet of Things and Cyber-Physical Systems.

Zhang, Y., Chen, D., Wang, S., & Tian, L. (2018). A promising trend for field information collection: An air-ground multi-sensor monitoring system. *Information Processing in Agriculture*, *5*(2), 224–233. doi:10.1016/j.inpa.2018.02.002

Chapter 3
Energy-Efficient and High-Performance IoT-Based WSN Architecture for Precision Agriculture Monitoring Using Machine Learning Techniques

Charles Rajesh Kumar J.
https://orcid.org/0000-0003-2354-6463
Effat University, Saudi Arabia

M. A. Majid
Effat University, Saudi Arabia

ABSTRACT

Traditional irrigation systems for agricultural lands are expensive, time-consuming, and labor-intensive. Utilizing cutting-edge technology like machine learning, the internet of things, and wireless sensor networks, smart farming addresses current issues with agricultural sustainability while boosting the quantity and quality of crop production from the fields to fulfill the rising food demand. Soil moisture and temperature sensors are used to create a low-cost, real-time IoT-based automatic irrigation system. Two groups have been formed with the sensor information such as "require water" and "not require water" and saved on the server. The device intelligently determines whether the field needs water and automatically turns "ON" or "OFF" the motor. Machine learning based models such as k-nearest neighbor, support vector machines, decision tree, and naive bayes are applied to decide irrigation requirements. Performance metrics show that the KNN classifier performs better than the other two models. The suggested framework allows for better field monitoring and visualization.

DOI: 10.4018/978-1-6684-7879-0.ch003

INTRODUCTION

Food needs have expanded due to the planet's rapidly increasing population of humans. Given the planet's finite resources, it is challenging to meet the world's food needs (Vootla et al., 2018). Technological innovations are applied in agriculture to boost productivity and overcome this problem. Precision agriculture (PA) uses IoT devices and near- and remote-sensing methods to keep track of crop conditions at various growing phases. PA entails collecting and processing a sizable volume of agricultural health data. Many factors, including water content and temperature, influence crops' health. With the help of PA, a farmer may accurately determine the conditions for a healthy crop and when, where, and in what quantity these conditions must exist. This necessitates gathering a considerable amount of data from various sources and areas of the farm, including soil nutrients, the existence of weeds and pests, plants' chlorophyll levels, and several environmental factors. To generate agronomic advice, all gathered data must be analyzed. For example, plants' chlorophyll content (amount of greenness) indicates the nutrients required depending on their development phase. These data are integrated with information about the soil characteristics where the crop will be grown, including climate prediction. The amount of a specific fertilizer that should be administered to that crop the following day is calculated using all the data that has been gathered. Farmers must receive agronomic advice properly and follow its suggestions to increase yields. WSN (wireless sensor network) consists of several nodes coupled to track and record the physical state of the environment, which is the primary driver of the PA. Figure 1 shows the various application of WSN-based IoT in agriculture monitoring.

Figure 1. Application of IoT-based WSN in agriculture monitoring

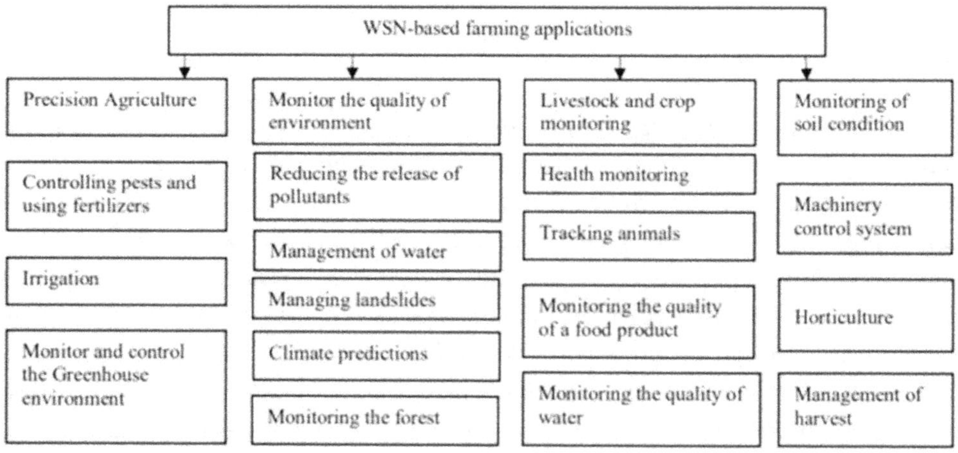

Every wireless node comprises a microcontroller, sensors, a radio transceiver, and additional circuitry that allows it to connect to a gateway and transmit data that the sensor has acquired (Kumar et al., 2022). Sensors gather data by measuring the physical parameters and sending it to the controller, who then sends it to a portable device or the cloud. The agriculture industry has a variety of needs, including those related to soil characteristics, crop varieties, climate, fertilizer classes, and water necessities. The requirements of crops vary based on the crops grown in the same area and the same plant grown on other lands with various climatic conditions. Sensors keep track of how these crop factors change over time. The size and expenditure of sensors have decreased due to the rapid development of WSN technologies, making it possible to use them in various fields, including agriculture. Sensors keep track of how these crop factors change over time. The size and expenditure of sensors have decreased due to the rapid development of WSN technologies, making it possible to use them in various fields, including agriculture.

Table 1 lists the most popular agriculture sensors to record crops' environmental factors. A WSN typically comprises one or more wireless nodes linked by sensors. These nodes, which are tiny units, are in charge of data collection (Kumar et al., 2023). Two kinds of nodes are source nodes that gather information and sink nodes or gateway that obtain data from source nodes. Compared to a source node, a sink node is more computationally powerful. When selecting wireless nodes, memory, area, energy, data rate, and cost restrictions exist. Table 2 compares various wireless nodes and lists their standard specifications. MICA2 is seen as more appropriate than the other wireless nodes due to its numerous extension connectors, which enable it to interface with various sensors. Cyber-physical systems typically integrate sensor networks with embedded computers to monitor and regulate the physical environment. Feedback loops enable these external stimuli to self-activate computing, control, and communication. The suggested system comprises three layers: the physical layer, where all sensory data is gathered; the network layer, where data is transported to the cloud; and the decision layer, where information is interpreted and processed to arrive at judgments based on observation. IoT-based systems face several difficulties due to the exponential growth of devices. The key issues underlying the IoT-based design are the requirement of minimal latency and minimum energy consumption, better use of bandwidth, and sporadic Internet access. Like a conventional IoT network, every node broadcasts information to a distant cloud, causing cloud congestion. The most advanced solutions to these problems, which lessen the cloud's computational load, are fog and computing (Kumar et al. 2021).

Table 1. List of the most popular agriculture sensors to record crops' environmental factors.

S.No	Acquired parameters	Name of the Sensor
1	Conductivity, The moistness of the soil, temperature of the soil	ECH2O-soil_moisture sensor
2	Temperature, Humidity	SHT71and SHT75-Humidity & temperature sensor
3	Conductivity, Soil Moisture, Soil Temperature, Salinity level	Hydra Probe-Water soil sensor
4	Air Temperature, Air Humidity, Air Pressure	XFAM_115_KPASR-Pressure_sensor.
5	, The moistness of the soil, The Temperature of the soil	MP406-Soil-moisture-sensor
6	Air Pressure, Wind Speed, Air Temperature, Air Humidity	Met Station One-(MSO)weather sensor
7	Conductivity, Soil Moisture, Salinity level, Soil Temperature,	ECE250-Electrical conductivity sensor
8	Air Pressure, Air Humidity, Air Temperature.	HMP45C-Temperature and Relative humidity sensor.
9	Soil Moisture, Soil Temperature	POGO-Wireless Soil sensor
10	Air Pressure, Wind Speed, Air Humidity, Air Temperature	CM-100-Weather Sensor
11	Plant Temperature, Air Temperature	107-L-BetaTherm100K6A1B Thermistor
12	Hydrogen level in Plant, Plant Temperature, CO2, Plant Wetness, Plant Moisture, Photosynthesis, Air humidity, Air temperature.	Cl-340-Photo synthesis system
13	Plant Temperature, Plant Wetness, Plant Moisture	237-L-Leaf Wetness sensor
14	CO2, Plant Temperature, Photosynthesis, Plant Moisture, Plant Wetness.	PTM-48A-photosynthesis monitor
15	Plant Temperature, CO2, Plant Wetness, Hydrogen.	NTM Sense H-R-Hydrogen sensor
16	Plant Temperature	LT-2M-Leaf temperature sensor
17	Photosynthesis	CM1000-Chlorophyll meter
18	CO2, Plant Moisture, Photosynthesis	TPS-2 portable photosynthesis
19	Photosynthesis	YSI6025-chlorophyll sensor
20	Plant Temperature, Plant Moisture	TT4-multi-sensor thermocouple
21	Plant Temperature, Plant Wetness, Plant Moisture	LW10-leaf wetness/Rainfall sensor

Fog computing's primary objective is to save bandwidth and power, which improves service quality for end-users. Energy efficiency is the essential factor to consider while creating any fog design. To extend the network's lifespan by evenly distributing the power among interconnected nodes, an energy-efficient protocol is required for a fog-supported WSN. PA has only been extensively applied in a few developed countries, even with all the IoT-related advancements. Due to a lack of resources, emerging economies like India do not frequently use remote sensing approaches to assess crops' health, reducing productivity (Thaher et al., 2020). Due to its vast fertile territory and climate changes, which allow for the cultivation of various crops, India is primarily an agricultural nation. Despite having access to many environmental assets, India struggles to produce large yields. Conventional farming methods employed for crop healthiness monitoring and yield analysis are the leading cause of poor productivity. These methods are entirely based on the instincts and knowledge of farmers. In the case of ample agricultural land, monitoring the crop requires frequent field visits, which is quite time-consuming and challenging. In this instance, the area being attacked by insects and pests is not correctly monitored, which may lead to excessive insecticide and pesticide application that harms crop nourishment.

Actuators and sensor systems are part of the field layers. Actuators and sensor systems are part of the field layers. The sensors measure light intensity, temperature, humidity, electrical conductivity, pH, soil moisture, etc. The actuators are capable of carrying out orders from higher-tier units. The short-range wireless communication modules are Wi-Fi and Zigbee, and the long-range wireless communication modules are cellular and LoRa. A fog computing layer's architecture is constructed to gather detected information from deployed sensor nodes and execute the needed data-processing tasks for intelligent collection, minimizing the traffic of the network and lowering power expenses. The fog layer, often referred to as a local gateway, develops the core elements of a given process model locally before transmitting them effectively and compactly to the cloud. As a result, there is a significant reduction in top layers' latencies and expenses. The gateway unit data obtained is constantly moved to a cloud computing environment. As a result, the layer of cloud computing allows farmers to utilize a knowledge-based directory to save various data types, including all the pertinent details about available farming and equipment possibilities. The user can see the outcomes of data analytics at the front-end layer. When triggered, an event can show prompt alerts on the user's mobile phone.

Table 2. Various wireless nodes and their standard specifications

S.No	Wireless Node	Microcontroller	Extension connectors (pins)	Data Speed/ rate	Existing sensors
1	MICA2 (MOTE-KIT4x0)	ATmega128L	51	38.4K Baud	Accelerometer, GPS, Magnetometer, Video, Sounder, Acoustic, Relative humidity (RH), Barometric Pressure, Humidity, Light, Temperature.
2	TelosB	TIMSP430	6, 10	250 Kbps	Humidity, Temperature, Light
3	IRIS	ATmega128L	51	250 Kbps	Microphone, Magnetometer, Sounder Video Sensor, Acoustic, Accelerometer, RH, GPS, Barometric Pressure, Temperature, Humidity, Light
4	MICA2DOT	ATmega128L	19	38.4 K Baud	RH, Acoustic, Accelerometer, Temperature, Barometric Pressure, Humidity, Light, GPS.
5	MICAz	ATmega128L	51	250 Kbps	Microphone, Magnetometer, Sounder, Video Sensor, Acoustic, Accelerometer, RH, GPS, Barometric Pressure, Temperature, Humidity, Light
6	Cricket	ATmega128L	51	38.4K Baud	Sounder, Magnetometer, Microphone, Sensor, Video, Ultrasonic, Barometric Pressure, Acoustic, RH, GPS, Humidity, Temperature, Light, Accelerometer.
7	Imote2	Marvell-XScalePXA271	40,20	250 Kbps	Accelerometer, Humidity, Temperature, Light.

Figure 2 shows the end-to-end communication for PA. In a smart irrigation system, the solenoid valves and IoT-based WSN regulate water flow based on soil moisture and provide real-time monitoring to farmers far from the farms.

Figure 2.Common architecture of PA system

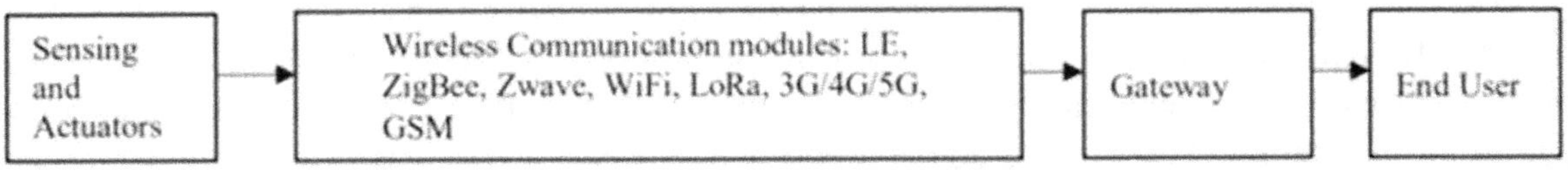

The system uses the soil sensor to measure the soil's humidity and temperature and then waters the crops accordingly. Farms with intelligent irrigation systems can check their irrigation systems automatically, eliminating the need for manual checking. The IoT-based WSN system enables crop monitoring. The sensor node connects the irrigation system to the mobile phone and regulates the time intervals between irrigation cycles and water flow from the pump. Artificial intelligence and Machine learning (ML) are changing farms into smart farmlands called Precision agriculture (PA). The next agricultural revolution will be driven mainly by ML and IoT-based WSN. One of the critical determinants of farm productivity and hydrological cycles is soil moisture, and accurate soil moisture forecasting is crucial for water-efficient use and management (Prakash et al., 2018). Farmers in the agricultural industry will benefit from forecasts of soil moisture. Soil moisture is important because of the following reasons: 1. Soil water acts as a solvent and a transporter for food nutrients to promote crop growth.2. The quantity of water that is available affects a crop's production.3. The water content of the soil itself is a nutrient.4. The flow of water in the soil controls the soil's temperature.5. The metabolic processes of microorganisms demand soil water.6. Soil water aids in the biological and chemical processes of the soil.7. Soil water is necessary for photosynthesis. Soil moisture can be predicted using ML approaches such as PCA, linear regression, Naive Bayes, and support vector machine regression. But because soil moisture is affected by many structural details and climatic variables, it is challenging to create the perfect mathematical model for predicting soil moisture. Using suitable ML approaches is crucial to increasing model performance and prediction accuracy. The following is a breakdown of the article's structure. Section 2 discusses the technologies for wireless communication, and Section 3 explains the related works. The proposed methods are elaborated in Section 4, and Section 5 covers the results and discussion. Section 6 concludes the paper.

WIRELESS COMMUNICATION TECHNOLOGIES

Table 3. Wireless communication techniques utilized in PA

S.No	Communication protocols	Topology	Data Speed/rate	Power	Physical Range	Standard
1	Wi-Fi	Star	1to 54 Mbps	Medium	50 m	IEEE 802.11
2	6LoWPAN	Mesh, Star	0.3 to50 Kbps	Low	2–5 km (urban), 15 km (suburban)	IEEE 802.15.4
3	RFID	Peer-to-Peer (P2P)	50 tags per second	Ultra-low	10–20 cm	RFID
4	ZigBee	Star, Tree, P2P, Mesh	250 Kbps	Low	10–100 m	IEEE 802.15.4
5	Bluetooth	Star, Bus	1to 2 Mbps	Low	30 m	IEEE 802.15.1
6	LoRaWAN	Peer-to-Peer, Star	27 to 50 Kbps	Very low	5km to 10 km	IEEE 802.11ah

Due to the explosive growth of IoT devices and WSN technology over the past few decades, numerous communication protocols have been developed (Al-Sarawi et al. 2017). Each protocol has unique specifications such as bandwidth, the number of available channels, data rate, batteries' timing, cost, etc. The wireless communication techniques are compared and shown in Table 3. Various wireless communication technologies include ZigBee, 6LoWPAN, Wi-Fi, RFID, BLE, ZigBee, ZigBee, and Cellular.

Cellular

Applications requiring a high data rate are best suited for cellular technologies. It may use GSM, 3G, 4G, and 5 G cellular connection technologies to deliver reliable high-speed Internet connections while using more energy. It requires the deployment of infrastructure, ongoing operational expenses, and support of people under a centralized management authority. Cellular technology is an excellent choice for underground WSNs, security techniques in agriculture, and projects involving smart homes (Zhang et al. 2017), even if it consumes more battery power. In (Patil et al. 2019), an integrated farm monitoring system that uses IoT and smartphone applications is presented. Farmers can remotely monitor their farms' soil moisture, pH level, leaf wetness period, humidity, and environmental temperature with this system. The device provides fresh insights to influence decision-making by instantly analyzing the climate and soil conditions in a specific location where the crop is cultivated.

Low-Power Wireless Personal Area Networks (6LOWPAN)

6LoWPAN is the first IP-based protocol utilized for IoT communication. With IPv6 capability for IEEE 802.15.4 connections, the 6LoWPAN protocol establishes itself as the first IoT standard. It is a standard protocol for implementing IPv6 on WSN made up of low-power wireless modules. Due to its small low power and bandwidth requirements, 6LoWPAN is inexpensive. Multiple topologies, including mesh and star topologies, are supported by 6LoWPAN. IoT-based smart agriculture is proposed in (Suryady et al.

2011) using 6LoWPAN. A 6LoWPAN-capable wireless sensor network was discussed, which is used to track the soil characteristics of crops.

ZigBee

The Zigbee protocol was created for personal area networks, a standard-based wireless technology. ZigBee is frequently used in PA to track temperature, pressure, oxygen, radiation, PH, pollutants, parasites, etc., affecting crops' health (Prakash et al. 2020). The sensor network is utilized in agriculture, where data is transferred via Zigbee and stored in microcontrollers. Farmers are using the data gathered from the sensors installed in the field. Zigbee is excellent for short- or medium-distance applications with high data needs because of its constrained range coverage and improved data rates. It supports tree, star, and mesh topologies and enables multi-hop data transfer. It supports many nodes and is simple to install, and it also has a flexible network configuration and a prolonged battery life. It has a narrow range, slow data throughput, and less safety compared to Wi-Fi-based technologies.

Bluetooth Low Energy (BLE)

The data collected from sensor nodes deployed throughout agricultural land can be observed utilizing Bluetooth low energy (BLE) mesh technologies (Shanbhag et al. 2017). The IoT will enable farmers to connect to their farms through the cloud. With relatively little human work, inexpensive, and less energy consumption, the system allows farmers to monitor various field conditions remotely and make decisions that would boost crop yield. BLE suits IoT applications that require reduced latency, low bandwidth, and short distances. Less setup time, Nodes in a star topology have complete support, and less energy consumption is among BLE's critical benefits over existing Bluetooth technologies. It only has a 10-meter range, which is pretty short. Moreover, its limitations on communication between devices, lack of security, and potential connection loss during transmission are negative.

Radiofrequency Identification (RFID)

The development of RFID sensing technologies has unleashed a new wave of agricultural applications (Motroni et al. 2021). The IoT technology uses sensors and RFID. Utilizing RFID communication protocols, the sensor data is gathered and delivered to the cloud. For instance, in an irrigation system, the sensor collects data regarding moisture and temperature. The sensor data is sent to the cloud using RFID communication protocol, and based on the soil's water content, the user can manage the water pump. Farmers can follow each step of the growing, harvesting, and distributing processes with the help of IoT sensors, RFID crop tagging, and network communications. Farmers may boost yields and spot any problem areas affecting their production using reliable data from each plant.

Wireless Fidelity (Wi-Fi)

The WiFi protocol supports IEEE 802.11n, 802.11g, 802.11b, and 802.11a. The system comprises a sensor to measure temperature, wetness in the soil, level of water and humidity, etc., along with a wireless fidelity module (WiFi). The sensors gather information about environmental changes and wirelessly transmit it to the cloud using a WiFi module. Over a WiFi connection, a web server receives the sensed information. The web server archives the data that has been received and offers up-to-the-minute reports on the weather and crops (Akram et al., 2021).

Long Range Wireless Area Network (LoRaWAN)

IoT applications may overcome bandwidth, power, and coverage limitations which are the primary shortcomings of traditional wireless communication technologies using the enhanced service provided by LPWAN technology. The LoRaWAN protocol, also known as LoRa in the LPWAN domain, offers additional benefits for creating IoT applications in the agriculture sector, such as robustness, security, and scalability. Due to its extensive range and low energy consumption, LoRaWAN is often utilized in agricultural applications (Singh et al. 2017).

RELATED WORKS

Broad definitions are given in the previous sections, and the other perspectives (literature reviews) are provided in the following section.

IoT-Based WSN Irrigation System

Agricultural activity automation can change the farm industry from laborious and static to intelligent and dynamic, resulting in increased productivity with less human supervision. The nodes of the WSN carry out the acquisition, gathering, and interpretation of data, such as soil temperature and moisture. Such information can automate agriculture irrigation while reducing water use, which has positive financial and environmental effects. Gutiérrez et al. 2014 proposed an automated irrigation system using WSN and the GPRS module. Appropriate sensors are used to measure the temperature and moisture of the root area. A gateway device manages sensor data, activates actuators, and transmits messages to a web application. A microcontroller-based gateway was designed with an algorithm based on temperature and moisture levels of soil threshold values to regulate water supply to the crops. The target plant is an organic sage. The sensors measure soil moisture and temperature, and a PIC24Fj64GB004 microcontroller and photovoltaic panels are utilized. Standards and protocols are ZigBee and GPRS. The programming languages are SQL server and C#. Sales et al. 2015 presented an innovative irrigation design using WSN, actuators, and the cloud. The target plant is a peach tree.

The sensors measure soil moisture and weather forecast, and an MSP430F2274 microcontroller and electro valve are utilized. Standards and protocols are IEEE802.15.4, ZigBee, GPS, GPRS, and HTTP. The programming languages are model view controller and JSON. (Monica et al. 2017) suggested an IoT-based self-regulating irrigation scheme using Bluetooth, GSM, sensors, and the cloud. The farmers can monitor the variables through the mobile app and cloud storage. The proposed method effectively identifies a technique to preserve water by studying and contrasting the data from the prior year with the current data. The target plant is groundnut. The sensors measure moisture, light, and temperatures, and an ATMEGA328 microcontroller and pump are utilized. Standards and protocols are Bluetooth, GSM, and Wi-Fi. The programming language used is Sparkfun.

Vaishali et al. 2017 designed an IoT-based intelligent irrigation monitoring and managing solution integrated with mobile devices. The primary goal of the proposed method is to use a smartphone to constantly monitor the water supply for the crops. The sensors measure soil moisture and temperatures, and a Raspberry Pi and pump motor is utilized. Standards and protocols are Bluetooth. The programming language used is Python. Haque et al. 2019 suggested an IoT-based agriculture irrigation tracking and control solution. The system uses node microcontrollers (ESP8266) for numerous methods that can be monitored and managed via the cloud. The microcontroller continuously tracks the states of multiple farm components and relays the information to the central control unit. After reviewing this data, the user can decide what steps to take, such as giving each microcontroller a specific task. The sensors measure soil moisture and temperatures, and a Raspberry Pi, solar cell, and pump are utilized. Standards and protocols are MQTT and Wi-Fi. The programming language used node.js. Balasooriya et al. 2020 developed an IoT-based intelligent irrigation system. The system uses a PH sensor and soil moisture sensors. Microcontrollers are used for real-time data processing in a cloud environment, making it possible to continuously check the pH and soil moisture levels.

Farmers can use the suggested technology to monitor and manage their watering system and plant environment through a smartphone app. The irrigation is regulated using Wi-Fi-enabled microcontrollers, sensors, and the cloud. Sah et al. 2021 devised an IoT-based irrigation system. The proposed system utilizes an Arduino UNO as its microcontroller and a GSM 900 module to interact with the farmers on a 2G phone. The automation process automatically determines whether the soil is dry or wet, applies the necessary irrigation, and notifies the farmer of the situation so that they may keep an eye on the farm even while not physically present. Tephila et al. 2022 presented an IoT-based automated irrigation system. The proposed system automatically controls time, steers clear of under- and over-irrigation problems, distribute water more efficiently, and monitors water reserves. The system uses open-source clouds, field-deployed sensors, sinks, and fusion centers for intelligent irrigation. In terms of the percentage of packets delivered, the number of packets sent to their destination, the length of the network's stability, and power use, the result is evaluated to that of other existing approaches.

Machine Learning (ML) Approach for Smart Decision

The various techniques for raising crops are still largely conventional and even outdated. Choosing the proper crop for the correct non-biological and biological components is still tricky for farmers. Thus, various AI techniques (ML methods) have been suggested by multiple researchers to increase crop productivity. Katarya et al. 2020 summarized the different ML techniques such as KNN, Neural networks, Ensemble-based models, and similarity-based models. They made the most significant suggestions, resulting in higher returns and minimal resources and capital usage. Zakzouk et al. 2021compared the various supervised ML algorithms for the automatic irrigation of rice plants. Multiple sensors measure soil moisture, humidity, temperature, and time.

The algorithms investigated are Random forest, logistic regression, decision tree, and KNN. These various models are trained and tested using approximately 100,000 datasets. The findings demonstrate that the Random Forest classifier has a 99.9% accuracy rate, which is the highest. Abuzanouneh et al. 2022 suggested an intelligent irrigation scheme based on IoT, for PA using ML algorithms. The suggested method enables sensing the farmland's parameters and making the proper irrigation selections. The presented concept uses various IoT-based sensors for soil moisture, light, humidity, and temperature. The sensor information is also sent to a cloud server for analysis and decision-making. The artificial algae algorithm (AAA) and the least-squares-support vector machine (LS-SVM) model were integrated to decide whether irrigation is necessary. The model's accuracy is 97.5%, with guaranteed more acceptable performance. Abdallah et al. 2022 proposed ML techniques for irrigation prediction. The ML approach utilized is Feature selection combined with various Regression techniques. The significance of the features is evaluated using SelectKBest, Recursive Feature, Random Forest, and Elimination (RFE) algorithms to accurately calculate the ideal amount of water required for a plant. Various regression techniques are developed based on the collection of useful features. The numerous models utilized in this approach are trained and tested employing a dataset that has been compiled on various crops like grapes, tomatoes, and lemons and includes multiple features such as crop data, irrigation data, soil data, and meteorological data.

Goldstein et al. 2018 presented a method to facilitate irrigation planning using ML algorithms. The sensors gather soil moisture data for monitoring crops in real-time. The study uses the data to forecast irrigation recommendations and monitor and control the yield. On this dataset, various regression (Gradient Boosted Regression Trees) and classification (Boosted Tree Classifier) techniques were used to create models that could forecast the weekly irrigation plan suggested by the agronomist. Bhoi et al. 2021 devised an automated precision irrigation strategy based on IoT for effective water use with minimal human involvement. Multiple ML-based regression and classification models are combined with intelligence. The testing results show that the support vector regression (SVR) and KNN classifier-trained systems are pretty efficient. Glória et al. 2021 devised a real-time irrigation technique based on IoT using ML algorithms. The developed method comprises wireless sensors and an actuators network. This mobile app allows the user to consult not only the information collected in real-time but also their history and act in accordance with the information it evaluates. ML-based adaptation approaches for agricultural irrigation systems were investigated to determine a suitable time to supply water to crops. Various ML algorithms studied were Support Vectors Machines, Decision Trees, Neural Networks, and Random forests. The Random forest approach showed the best result of 84.6% compared to other ML algorithms considered in the study. The system's adoption discovered the designed solution's effectiveness and ability to save up to 60% of water.

PROPOSED DESIGN

The following section discusses the proposed IoT-based irrigation system and the various modules utilized to build the system. Furthermore, the Machine Learning approach for smart decision is presented.

IoT-Based WSN Irrigation System

The agricultural industry currently uses WSNs for a variety of purposes. Smart fertilization, Smart irrigation, greenhouse monitoring, and intelligent pest control are prevalent uses. Once the WSN integrates with the IoT, its full potential will be realized. An intelligent irrigation design was developed to maximize water utilization for crops. The smart irrigation strategy can reduce the amount of water entering the crop field by turning the motor off automatically without human interference, depending on the soil's water content. The process can also be manual. The user can observe the surface soil moisture, and a signal is produced when a particular threshold is reached. A smartphone app allows the user to "OFF" or "ON" the motor. Figure 3 shows the proposed irrigation system. A real-time sensor node system based on IoT was created to regulate the water amount utilized in PA.

Figure 3. Proposed architecture of the intelligent irrigation system

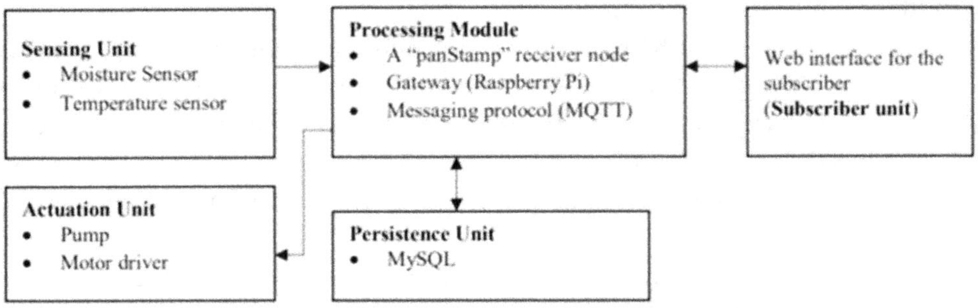

Sensing Module

Soil moisture sensors (SEN-13637), soil temperature sensor (DS18B20), Wireless node/mote (panStamp NRG 2.0). Panstamp is a compact and energy-efficient wireless sensor mote that can be programmable from Arduino IDE. It employs a transceiver (CC1101 RF) integrated into an MSP430 core to create the System on a Chip (CC430F5137). SEN-13637 comprises two pads and works as a variable resistor when combined. The conductivity between the pads improves with soil moisture, lowering resistance. The operating range of DS18B20's operating temperature range is between -55 to 125 Celcius. The devices are powered using a 5V DC battery. Figure 4 shows the block diagram of the sensing unit.

Figure 4. Block diagram of the sensing unit of the proposed architecture

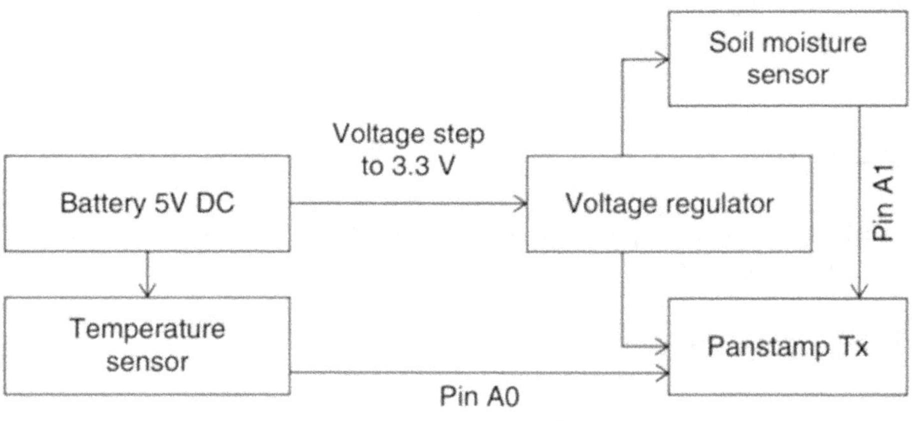

Processing Module

A panStamp receiver node in the processing unit collects sensory information from farms that may be far away from the user and delivers the data to the gateway. Here the Raspberry Pi is the gateway. A low-cost, low-energy embedded board is utilized in the suggested IoT-based system. The Raspberry Pi should have sufficient speed to execute ML algorithms to facilitate cloud transmission and gather information from several sensor nodes. Selecting the suitable communication protocol from the many already in use is crucial, including HTTP, MQTT, AMQP, and CoAP. We choose the MQTT protocol because of its ability to transmit brief messages and minimal bandwidth consumption, making it perfect for Machine to Machine (M2M) connections. MQTT is an IoT device's communication standard that deals with messages, clients, servers, brokers, and topics.

Clients are any applications or devices that make use of MQTT. The client performs the "publish" and "subscribe" operations. Whenever the client transmits the information to the server, the technique is called a "publish." The data sent by the server to the client is called a "subscription." The client's ability to "publish" and "subscribe" to messages is made possible by the server. A server handles the client's network connection, receives the client's messages, handles unsubscribe and subscribe requests, sends the client's application messages, and terminates the client's network connection. The term "topic" refers to the broker's method of filtering messages for every connected client. Messages are sent via a topic (a provided information channel). The broker obtains published messages, and customers who have previously subscribed receive the information from the broker. An open-source message broker Mosquitto is utilized, which implements the MQTT protocol. The Mosquitto, like the Raspberry Pi, is appropriate for single-board computers with minimal power consumption.

When deciding which Web servers are best for the suggested system, Flask, Nginx, and Lighttpd are considered. Lighttpd stands out as a high-performance web server, an excellent option for embedded devices. Lighttpd provides auto-indexing, load balancing, and tolerance for faults capabilities when "Flask" and "Nginx" are compared. Flask is a lightweight structure with essential features, including a quick debugger and an integrated development server. Flask is the best choice because the Web server won't do severe load tasks in our situation. The capacity to communicate directly with the broker is the most critical component of a web interface. Any interaction requires a client to broker a connection established via the web server (Raspberry Pi). The Raspberry Pi hosts the app's web pages, and the Raspberry Pi serves as a Mosquitto/Broker and a web server. Figure 5 shows the transfer of information within the processor unit. The server provides web pages to the client, and the client can use the standard HTTP request technique to get them. The web page uses the MQTT protocol to communicate with the broker. The "Eclipse Paho MQTT" JavaScript library creates the "MQTT client." The browser uses web sockets to interface with MQTT.

Figure 5. Transfer of information within the processor unit

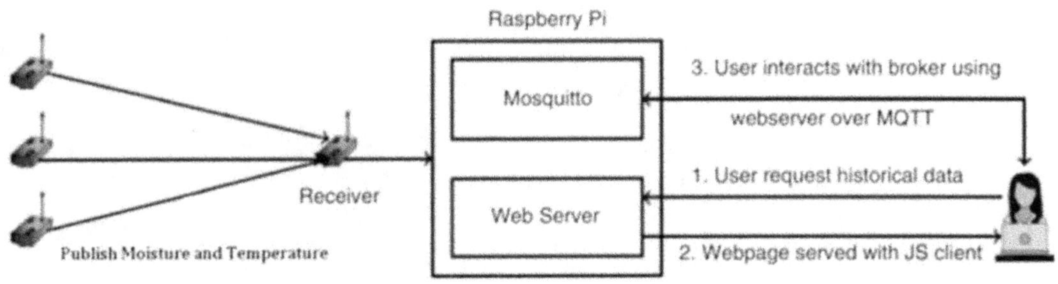

The Subscriber Module

The subscriber module involves designing and implementing an online user interface for subscribers. The farmer would have access to a web interface that would allow them to "subscribe" to "Topics" or to the act of "publishing" that involves transmitting messages. The online web user interface supports log monitor format and data visualization with historical data utilizing the tabular form. Python was chosen to run tasks on the Web server, and the CSS and HTML form the basis of the Web interface. The feature of Web sockets in the broker (Mosquitto) made use of using JavaScript. The libraries are available in JavaScript, making it easier to interact with Mosquitto. After inputting the broker's username, port, and login information, the user's data is provided to the Mosquitto broker to verify the subscriber's authenticity.

Actuation Module

Figure 6 shows the actuation module of the proposed system. The Raspberry Pi is connected to an L298N dual H-bridge motor driver and a DC motor (12V, 200-300mA). One silicone tube is employed to draw water from the tank, while another, with a flow rate of 100 mL per minute, is operated to water the plants. The motor driver uses a 5 to 46 V DC power supply with a maximum 2A peak current. Using Raspberry Pi, the sensed data are sent to the server. The real-time data from the sensors that measure the temperature and moisture content of the soil are displayed in real-time graphs when the client login into the mobile application. The user can switch "ON" the motor remotely using the mobile application if they see that the moisture in the soil measurement is under the threshold or that the temperature is higher than the threshold value. The motor status is changed to "ON" when the server receives this instruction. The motor is switched "ON" to pump water to the crops when the motor driver receives an "ON" status signal from the controller (Raspberry Pi). The user can "OFF" the motor manually or automatically whenever the ground is sufficiently moist. The motor status is switched to "OFF" when the server obtains the instruction. The motor driver gets a signal from the gateway to stop the motor when it notices the "OFF" status.

Figure 6. Actuation module of the proposed system

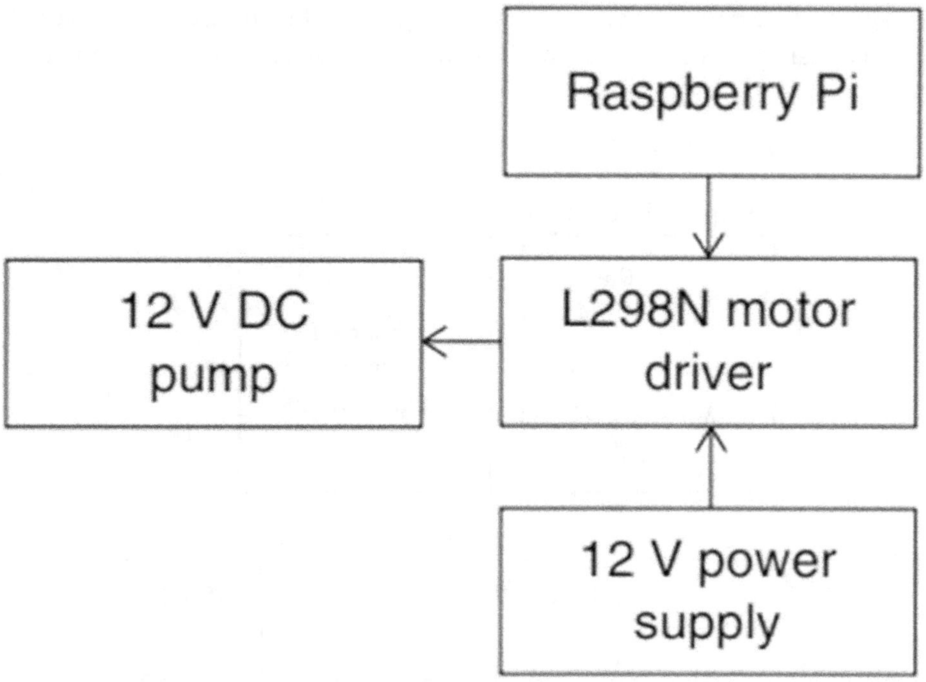

Persistence Module

The data is transmitted to the subscriber node from wireless nodes over MQTT. The data packets are stored in a MySQL database so they may be accessed later. MySQL is an open-source relational and large-scale database. Because MySQL is built on a client-server model, an SQL server that supports multithreading and multiprocessing is necessary. Due to its scalability and support for multiuser characteristics, MySQL is an ideal choice for distributed applications. The persistence module is shown in Figure 7.

Figure 7. Persistence module of the proposed architecture

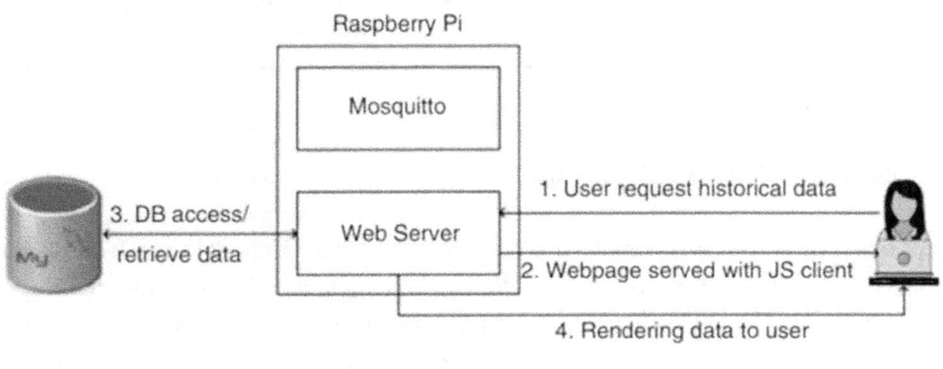

Machine Learning (ML) Approach for Smart Decision

The proposed architecture based on ML is shown in Figure 8. It comprises three layers 1. A layer for data collection and management, 2. Layer for data processing, and 3. An application layer.

Figure 8. Proposed architecture based on ML

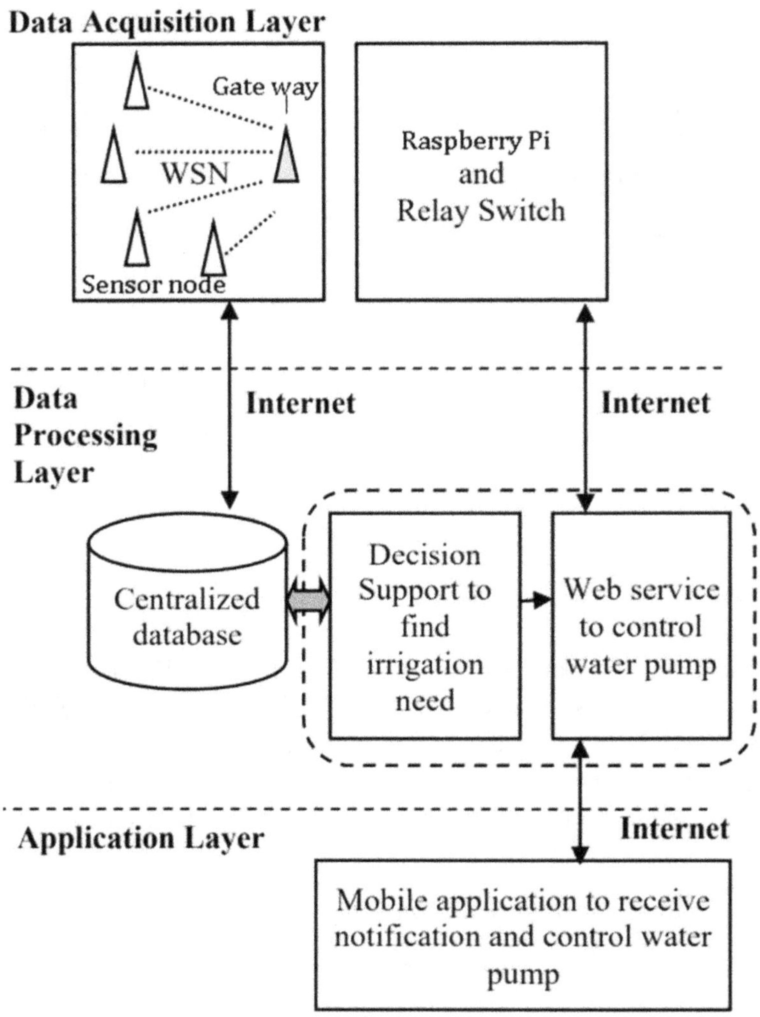

Layer for Data Collection and Management

The IoT components are primarily located in the collection and management Layer. It consists of two parts, one for motor control and the other for data collection. WSN and wireless communication technologies such as ZigBee are implemented with moisture and temperature sensors. These sensors' outputs are read by a Raspberry Pi attached to a gateway node by ZigBee, which then utilizes web services to transfer the information to the server. A Raspberry Pi or Arduino Uno node controls a relay switch attached to a water pump. The categorization model triggers a web service that controls the node. PHP was utilized to create the web service and Set up Windows' Apache_Web_Server to execute PHP.

Layer for Data Processing

The process flow diagram for decision support is shown in Figure 9. Three elements needed for data processing are 1. A centralized database, 2. A decision support system to forecast irrigation requirements, and 3. A web service to manage the water pump. The server receives the data from the data collection and management layer, which is then sent and placed into the MySQL database. Two groups have been formed with the sensor information, such as "require water" and "not require water," and saved on the server. The device intelligently determines whether the field needs water and automatically turns "ON" or "OFF" the motor, saving the farmer's time and human labor. The classification model is built with the clustered data using classification models such as KNN, Naive Bayes, and SVM. The water motor can be started and stopped using a web service. This PHP-based web service transmits signals to the Raspberry Pi or Arduino-Uno to operate the relay switch, which starts and stops the water motor. Additionally, it notifies the user when irrigation is necessary.

Figure 9. Process flow diagram for decision support

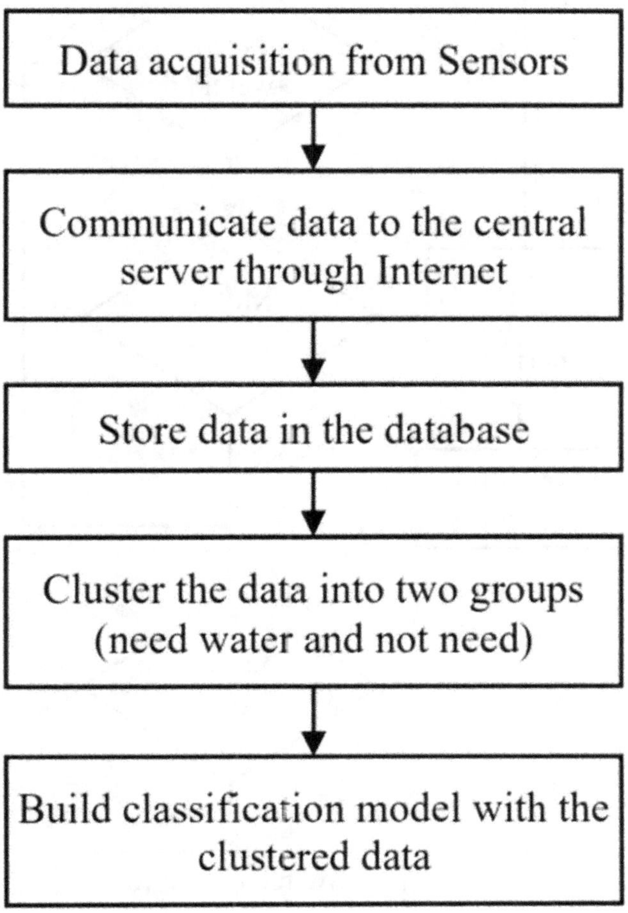

Application Layer

A mobile application for a user's smartphone is running in the application layer. When the web application notifies the user, the notification outlining the field's requirement for irrigation is displayed, and the motor is turned ON. The user can manually manage the motor utilizing the mobile app as well. This relay switch is attached to a Raspberry Pi or an Arduino Uno, and the mobile app furnishes information to the web service, which controls it. The flow diagram for controlling the relay switch is shown in Figure 10.

Figure 10. Flow diagram for controlling the relay switch

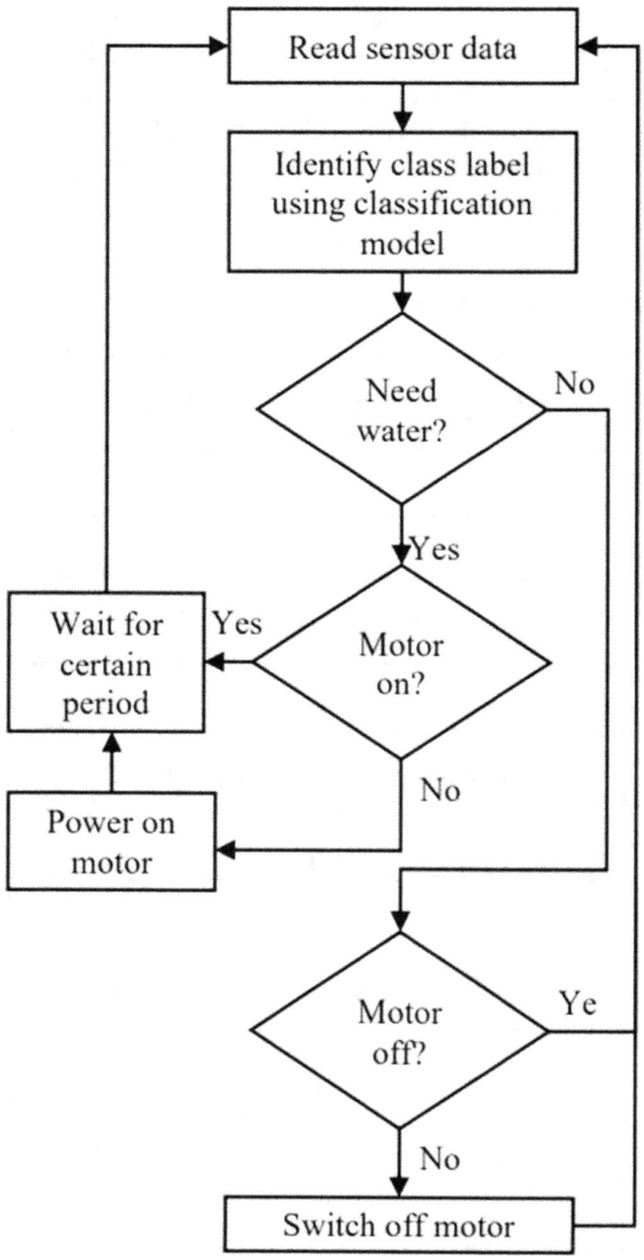

RESULTS AND DISCUSSION

This section discusses the IoT-based WSN irrigation system, a real-time temperature and moisture monitoring system. Furthermore, it explains the machine learning approach for predicting soil moisture.

IoT-Based WSN Irrigation System

The Web interface gives users access to any node's tabular and plotted historical information. The Web server receives a request via the HTTP protocol, consults the databases, and then returns the user's responses. Since this capability deals directly with the Web server, the user need not be linked to the Mosquitto broker to fetch data. The Web application allows frequent and real-time temperature and soil moisture data monitoring, allowing users to remotely switch "ON" and "OFF" the motor in the field as needed. After continuously updating the device on the Web server and checking it, the motor is switched "ON" or "OFF" using the control signal generated by the Raspberry Pi. The location of the node can be fetched through Google Maps. Real-time monitoring of temperature and moisture is shown in Figure 11.

Figure 11. Real-time monitoring of temperature and moisture

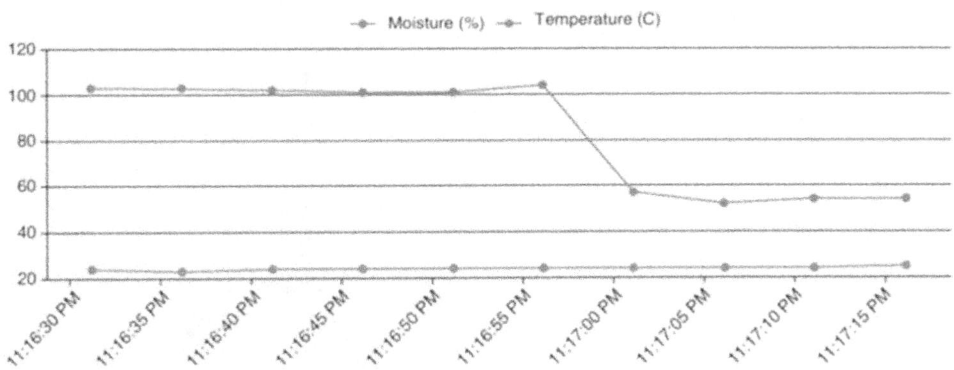

Machine Learning (ML) Approach for Prediction of Soil Moisture

The suggested method offers an automated decision-making system that sends water to the field using IoT-based WSN and data mining approaches when required. The gathered information is transmitted to the server over the internet. The server utilized in this configuration includes an Intel i7 processor clocked at 3.40GHz, 4GB of RAM, and 500GB of hard drive space. The server machine houses a database for storing received data, a decision support model for processing it, and a web service for controlling a relay switch. The user's Android-based smartphone is running an app. The user can physically operate the motor while still receiving the notification on their smartphone from wherever they are. Accuracy, recall, and precision, which are provided in equations (1), (2), and (3), are the performance measures taken into account to assess the classification model in the decision support procedure. The suggested method would make it easier for the user to drive when he is out of stations instead of going to the farm. Additionally, the user can manually regulate the motor, which would be helpful for him in critical instances. The created system lessens the user's human labor and aids in irrigating the field only when necessary, ensuring that the required volume of water is utilized efficiently. The suggested method uses ML data mining approaches, allowing the model to learn from primary information before making decisions based on new data. It avoids integrating intelligent irrigation systems with just thresholds. ML Classifiers such as KNN, Naive Bayes, and SVM decide if watering is required. ML classification performance measures demonstrate that the K-NN classifier outperforms the other two models considered for this investigation, as shown in Figure 12. The K-NN aids farmers in the timely irrigation of their fields and efficient water utilization.

$$\text{Accuracy}=\frac{\text{True Positive}+\text{True Negative}}{\text{The total\_number of positive\_samples}+\text{The total\_number of negative\_samples}} \tag{1}$$

$$\text{Recall}=\frac{\text{True Positive}}{\text{True Positive}+\text{False Negative}} \tag{2}$$

$$\text{Precision}=\frac{\text{True Positive}}{\text{True Positive}+\text{False Positive}} \tag{3}$$

Figure 12. ML classification performance measures

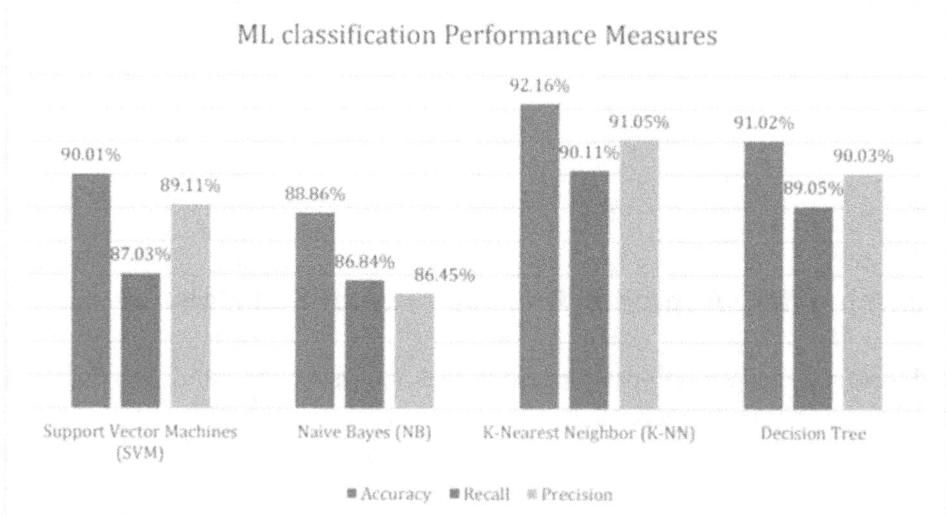

It demonstrates that the K-NN classification algorithm has the highest accuracy compared to other classification models (NB, SVM). The findings show that the KNN classification method yields more precise results. KNN does not create a model because it is a lazy learner. To calculate similarity, all accessible data are used. Thus, the KNN algorithm could be applied to the decision-support process of determining the requirements for irrigation. It aids farmers in the timely irrigation of their fields and efficient water utilization.

SOLUTIONS AND RECOMMENDATIONS

1. Precision agricultural technologies increase output while reducing needless spending on inputs like seeds, fertilizer, herbicides, and gasoline, and this is accomplished by employing GPS data.
2. With more accuracy than traditional image processing methods, deep learning can be heavily utilized in digitizing agricultural lands.
3. Crop selection, fertilizer suggestion, and plant disease prediction applications can all be created using machine learning algorithms.
4. The user can input their soil data into the crop suggestion application, and the application will then forecast which crop the farmer should produce.
5. The user can enter soil information and the crop type they are growing into the fertilizer suggestion application, and the application will forecast what the soil lacks or has an abundance of and suggest changes.
6. In the plant illness prediction application, the farmer can upload a picture of a sick plant leaf, and the application will identify the disease and provide some historical context and treatment recommendations.
7. The difficulties of IoT devices and ML, the functions of AI, ML, and agriculture sensors need to be thoroughly examined.
8. The ML-based methods assist in identifying the best practices for crop management and harvesting in light of the current environmental changes.
9. Using ML, smart farming enables individuals to cultivate crops effectively, resulting in great yields at a cheap cost.
10. ML-based techniques aid in estimating the whole cost of farming. This would enable a person to plan their production tasks, leading to an integrated agricultural solution.
11. By offering detailed advice and information on the crops, ML technology helps farmers reduce their farming losses.

FUTURE RESEARCH DIRECTIONS

Future research should focus on predicting soil variables, including organic carbon, predicting agricultural yields, identifying diseases and weeds in crops, and identifying species. The classification of diverse sets of crop images using ML and computer vision will be investigated to assess crop quality and production. This strategy can be applied to improve livestock output by identifying reproductive trends, identifying eating disorders, and predicting the behavioral patterns of cattle using data from collar sensors. Drip irrigation and other smart harvesting methods are also included in the research, greatly decreasing the need for human effort in irrigation. Future studies will show improvements in knowledge-based farming which can raise the productivity and quality of products over the long term.

CONCLUSION

Since water availability will continue to be a global problem in the future decades, managing water in agriculture presents significant challenges. Therefore, precision irrigation(PA) is presented as a novel approach to boost agricultural productivity and control water excess utilizing cutting-edge technologies like cloud computing, WSN, and IoT. We have explored the most popular wireless technologies and the overall design of IoT-based irrigation methods. The main goal of developing intelligent irrigation systems based on the IoT and WSN is to reduce labor costs, time spent on tasks, and water usage. A real-time, low-cost automatic watering design based on IoT and WSN is created. With the advancement of technology, rule-based irrigation systems have been created. In rule-based systems, the system verifies a threshold before turning "ON" or "OFF" the motor. Sensing information is sent to the Raspberry Pi, a system's computing unit, and MQTT protocol was employed to gather data from several sensor nodes. The Mosquitto serves as a message broker. Suppose the measured moisture content of the soil is below the appropriate threshold. In that case, the farmer can immediately control the motor using a mobile application, minimizing the amount of water and energy wasted. The sensor information is then grouped into two groups and saved on the server. ML Classifiers such as KNN, Naive Bayes, and SVM decide if watering is required. ML classification performance measures demonstrate that the K-NN classifier outperforms the other two models considered for this investigation. The created system lessens the user's human labor and aids in irrigating the field only when necessary, ensuring that the required volume of water is utilized efficiently.

ACKNOWLEDGMENT

The authors would like to thank Bir Tikendrajit University, India, for their essential advice and assistance during the research process. The knowledge and insights of the Faculty of Engineering greatly influenced the direction and focus of our research. We also acknowledge the Department of Electrical and Computer Engineering at Effat University, Saudi Arabia, for giving us the resources and support required to finish this study. Their knowledge, perceptions, and encouragement were crucial in assisting us to finish this research.

REFERENCES

Abdallah, E. B., Grati, R., Fredj, M., & Boukadi, K. (2022). A Machine Learning Approach for a Robust Irrigation Prediction via Regression and Feature Selection. In *International Conference on Advanced Information Networking and Applications [AINA 2022: Advanced Information Networking and Applications],* (*vol. 449*, pp.491-502). Springer. 10.1007/978-3-030-99584-3_43

Abuzanouneh, K. I. M., Al-Wesabi, F. N., Albraikan, A. A., Duhayyim, M. A., Al-Shabi, M., Hilal, A. M., Hamza, M. A., Zamani, A. S., & Muthulakshmi, K. (2022). Design of machine learning based smart irrigation system for precision agriculture. [Tech Science Press]. *Computers, Materials & Continua, 72*(1), 109–124. doi:10.32604/cmc.2022.022648

Akram, V. K., & Challenger, M. (2021). A Smart Home Agriculture System Based on Internet of Things. In *2021 10th Mediterranean Conference on Embedded Computing (MECO)* (pp. 1-4). IEEE. 10.1109/MECO52532.2021.9460276

Al-Sarawi, S., Anbar, M., Alieyan, K., & Alzubaidi, M. (2017). Internet of Things (IoT) communication protocols: Review. In 2017 8th International Conference on Information Technology (ICIT), (pp. 685-690).IEEE.

Balasooriya, T. N., Mantri, P., & Suriyampola, P. (2020). IoT-Based Smart Watering System Towards Improving the Efficiency of Agricultural Irrigation. In *2020 IEEE Global Conference on Artificial Intelligence and Internet of Things (GCAIoT)* (pp. 1-7). IEEE. 10.1109/GCAIoT51063.2020.9345902

Bhoi, A., Nayak, R. P., Bhoi, S. K., & Sethi, S. (2021). Automated Precision Irrigation System Using Machine Learning and IoT. In Intelligent Systems, Lecture Notes in Networks and Systems, (vol 185, pp. 275-282). Springer. doi:10.1007/978-981-33-6081-5_24

Glória, A., Cardoso, J., & Sebastião, P. (2021). Sustainable Irrigation System for Farming Supported by Machine Learning and Real-Time Sensor Data. *Sensors (Basel)*, *21*(9), 3079. doi:10.339021093079 PMID:33925142

Goldstein, A., Lior Fink, L., Amit Meitin, A., Shiran Bohadana, S., Oscar Lutenberg, O., & Ravid, G. (2018). Applying machine learning on sensor data for irrigation recommendations: Revealing the agronomist's tacit knowledge. *Precision Agriculture*, *19*(3), 421–444. doi:10.100711119-017-9527-4

Gutiérrez, J., Villa-Medina, J. F., Nieto-Garibay, A., & Porta-Gándara, M. Á. (2014). Automated Irrigation System Using a Wireless Sensor Network and GPRS Module. *IEEE Transactions on Instrumentation and Measurement*, *63*(1), 166–176. doi:10.1109/TIM.2013.2276487

Haque, M. S. T., Rouf, K. A., Khan, Z. A., Emran, A., & Zishan, M. S. R. (2019). Design and Implementation of an IoT based Automated Agricultural Monitoring and Control System. In *2019 International Conference on Robotics, Electrical and Signal Processing Techniques (ICREST)* (pp. 13-16). IEEE. 10.1109/ICREST.2019.8644212

Katarya, R., Raturi, A., Mehndiratta, A., & Thapper, A. (2020). Impact of Machine Learning Techniques in Precision Agriculture. In *2020 3rd International Conference on Emerging Technologies in Computer Engineering: Machine Learning and Internet of Things (ICETCE),* (pp. 1-6). IEEE. 10.1109/ICETCE48199.2020.9091741

Kumar, J. C. R., & Kumar, D. V. (2022). Energy-efficient, high-performance and memory efficient FIR adaptive filter architecture of wireless sensor networks for IoT applications. *Sadhana*, *47*(248), 248. doi:10.100712046-022-02013-y

Kumar, J. C. R., Kumar, D. V., Arunsi, M., & Baskar, D. (2021). Energy-Efficient and Secure IoT architecture based on a Wireless Sensor Network Using Machine Learning to Predict Mortality Risk of patients with CoVID-19. In *2021 6th International Conference on Communication and Electronics Systems (ICCES)* (pp. 1853-1861).IEEE.

Kumar, J. C. R., & Majid, M. A. (2023). Optimizing Energy Efficiencies of IoT-based Wireless Sensor Network Components for Metaverse Sustainable Development using Carry Resist Adder based Booth Recoder (CRABRA). In *2023 20th Learning and Technology Conference (L&T)* (pp. 91-96). IEEE.

Monica, M., Yeshika, B., Abhishek, G. S., Sanjay, H. A., & Dasiga, S. (2017). IoT based control and automation of smart irrigation system: An automated irrigation system using sensors, GSM, Bluetooth and cloud technology. In *2017 International Conference on Recent Innovations in Signal processing and Embedded Systems (RISE)* (pp. 601-607). IEEE. 10.1109/RISE.2017.8378224

Motroni, A., Rosati, F., Nepa, P., Buffi, A., Pirozzi, M., Donato, L. D., Tomassini, L., & Ferraro, A. (2021). An RFID Tracking System for Agricultural Safety. In *2021 IEEE International Conference on RFID Technology and Applications (RFID-TA)* (pp. 28-31). IEEE. 10.1109/RFID-TA53372.2021.9617356

Patil, M. A., Adamuthe, A. C., & Umbarkar, A. J. (2019). Smartphone and IoT Based System for Integrated Farm Monitoring. In *Proceedings of the 2nd International Conference on Advanced Technologies for Societal Applications* (vol. 1, pp. 471-478).Springer.

Prakash, S. (2020). Zigbee based Wireless Sensor Network Architecture for Agriculture Applications. In *2020 Third International Conference on Smart Systems and Inventive Technology (ICSSIT)* (pp. 709-712).IEEE. 10.1109/ICSSIT48917.2020.9214086

Prakash, S., Sharma, A., & Sahu, S. S. (2018). Soil Moisture Prediction Using Machine Learning. In *2018 Second International Conference on Inventive Communication and Computational Technologies (ICICCT)* (pp. 1-6).IEEE.

Sah, R. R., Khatri, R., Basotra, T., & Chauhan, A. (2021). IoT Based Irrigation System -A Farmer Friendly Version. In *2021 3rd International Conference on Advances in Computing, Communication Control and Networking (ICAC3N)* (pp. 677-681). IEEE.

Sales, N., Remédios, O., & Arsenio, A. (2015). Wireless sensor and actuator system for smart irrigation on the cloud. In *2015 IEEE 2nd World Forum on Internet of Things (WF-IoT)* (pp. 693-698). IEEE. 10.1109/WF-IoT.2015.7389138

Shanbhag, N., Tambosker, S., Bhagat, R., Gaonkar, V., Rodrigues, D., & Leitao, F. (2020). Design of Sensor Nodes for Agricultural field monitoring Using IoT and BLE mesh technology. In *2020 IEEE 7th Uttar Pradesh Section International Conference on Electrical, Electronics and Computer Engineering (UPCON)* (pp. 1-6). IEEE. 10.1109/UPCON50219.2020.9376426

Singh, D. K., Sobti, R., Jain, A., Malik, P. K., & Le, D. (2022). LoRa based intelligent soil and weather condition monitoring with internet of things for precision agriculture in smart cities. *IET Communications*, *16*(5), 604–618. doi:10.1049/cmu2.12352

Suryady, Z., Shaharil, M. H. M., Bakar, K. A., Khoshdelniat, R., Sinniah, G. R., & Sarwar, U. (2011). Performance evaluation of 6LoWPAN-based precision agriculture. In *the 2011 International Conference on Information Networking (ICOIN2011)* (pp. 171-176).IEEE.

Tephila, M. B., Sri, R. A., Abinaya, R., Lakshmi, J. A., & Divya, V. (2022). Automated Smart Irrigation System using IoT with Sensor Parameter. In *2022 International Conference on Electronics and Renewable Systems (ICEARS)* (pp. 543-549).IEEE 10.1109/ICEARS53579.2022.9751993

Thaher, T., & Ishaq, I. (2020). Cloud-based Internet of Things Approach for Smart Irrigation System: Design and Implementation. In *International Conference on Promising Electronic Technologies (ICPET)* (pp. 32-37). IEEE. 10.1109/ICPET51420.2020.00015

Vaishali, S., Suraj, S., Vignesh, G., Dhivya, S., & Udhayakumar, S. (2017). Mobile integrated smart irrigation management and monitoring system using IOT. In *2017 International Conference on Communication and Signal Processing (ICCSP)* (pp. 2164-2167). IEEE. 10.1109/ICCSP.2017.8286792

Vootla, P., Remeithi, F. A., Bariaghabr, S. A., & Mansoori, F. A. (2018). Food waste - A global challenge to sustainability. In *2018 Advances in Science and Engineering Technology International Conferences (ASET),* (pp. 1-5). IEEE.

Zakzouk, S., Ehab, M., Atef, S., Yousri, R., Tawfik, R. M., & Darweesh, M. S. (2021). A Comparative Study of Machine Learning Techniques for Automatic Rice Crop Irrigation. In *2021 17th International Computer Engineering Conference (ICENCO)* (pp. 46-50). IEEE. 10.1109/ICENCO49852.2021.9698935

Zhang, X., Andreyev, A., Zumpf, C., Negri, M. C., Guha, S., & Ghosh, M. (2017). Thoreau: A subterranean wireless sensing network for agriculture and the environment. In *2017 IEEE Conference on Computer Communications Workshops (INFOCOM WKSHPS),* (pp. 78-84). IEEE. 10.1109/INFCOMW.2017.8116356

Chapter 4
Cyber-Physical System Framework for Efficient Management of Indoor Farming Production

Ahmed Mubarak Mekki Awouda
Politecnico di Torino, Italy

Benedetta Fasciolo
Politecnico di Torino, Italy

Giulia Bruno
Politecnico di Torino, Italy

Valentino Razza
Politecnico di Torino, Italy

ABSTRACT

World population is expected to reach 9.7 billion by 2050. Thus, a significant increase in agricultural production is required to accommodate population growth. However, environmental crises and urbanization pose a threat to agriculture. Vertical farming can partially mitigate these effects by growing plants while optimizing space and maximizing several outdoor resources. However, this approach necessitates thoroughly re-evaluating cultivation techniques and environmental factors. This chapter proposes a holistic, sustainable agricultural framework for developing indoor farming systems, which consists of a methodology for creating crop growth procedures from experimental data using statistical analysis and artificial intelligence algorithms. The proposed method aims to balance resource utilization and productivity in vertical farming. Furthermore, the authors propose a design framework to create a sustainable aeroponic system structure. This structure aims to provide a test bench to prove the effectiveness of the said methodology.

DOI: 10.4018/978-1-6684-7879-0.ch004

INTRODUCTION

To meet the food demand from the world's growing population, food production must double by 2050 (United Nations, 2022). A significant increase in agricultural production is essential to meet the food demand growth. Such an increase can be challenging to achieve, mainly due to the climate change occurring in the recent years.

One of the main effects of climate change is the temperature increase. Higher temperatures lead to higher crop respiration and evapotranspiration rates, wider pest infestation, a shift in weed flora, and reduced crop duration (Malhi et al., 2021). The higher temperatures also increase the microbial population present in the soil and their enzymatic activities. If the enzymes rise above a critical range, they denature (Tudi et al., 2021). Also, soil salinity is affected by the temperature increase and the consequent water reduction (Kumar et al., 2020). The increase in salt concentration harms plants due to osmotic stress and ion toxicity (Safdar et al., 2019). The soil pH, influenced by soil conditions and anthropogenic activities, impacts plant growth and development by affecting nutrient uptake, root diameter, and root length. For instance, acidic pH inhibits mineral translocation, while basic pH reduces plant zinc accumulation (Rahman et al., 2021). Climate change also affects the concentration of nutrients in the soil. Particularly, it leads to shortages of macro and micronutrients, such as phosphorus and nitrogen (Young et al., 2021). Soil usually contains a considerable amount of total phosphorus, but the plants can absorb only orthophosphate ions found in low concentrations in the soil solution (Alotaibi et al., 2021). Even if organic and inorganic fertilizers increase phosphorus availability, orthophosphates in the soil solution react rapidly with other soil components and become unavailable for plants.

On the other hand, the increase in current agricultural activities also negatively impacts the environment. The first negative effect is that, due to deforestation and intensive agriculture, a large amount of CO_2 is released into the atmosphere (Lal, 2004). Another adverse effect is due to the use of fertilizers and pesticides. Even if they are essential to provide plants with the necessary nutrients, they cause water and air pollution (Lask et al., 2021). Crops absorb only one-third of the applied fertilizer, while the leftover gets lost in the environment (Kumar et al., 2021). The use of pesticides also results in a loss of biodiversity (Young et al., 2021). About one-third of agricultural products are produced based on the application of pesticides. However, their use affects environmental pollution of water, air, soil, and biodiversity (non-target organisms) through adsorption, leaching, volatilization, spray drift, and runoff (Tudi et al., 2021). Lastly, some agricultural techniques hurt the environment. For example, continuous cropping has decreased rhizosphere soil pH, nutrient imbalance, reduced enzyme activity, and modified microbial community composition, causing an increase in pathogenic microorganisms (Zhao et al., 2021).

For these reasons, it is evident that there is a significant mutual influence between agriculture and the environment. Adopting sustainable agricultural techniques can decrease the negative effects of this influence. Several methods have been developed to mitigate the impacts described above. Some involve smart agriculture technologies such as irrigation and nutrition management or smart controlled agriculture (Malhi et al., 2021; Ragaveena et al., 2021). Others can be grouped into traditional agricultural techniques such as crop residue retention, cover cropping, no-tillage, etc. (Murphy et al., 2016; Young et al., 2021; Lal, 2004). Moreover, various benefits of using biochar have been studied, such as its positive influence on phosphorus availability in arid soils (Alotaibi et al., 2021) and its better-performing fertilizer behavior than traditionally used chemicals (Kumar et al., 2021). Additionally, the beneficial role of Plant-Growth Promoting Bacteria (Kumar et al., 2020) and seedlings primed with humic acids help plants to grow in high salinity conditions (Souza et al., 2021). Lastly, nano fertilizers have been studied to increase the efficiency of nutrient delivery to plants (Mejias et al., 2021).

This chapter considers one of these sustainable agriculture technologies, the aeroponic system, a vertical farming technique. Vertical farming is a method that, by exploiting vertical surfaces, allows for saving soil and natural resources, and it is especially used in urban contexts. Vertical farming mainly includes techniques such as hydroponics, aquaponics, bioponics, and aeroponics. Hydroponics is a technology to grow plants without soil by providing nutrients or fertilizers in a nutrient solution (Richa et al., 2020). Aquaponics is a technology that allows the mutual growth of plants and fish, where the waste from fish is used as fertilizer for crops. Bioponics combines aquaponics and hydroponics techniques, where the nutrients released by the biological activity of microorganisms are derived from multiple sources, such as natural substances of plant, animal, and mineral origin (Ragaveena et al., 2021). Aeroponics is a technology that provides the nutrient solution by nebulization onto the plant's roots. Also, some plants live naturally in aeroponic systems, collecting nutrients from mist aerosol droplets through the leaves and roots, such as epiphytic orchids and bromeliads (Bethany et al., 2020). Figure 1 shows the architecture of the aeroponic technology. The plant roots are exposed to aerosol droplets that contain the nutrient solution. One of the most used techniques for nebulization is high-pressure atomization, where a high-pressure nutrient solution is forced through an orifice turning the liquid into droplets. The aeroponic structure usually consists of a tank where plant roots and nozzles are placed and pipes that recover or supply the nutrient solution. This tank is contained within a thermally insulated chamber. In the aeroponic systems, the following actuators are used: (i) pumps and nozzles to provide the nutrient solution to the roots, (ii) LED lights to provide artificial lighting to the system, (iii) air conditioning system and fans to regulate air circulation, temperature, and humidity, and (iv) piping system for the waste recovery of the nutrient solution.

Even if, compared with the hydroponics system, the aeroponics system requires more extensive farm infrastructure and actuators, the present work considers aeroponics due to its potentiality to solve several plant physiological constraints occurring during hydroponic cultivation, such as a greater oxygen availability within the root bed and an enhanced water use efficiency. Furthermore, it is important to integrate IoT and AI into vertical farming to make aeroponic systems more efficient and ensure less environmental impact. The system can acquire data on plant physiology and growth in an agricultural-controlled environment through IoT sensors and connections. Then, the collected data can be analyzed by AI algorithms to monitor plant development by optimizing resources. In recent years, several papers have been published describing soilless techniques and comparing them to traditional agriculture. Moreover, several works study the integration of IoT and AI networks into these systems. However, a gap in the literature is found on how these integrated systems can be optimized to improve indoor productivity and resource utilization. This chapter presents a methodology to develop a crop growth procedure to maximize plant production and the utilization of resources within the system.

Figure 1. Aeroponic system

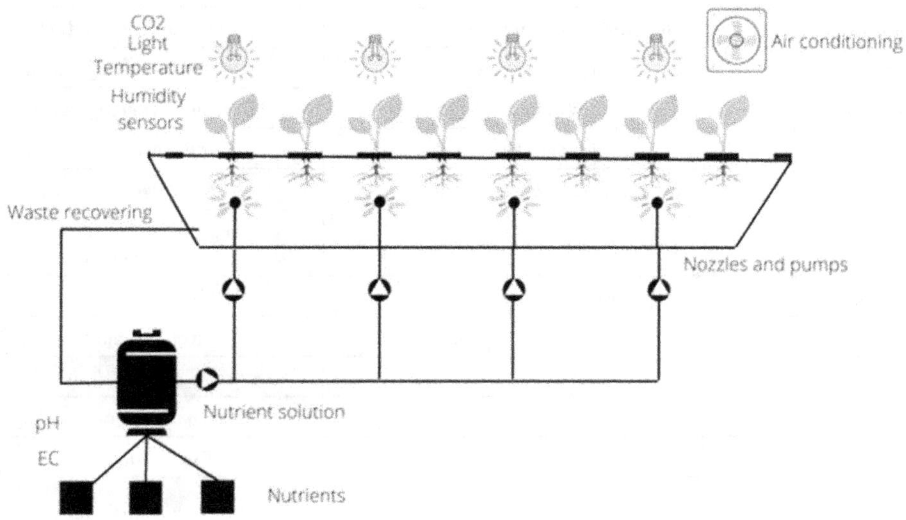

Literature Review

The first experiments of vertical farming with hydroponics technology were done in the mid-17th century; then, in the early 1900s, the technique was improved, and the term "hydroponics" was coined (El-Kazzaz, 2017). Given the pressure of the environmental crisis in recent decades, soilless agriculture obtained increasing interest.

A literature search was conducted to highlight this growing trend. The applied procedure is structured in the following steps.

1. Selection of keywords
2. Extraction and selection from the Scopus database
3. Combination of queries
4. Analysis of results

Three queries with different keywords were constructed on specific topics summarized in Table 1. In the first query, synonyms for vertical farming were selected to find papers on soilless technologies. The objective of the second query was to find articles related to IoT implementation in agriculture. Finally, the last query is used to find papers related to the use of AI in agriculture. By executing the queries separately and searching papers by the terms in the title, keywords, and abstract, a total number of results, 27091, 22422, and 9208, respectively, have been obtained.

Table 1. Queries and Keywords

Vertical farming query	IoT and agriculture query	AI and agriculture query
vertical *and* farming*	IoT *w/2* agriculture	AI *w/2* agriculture
soilless *w/2* system*	Internet *and* of *and* things *w/2* agriculture	Artificial *and* intelligence *w/2* agriculture
hydroponic*	cps *w/2* agriculture	Neural network *w/2* agriculture
aeroponic*	cyberphysical *w/2* agriculture	AI AND algorithm* *w/2* agriculture
aquaponic*	cyber-physical *w/2* agriculture	Artificial *and* intelligence *and* algorithm* *w/2* agriculture
urban *and* farming*	precision *and* farming*	Machine *and* learning *w/2* agriculture
soilless *w/2* agriculture	automated *and* agriculture	Deep *and* learning *w/2* agriculture
		smart *and* agriculture
		Data *and* analytics* *w/2* agriculture
		Big *and* data *w/2* agriculture

The query on vertical farming was then combined with the other queries to explore how soilless systems are integrated with IoT and AI. To conduct this analysis, the word "agriculture" was removed from the keywords in the second and third queries, as it might have been redundant. The intersection of vertical farming and IoT produced 717 articles. This means that, although the two topics are highly researched, there are limited combined results. However, in Figure 2, there is evidence that the intersection between the two subjects has been growing strongly in recent years.

Figure 2. Documents by year: the intersection of vertical farming and IoT

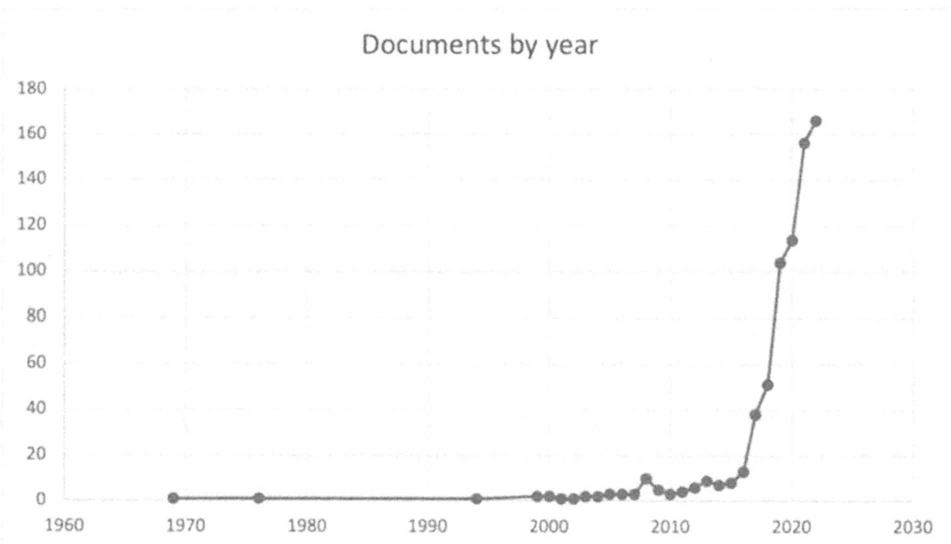

Similarly, the intersection of vertical farming and AI produced a total of 691 items. Figure 3 shows the growing trend of this subject.

Figure 3. Documents by year: the intersection of vertical farming and AI

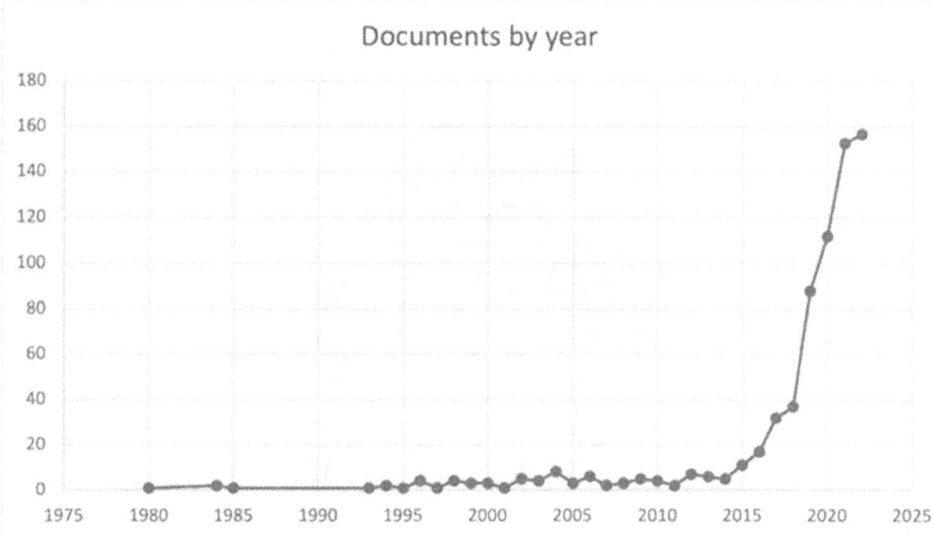

This analysis also shows that vertical farming is well spread across different disciplines. Indeed, Figure 4 and Figure 5 show that AI and IoT integrated into soilless agriculture is a topic of interest not only in engineering and computer science disciplines but also in other fields such as environmental, social, or agricultural.

Figure 4. Disciplines: soilless agriculture and AI

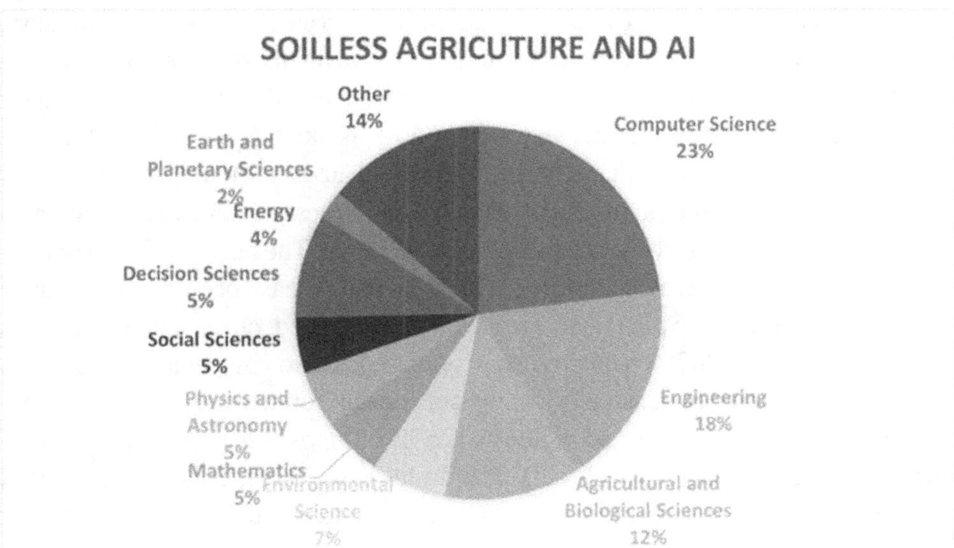

Figure 5. Disciplines: soilless agriculture and IoT

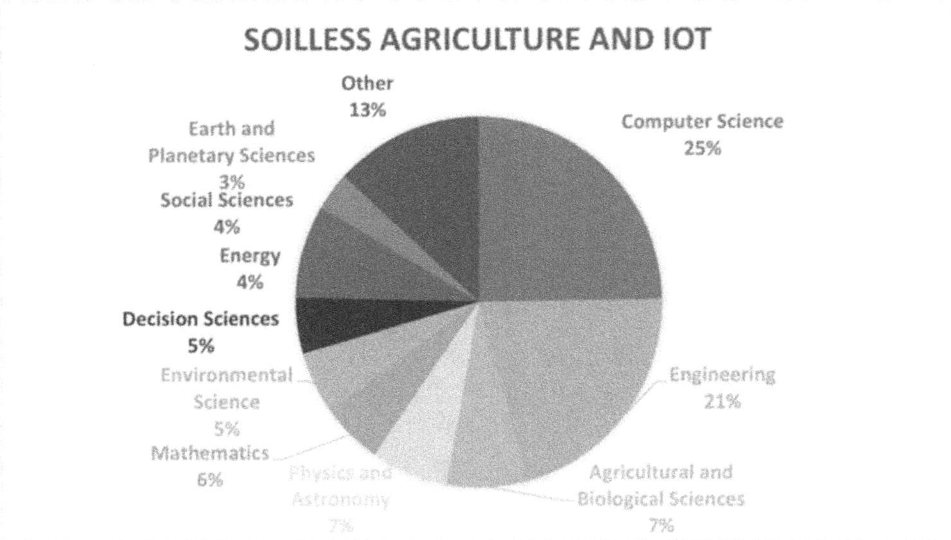

Several publications outlining soilless techniques and contrasting them with traditional agriculture have recently been published. Additionally, efforts are being made to incorporate IoT and AI networks into these systems. However, this merger is yet to be consolidated and properly evaluated.

Utilizing vertical farming methods to grow plants can have several benefits and reduce some drawbacks that impact open-field agriculture's efficacy and productivity. First, the decrease in CO_2 emissions from transporting the crop to metropolitan areas is an advantage worth highlighting. Abandoned structures can be reclaimed and exploited to practice vertical farming in urban and suburban regions (Avgoustaki & Xydis, 2020; Eldridge et al., 2020; Tunio et al., 2020). Furthermore, crop waste and harvest quality can both be affected by transportation. Even at extremely low refrigeration temperatures, the nutritional content of fresh fruits and vegetables is reduced during transport, and the food supply chain results in the loss or waste of 40 to 50 percent of fruits and vegetables (Avgoustaki & Xydis, 2020). Moreover, vertical farming can use up to 98% less water than an open field. The air conditioner's evaporator compresses and collects 95% of the water from plant evapotranspiration in these systems. After being sterilized, it is stored in the water tank and reused (Avgoustaki & Xydis, 2020). The decrease in pesticides and fertilizers is yet another reason for significant savings. This is so that disease occurrence is reduced, and the amount of nutrition delivered to the plant is maximized in a controlled, closed environment (Avgoustaki & Xydis, 2020; Eldridge et al., 2020). The productivity of yields also rises in a vertical farming system since production is not impacted by seasonality or the outside environment. Due to these elements, vertical systems can be built even in nations and regions where it is impossible to cultivate some crops due to climatic circumstances (Avgoustaki & Xydis, 2020; Eldridge et al., 2020; Tunio et al., 2020). Crop yield without affecting soil health is a furthermore advantage. This could lead to converting certain agricultural lands back into forests, enhancing global soil health and soil carbon sequestration (Hati & Singh, 2021). However, modern vertical farming systems do face various restrictions and difficulties. In fact, these systems require electricity to run the air conditioning and artificial lighting (Avgoustaki & Xydis, 2020), and the system's components and materials may be pricey. For instance, sensors, high-pressure pumps, and LEDs can be more expensive than standard greenhouse equipment (Tunio et al., 2020). Finally, to run a vertical farming system effectively, the user needs a proper knowledge (Tunio et al., 2020).

Based on the results of past studies, the three main resource utilization indicators are (i) water consumption, (ii) land consumption, and (iii) energy consumption. Vertical farming performs significantly better than greenhouse and open-field agriculture in terms of water and land consumption. On the other hand, the energy consumption of vertical farming is usually higher than that of other systems.

Regarding water consumption, according to Orsini et al. (2020), only 3 to 20 grams of lettuce can be produced with 1 liter of water in the open field, compared to 5-60 gr/L in a greenhouse and 45-80 gr/L in a vertical farm. This is because the water use efficiency in vertical farms is over 0.95, while it is about 0.02 in a greenhouse (Toyoki Kozai et al., 2019). This high efficiency is due to optimized water use during crop growth and avoiding the use of water to wash crops before eating to remove pesticides, insects, and foreign particles (Toyoki Kozai et al., 2019). Considering that the agricultural industry uses over 70% of freshwater worldwide, it is clear that there is an urgent need to find cropping systems that minimize water consumption (FAO, 2018).

Regarding land exploitation, in one square meter of soil, 10-15 grams of lettuce are produced per day in the open field, 100-300 g/m2d in a greenhouse, and 1300-3000 g/m2d in vertical farming (Barbosa G. et al., 2015). Soil fertility is one of the main crop production requirements in the open field. However, up to 33% of global soils have reduced or lost fertility due to degradation processes caused by unsustainable soil management practices (FAO, 2022). Moreover, more than two-thirds of the world is expected to be urban by 2050, which will lead to a large amount of soil being converted into urban areas (United Nations, 2019). Soil degradation and urbanization are the main factors that necessitate finding a cropping method to minimize soil exploitation and consumption.

The main drawback of vertical farming is that energy consumption is usually higher than that of greenhouses and open fields. This is mainly due to the consumption of the lighting system and climate control (Yokoyama, R., 2019). According to Orsini et al. (2020), 1 kg of fresh weight of lettuce can be produced with only 0.5 kWh in the open field, while it requires 9 kWh in a greenhouse and around 13.3 kWh in a vertical farming system. The higher energy requirements can lead to increased economic costs and environmental impact, although this depends on how the energy is produced. For instance, the vertical farming life cycle assessment reported by Romeo D. et al. (2018) found that the CO2 emissions of a vertical hydroponic farm are very similar to those of the open field (0.39 kg CO2eq and 0.29 kg CO2eq, respectively) compared to the greenhouse (7.1 kg CO2eq). This is because the considered system used an energy mix made of more than 70% nuclear energy. If renewable energy sources are used, the environmental impact can be further reduced. Another way to reduce the economic cost and the environmental impact of energy consumption is to optimize the lighting system, which is responsible for 50% of the energy costs (Yokoyama, R., 2019). To this aim, some works suggest using optimized LEDs and identifying the optimal light spectrum (Kozai T. et al., 2015). Finally, concentrating crop production in the less cold season could reduce emissions and energy costs due to their seasonal dependence (Barla S. et al., 2020).

FRAMEWORK FOR SMART INDOOR FARMING SYSTEM

The authors propose a novel approach to tailor the best crop growth procedure in the latter, optimizing some metrics like productivity or sustainability. The design of an intelligent IoT-based aeroponic system is also reported in the second part of this section.

Methodology

The proposed methodology aims at defining a crop growth procedure (CGP) to balance resource utilization and productivity in an aeroponic greenhouse, i.e., derive the optimal values of greenhouse parameters to achieve a sustainable and efficient system. The methodology is divided into three phases, as shown in Figure 6. Such phases are repeated for each plant growth stage within the soilless system (seedling and adult plant). For each step, the arrows entering from the left side represent the inputs of the phase, the arrows exiting from the right side represent the outputs, the incoming arrows from the upper side represent the controls used to implement the stage, and the incoming arrows from the lower side represent the resources used. Each phase is described in detail in the following paragraphs.

Figure 6. Proposed methodology

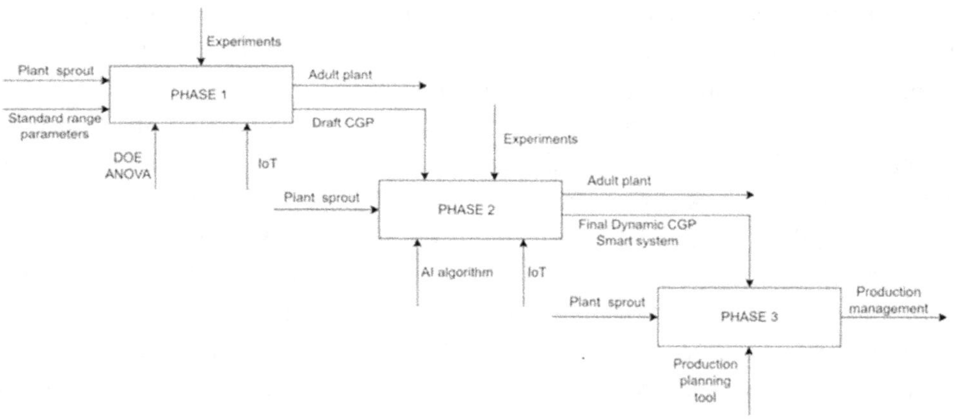

In the first phase, the authors study the cause-and-effect relationships between the input and the output parameters. Initially, the primary and essential parameters for plant development, such as nutrients, water, light, temperature, etc., are studied. Then, the same analysis is done for the secondary parameters, which include the structure of the aeroponic system, such as nozzle size, nutrient pressure, and droplet size. To generate a considerable amount of data, several experiments will be carried out on a specific type of crop in the aeroponic greenhouse by changing the input values and measuring the plant's response. As described later, the system is monitored and controlled by sensors and actuators. The collected data are used to perform statistical analyses to identify cause-and-effect relationships between the inputs and outputs of the system by using the Design of Experiment and ANOVA. At the end of the first step, there will be an initial understanding of the cause-effect relationships between plant growth and primary and secondary parameters. Moreover, a draft crop growth procedure is defined to get the basic parameters for implementing the second phase.

The data from the initial crop growth procedure and the new experiments are used to train AI algorithms to derive more knowledge. Indeed, a trained AI algorithm can be used as a model to evaluate new strategies not tested in previous experiments. The final crop growth procedure can be considered a dynamic recipe since the recurrent learning phase of the AI algorithm allows continuous improvement of the process. In addition, the combination of AI and IoT leads to a smart aeroponic system that will optimize crop production.

The last phase of the methodology addresses production planning according to demand. Nowadays, in industrial scenarios, food production must cope with volumes and delivery times stated by the consumers, thus forcing the indoor plant to produce a set of mixed crops, possibly with different compositions in different periods. Crop production in aeroponic systems should follow a just-in-time logic. This is essential to ensure good product quality and avoid food waste. In particular, this happens when the production is off-season, as the crop could quickly deteriorate once out of the aeroponic system due to hostile climatic conditions. Therefore, with the specific receipt and crop growth time obtained from the previous phases, the production can be scheduled according to demand minimizing waste and resources through a production planning tool. This phase will also involve several experimental activities on mixed crops grown to suggest the best crop mixing and timing for an optimal trade-off among quality, yield, and resource utilization.

IoT/CPS Architecture

An intelligent aeroponic growth system has been designed and developed to implement the described methodology. The overall system architecture is shown in Figure 7. It consists of sensors, actuators, and various informatics systems. In the following, each part of the architecture is explained in more detail.

Figure 7. System architecture

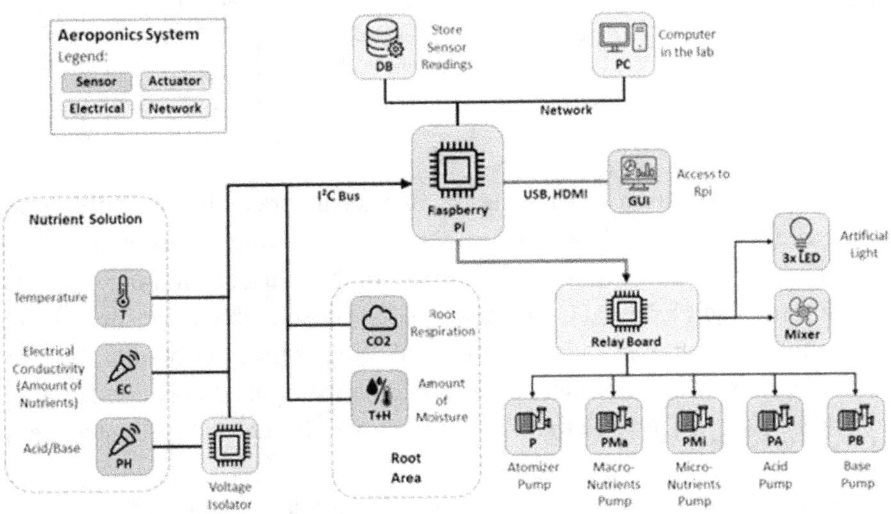

Hardware

Several hardware components have been used in this work, whose main objective is to provide monitoring, control, and automatic data acquisition from the system. These components include sensors, actuators, and controllers.

Sensors

Depending on the sensed environment, the sensors operate in the nutrient environment or the plant's climate. The detailed technical specifications of the sensors considered in this work are reported in Table 2. Nutrient sensors are placed inside the solution reservoir to acquire the main solution parameters, i.e., pH and electrical conductivity. Climate sensors measure the temperature, humidity, and CO_2 in the plant's environment. They have been installed in the growth chamber.

Table 2. Sensors used in the aeroponic system

Designation	Sensor	Transmitter	Brand Source	Specifications
PH	Double junction silver/ silver chloride Lab Grade pH Probe	EZO pH Circuit Protocol: I2C	Atlas Scientific (PH Kit)	Range 0-14 Accuracy: 3mm Conformity: ISO 10523
Electrical Conductivity	Mini Conductivity Probe K 1.0	EZO EC Embedded conductivity circuit Protocol: I2C	Atlas Scientific (Conductivity Kit)	Range: 5 – 200,000 µS/cm Accuracy: +/- 2% Conformity: NSF/ANSI 51
Solution Temperature	Waterproof DS18B20 Digital temperature sensor	Protocol: 1-Wire	Adafruit (Digital temperature sensor)	Range: -55-125°C Accuracy: ±0.5°C
CO2	Embedded NDIR Carbon Dioxide sensor	Protocol: I2C	Atlas Scientific (Ezo-CO2)	Range: 0-10,000 ppm Accuracy: (+/- 5%) + (+/- 50 ppm)
Light Intensity	GY-30 with AD converter	Protocol: I2C	Tecnoiot	Range: 0-65535 lux Accuracy: +/- 1 lux
Air Temperature/ Humidity	AM2315C enclosure contains: DS18B20 Temperature sensor +Capacitive Humidity sensor	Protocol: I2C	Adafruit (Temperature Humidity sensor)	Range T: -40– 80°C Accuracy: T: +/- 0.3°C H: +/- 2%

The pH sensor provides information about the acidic or basic nature of the nutrient solution. Nominal pH must be kept between 5.5 and 6.5, depending on the crop type (Ragaveena et al.,2021). pH values far from this specific range could prevent the plant's uptake of the nutrient solution, leading to suffering plants' growing conditions. The Electrical Conductivity (EC) sensor, which measures the total concentration of salts in the nutrient solution, indicates the total ionic concentration and the strength of the nutrient solution. Its measure, whose optimal range should be maintained between 1.5 and 2.5 dS m-1 (Richa et al., 2021), is necessary to provide the exact amount of nutrients and guarantee the repeatability of the experiments.

The growth chamber environment is monitored through a temperature sensor to capture temperature fluctuations that can significantly affect the root growth, respiration, transpiration, flowering, and dormant period (Lakhiar et al, 2018). A humidity sensor is used to maintain the moisture of the aeroponic chamber at a specific range. In the literature, humidity is recognized as a key indicator of the number of nutrients available in the chamber's root area (e.g., Lakhiar et al, 2018) since nutrition is provided via tiny droplets in the air. Humidity readings can be substituted with Vapour pressure deficit (VPD), VPD is a more accurate humidity measurement because it considers the effect of temperature on the water-holding capacity of air, which is what ultimately drives transpiration of the leaf surface (Amitrano et al, 2019). Light intensity sensors are used to measure Lux, which is a measurement of light intensity as perceived by humans; this can then be converted into PPFD, which measures the number of photons within the PAR spectrum reaching the leaf surface. Light is an essential element for plant growth because it influences vital processes such as photosynthesis (Ragaveena et al., 2021). Finally, a CO2 sensor is needed to check the air quality; carbon dioxide concentration fluctuations influence roots' respiration, nutrient absorption, and overall plant growth (Gour et al. 2020).

Actuators

The primary actuators used for this aeroponic system are pumps. They're used for irrigation and nutrient solution management. The main pump delivers the solution to the plants via the atomization nozzles. Two peristaltic pumps have been used to control the PH of the solution by controlling the amount of acid or base added to the solution. Another pair of peristaltic pumps are used to add Macronutrients & Micronutrients. Furthermore, a magnetic mixer is used to ensure the homogeneity of the solution.

Lighting is essential for plants as it plays a vital role in photosynthesis. Lack of light significantly impacts the photosynthesis process, which, in turn, impedes plant growth. In this work, the type of light used was a 24V LED lamp. It has been selected because it has high blue-light intensity, which makes it suitable for growing leafy vegetables. The specifications of the actuators used are shown in Table 3.

Table 3. Actuators used in the aeroponic system

Designation	Component	Specifications
Atomizer Pump	Booster Pump	Flow rate: 50 GPD Pressure: 125 psi Rating: 24V AC
Macro-Nutrients Pump Micro-Nutrients Pump Acid Pump Base Pump	Embedded Peristaltic Dosing Pump (Ezo-pump)	Flow rate: 0.5-105ml/min Accuracy: +/- 1% Protocol: I2C
Artificial Light	LED	Size: 70 cm Rating: 24V
Mixer	Stir Bar	Material: neodymium magnet Dimensions: 1x5 cm

Controllers and Boards

Raspberry Pi was used, a single-board microcomputer equipped with a 40-pin GPIO header to connect sensors, actuators, and other peripherals (Upton et al, 2016). The model used in this work is a Raspberry

Pi 4 Model B, running a Broadcom BCM2711 (1.5GHz) processor, together with a few peripherals to connect the board to the network (i.e., wireless, Bluetooth, and Gigabit Ethernet) for the communication with a remote database. To let the Raspberry Pi drive the non-I2C-based actuators, a relay board is used for an ON-OFF control of such devices.

Software and System

Node-Red was chosen as a development environment for programming the system logic and managing the various data flows. Node-Red is a flow-based open-source programming tool built upon Node.js. More specifically, it is a process-based virtual programming environment that creates data streams from the sensor to the cloud by connecting hardware and software (Node-RED Features). It provides a web browser-based flow editor, which can be used to create JavaScript functions. It has been chosen because of its ability to run on the Raspberry Pi and the abundance of libraries available, together with a graphical programming tool that allows the user to focus on logic development rather than fixing code syntax and debugging. Finally, a remote database is designed to store the acquired input and output signals for post-processing analysis. The database system, based on PostgreSQL, is reachable through the local network.

Structure

The growth chamber makes an isolated and controlled environment. The chamber is a 150x150x200cm grow tent made from 1680D waterproof polyester, while the interior is lined with reflective mylar. The structure of the chamber is made from 16mm steel tubes to ensure structural stability and rigidity. Inside the chamber, the plant bed consists of jute dowels suspended on a plastic mesh of 60x40cm that allows 228 plants to be grown simultaneously.

Functions Supported

The designed setup will allow for validating the proposed approach. In particular, the overall system can regulate the main crop parameter to desired values and collect information for later analysis. The growth chamber can be monitored and controlled remotely, allowing an intelligent system to develop and apply an automatically generated growth recipe. Thanks to the full automatization of the growth chamber, it is possible to manage several aspects of the crop, such as nutrition management, environment control, and climate monitoring, continuously during the day.

The control logic manages crop nutrition by providing the correct amount of nutrients delivered at the right time to the plant. The nutrition solution is kept at the desired conditions thanks to the control performed with pH and electrical conductivity. A feedback control law is designed to maintain these values at a certain threshold via feedback control using peristaltic pumps that inject the correct amount of minerals into the solution. Furthermore, the system operates the main pump to transfer the obtained solution to the plants via the atomization nozzles only when the desired nutrient parameters are achieved. All operations are unattended, letting the system operate autonomously 24h per day to maintain the crop condition at the suitable value for the experiments. An automatic error messaging system warns the user if the water level in the nutrient solution reservoir is below a given threshold.

Another aspect is environmental control, especially related to incident lighting. Light is essential for plant growth as it is needed for photosynthesis. Several LED lights can be switched on and off within the growth chamber to combine colors and luminance frequency.

The user can continuously monitor the state of the aeroponic chamber by remote. All the sensor data is available in real time, and an error signaling system warns the user when parameters are outside the safety range. This should let the user fix problems without compromising the experiments. Furthermore, a remote database stores data throughout the growth cycle to allow a-posteriori data processing.

CASE STUDY

Since the described system is currently under development, in order to show the application of the framework, a public dataset on aeroponic culture was used (Torres Tello et al., 2020). The dataset description and the methodology application are reported in the rest of this section.

Dataset Structure

The dataset included six different crops: garlic chives, basil, red chard, rainbow chard, arugula, and mint. In total, 200 samples have been collected. Two kinds of features are available for each sample: static inputs and sensor data. The sensor data are further divided into nutrient variables and environmental variables. The sensor data were collected hourly for each crop; then, the average value was computed for the time frame in which each crop was in the aeroponic system. Table 4 reports the list of sensor data and static inputs available. In addition, for each crop, the crop yield measured in weight per spot was recorded at the end of the growing period. This is the target variable to predict.

Table 4. Features available in the aeroponic culture dataset

Nutritional variables	Environmental variables	Static inputs
Solution pH	Temperature	Type of crop
Solution TDS (total dissolved solid)	Humidity	Number of days in the tray
Reservoir temperature	CO_2 level	Number of days in the tower
	Room light level	Grow number (how many times a plant has been transplanted)
	Vapor pressure deficit	Harvest number (how many times the plant has been harvested)

Machine Learning Application

It is assumed that the first phase of the methodology was already applied to define the experiments described by (Torres Tello et al., 2020). In fact, they have chosen the inputs to collect and the range of values of the sensors. Thus, the authors proceeded with the second phase, in which different machine learning algorithms are exploited to predict crop yield.

The following machine-learning algorithms were applied in order to find the best regressor model.

- Linear regression (LR), which identifies the best fitting model to represent the connection between an outcome variable and a group of independent variables as a straight line or a hyperplane (Rencher and Christensen, 2012).
- Decision tree regression (DT), which builds a model as a tree structure with decision nodes and leaf nodes (Rokach and Maimon, 2014). A decision node has two or more branches, each representing values for the attribute tested. A leaf node represents a decision on the numerical target. The root node is the topmost decision node in a tree, corresponding to the best predictor.
- Support vector regression (SVR), which builds a hyperplane or set of hyperplanes in a high dimensional space and tries to fit the best line within a threshold value, i.e., the distance between the hyperplane and boundary line (Awad and Khanna, 2015).
- Artificial neural network (ANN), which builds a network of nodes, and learning occurs by changing connection weights after each piece of data is processed, based on the amount of error in the output compared to the expected result (Haykin, 1998).

Before applying the algorithms, the features are standardized by subtracting the mean and scaling to unit variance (Han et al., 2011). The leave-one-out cross-validation has been adopted to divide samples into training and test sets. It is a particular type of cross-validation where the number of folds is equal to the number of instances in the dataset. It means that the algorithms are applied using, for every run, one sample as test set and all the others as training set (Sammut and Webb, 2011).

The performance metric used to compare the models is the mean average error (MAE), i.e., the average absolute difference between the real and predicted values (Willmott and Matsuura, 2005). The Python open-source library Scikit-learn (https://scikit-learn.org/stable/) has been used to implement the machine learning algorithms. The obtained results are reported in Table 5. DT shows the minimum error; thus, for this dataset, it is the most accurate algorithm to predict crop yield. Figure 8 shows the graphical comparison between the actual and predicted crop yields by the DT algorithm.

Table 5. Results of the ML algorithms

	LR	**SVR**	**ANN**	**DT**
MAE	0.34	0.31	0.28	**0.21**

Figure 8. Comparison between the actual crop yield (blue) and predicted crop yield by the DT algorithm (red)

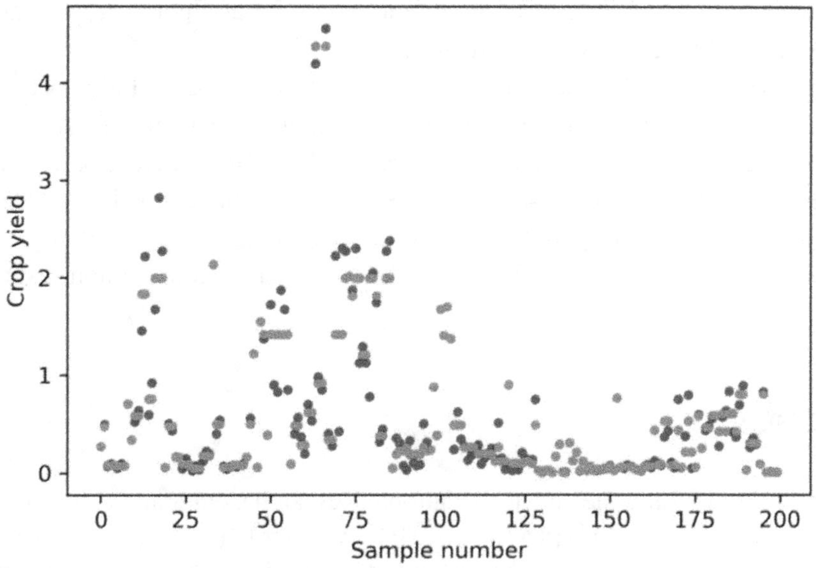

Generalization

The proposed methodology is a general approach for defining a tailored recipe for crop growth. Depending on the available data, the user can select different objectives for production, e.g., increase product quality, have sustainable production, reduce costs, or increase productivity. In some cases, these objectives contrast with each other (e.g., a high-quality product may be more expensive and with a lower production rate). For this reason, when the user tries to select more than one objective, the optimal solution cannot be unique. A trade-off, defined as the weighted sum of each objective, should be defined time by time to meet the user target.

CONCLUSION

This chapter examined how agriculture affects the environment and how the climate crisis threatens crop production. Systems for vertical farming can reduce some of the negative effects of harvesting on the environment. Nevertheless, urban farming methods alone cannot revolutionize agriculture and satisfy production demands; hence, they need to be coupled with technological innovations to reach their full potential and benefits. Enabling technologies such as the Internet of Things (IoT) and machine learning (ML) can enhance urban farming greatly if utilized correctly. Furthermore, the combination of these technologies with urban farming provides a much-needed solution for the issue of urbanization by allowing people to cultivate independently of external climatic conditions (Anytime, anywhere cultivation).

The methodology and system design framework presented in this work can optimize the production of soilless crops. It can also be used to strengthen the system's sustainability. A crop growth procedure could be created using the findings from earlier experiments to strike a balance between maximizing productivity and minimizing resource use. An example of using the methodology has also been presented in this work, specifically a case study for the application of phase two. Various machine learning algorithms have been used to create a yield prediction model for aeroponic crops. The best result was obtained using Decision Tree Regression with an MAE of 0.21

Even though it can significantly reduce the amount of resources used, including water, soil, and fertilizer, vertical farming is more expensive and requires more expertise than open-field agriculture. Because of this, it is crucial to conduct a socioeconomic analysis to determine whether or not this technique can supplement or integrate current crop production. The energy efficiency of the system is an additional crucial factor. Both artificial lighting and air conditioning systems are power consumers in aeroponic greenhouses. Therefore, energy consumption must also be considered in economic and environmental impact studies.

ACKNOWLEDGMENT

This study was carried out within the Agritech National Research Center and received funding from the European Union Next-GenerationEU (PIANO NAZIONALE DI RIPRESA E RESILIENZA (PNRR) – MISSIONE 4 COMPONENTE 2, INVESTIMENTO 1.4 – D.D. 1032 17/06/2022, CN00000022) and PIANO NAZIONALE DI RIPRESA E RESILIENZA (PNRR) – Decreto Ministeriale n. 1061. This manuscript reflects only the authors' views and opinions, neither the European Union nor the European Commission can be considered responsible for them.

REFERENCES

Alotaibi, K. D., Arcand, M., & Ziadi, N. (2021). Effect of biochar addition on legacy phosphorus availability in long-term cultivated arid soil. *Chemical and Biological Technologies in Agriculture*, *8*(1), 47. Advance online publication. doi:10.118640538-021-00249-0

Amitrano, C., Arena, C., Rouphael, Y., De Pascale, S., & De Micco, V. (2019). Vapour pressure deficit: The hidden driver behind plant morphofunctional traits in controlled environments. *Annals of Applied Biology*, *175*(3), 313–325. doi:10.1111/aab.12544

Avgoustaki, D. D., & Xydis, G. (2020). Indoor vertical farming in the Urban nexus context: Business growth and resource savings. *Sustainability (Basel)*, *12*(5), 1965. doi:10.3390u12051965

Awad, M., & Khanna, R. (2015). Support Vector Regression. In *Efficient Learning Machines*. Apress. doi:10.1007/978-1-4302-5990-9_4

Barbosa, G., Gadelha, F., Kublik, N., Proctor, A., Reichelm, L., Weissinger, E., Wohlleb, G., & Halden, R. (2015). Comparison of land, water, and energy requirements of lettuce grown using hydroponic vs. conventional agricultural methods. *International Journal of Environmental Research and Public Health*, *12*(6), 6879–6891. doi:10.3390/ijerph120606879 PMID:26086708

Barla, S., Salachas, G., & Abeliotis, K. (2020). Assessment of the greenhouse gas emissions from aeroponic lettuce cultivation in Greece. *Euro-Mediterranean Journal for Environmental Integration*, *5*(2), 29. doi:10.100741207-020-00168-w

Adafruit. (n.d.). *Digital temperature sensor DS18B20*. Adafruit. https://www.adafruit.com/product/381

El-Kazzaz, A. A.El-Kazzaz. (2017). Soilless Agriculture a New and Advanced Method for Agriculture Development: An Introduction. *Agricultural Research & Technology: Open Access Journal*, *3*(2). doi:10.19080/ARTOAJ.2017.03.555610

Eldridge, B. M., Manzoni, L. R., Graham, C. A., Rodgers, B., Farmer, J. R., & Dodd, A. N. (2020). Getting to the roots of aeroponic indoor farming. *The New Phytologist*, *228*(4), 1183–1192. doi:10.1111/nph.16780 PMID:32578876

Atlas Scientific. (2022). Ezo-CO2. *Embedded Carbon Dioxide Sensor*. Atlas Scientific. https://atlas-scientific.com/probes/co2-sensor/

Atlas Scientific. (2022). Ezo-Pump. *Embedded Dosing Pump*. Atlas Scientific. https://atlas-scientific.com/peristaltic/ezo-pmp/

FAO. (2018). *Water for sustainable food and agriculture: A report produced for the G20 Presidency of Germany*. Food & Agriculture Org.

FAO. (2022). *Soils for nutrition: state of the art*. FAO EBooks. doi:10.4060/cc0900en

Gour, V. Reddy, V. M., S. N., Vishuvardhan, & Ram, V. T. (2020). IoT based Farming Techniques in Indoor Environment: A Brief Survey. *2020 5th International Conference on Communication and Electronics Systems (ICCES)* (pp. 790-795). IEEE. 10.1109/ICCES48766.2020.9137950

Han, J., Kamber, M., & Pei, J. (2011). Data Transformation and Data Discretization. In *Data Mining: Concepts and Techniques* (pp. 111–118). Elsevier.

Hati, A. J., & Singh, R. R. (2021). Smart Indoor Farms: Leveraging Technological Advancements to Power a Sustainable Agricultural Revolution. In AgriEngineering (Vol. 3, pp. 728–767). MDPI. doi:10.3390/agriengineering3040047

Haykin, S. (1998). Neural Networks: A Comprehensive Foundation (2 ed.). Prentice Hall.

Kozai T., & Niu G. (2015). *Challenges for the Next-Generation PFAL, Plant Factory: An Indoor Vertical Farming System for Efficient Quality Food Production*, 387-393. Academic Press. doi:10.1016/B978-0-12-801775-3.00027-5

Kozai, T., & Niu, G. (2019). Water use efficiency, In Plant Factory: An Indoor Vertical Farming System for Efficient Quality Food Production, pp. 97-99. Research Gate.

Kumar, A., Joseph, S., Graber, E. R., Taherysoosavi, S., Mitchell, D. R. G., Munroe, P., Tsechansky, L., Lerdahl, O., Aker, W., & Sæbø, M. (2021). Fertilizing behavior of extract of organomineral-activated biochar: Low-dose foliar application for promoting lettuce growth. *Chemical and Biological Technologies in Agriculture*, *8*(1), 21. doi:10.118640538-021-00222-x

Kumar, A., Singh, S., Gaurav, A. K., Srivastava, S., & Verma, J. P. (2020). Plant Growth-Promoting Bacteria: Biological Tools for the Mitigation of Salinity Stress in Plants. In Frontiers in Microbiology (Vol. 11). Frontiers Media S.A. doi:10.3389/fmicb.2020.01216

Lakhiar, I. Gao J., Syed, T., Chandio, F., Buttar, N., & Qureshi, W. (2018). Monitoring and Control Systems in Agriculture Using Intelligent Sensor Techniques: A Review of the Aeroponic System. *Journal of Sensors*. doi:10.1155/2018/8672769

Lal, R. (2004). Soil carbon sequestration to mitigate climate change. In Geoderma (Vol. 123, pp. 1–22). doi:10.1016/j.geoderma.2004.01.032

Lask, J., Kam, J., Weik, J., Kiesel, A., Wagner, M., & Lewandowski, I. (2021). A parsimonious model for calculating the greenhouse gas emissions of miscanthus cultivation using current commercial practice in the United Kingdom. *Global Change Biology. Bioenergy*, *13*(7), 1087–1098. doi:10.1111/gcbb.12840

Malhi, G. S., Kaur, M., & Kaushik, P. (2021). Impact of climate change on agriculture and its mitigation strategies: A review. In Sustainability (Switzerland) (Vol. 13, pp. 1–21). MDPI. doi:10.3390u13031318

Mejias, J. H., Salazar, F., Pérez Amaro, L., Hube, S., Rodriguez, M., & Alfaro, M. (2021). Nanofertilizers: A Cutting-Edge Approach to Increase Nitrogen Use Efficiency in Grasslands. In Frontiers in Environmental Science, 9. Frontiers Media S.A. doi:10.3389/fenvs.2021.635114

Murphy R., Montes-Molina J., Govaerts B., Six J., van Kessel C., Fonte S. (2016). Crop residue retention enhances soil properties and nitrogen cycling in smallholder maize systems of Chiapas, Mexico. Applied Soil Ecology, (2016), 110-116, 103. doi:10.1016/j.apsoil.2016.03.014

Node-RED. (n.d.). *Features*. Node-RED. https://nodered.org/about/

Orsini (2020). Sustainable use of resources in indoor farms with artificial lighting. *European Journal of Horticultural Science*, *85*, 5. doi:10.17660/eJHS.2020/85.5.1

Ragaveena, S., Shirly Edward, A., & Surendran, U. (2021). Smart controlled environment agriculture methods: a holistic review. In Reviews in Environmental Science and Biotechnology (Vol. 20, Issue 4, pp. 887–913). Springer Science and Business Media B.V. https://doi.org/ doi:10.100711157-021-09591-z

Rahman, S. U., Xuebin, Q., Riaz, L., Yasin, G., Shah, A. N., Shahzad, U., Jahan, M. S., Ditta, A., Bashir, M. A., Rehim, A., & Du, Z. (2021). The interactive effect of pH variation and cadmium stress on wheat (Triticum aestivum L.) growth, physiological and biochemical parameters. *PLoS ONE, 16*. doi:10.1371/journal.pone.0253798

Rencher, A. C., & Christensen, W. F. (2012). Chapter 10, Multivariate regression – Section 10.1, Introduction. Methods of Multivariate Analysis, Wiley Series in Probability and Statistics (3rd ed.), John Wiley & Sons, p. 19

Richa, A., Fizir, M., & Touil, S. (2021). Advanced monitoring of hydroponic solutions using ion-selective electrodes and the internet of things: A review. *Environmental Chemistry Letters*, *19*(4), 3445–3463. doi:10.100710311-021-01233-8

Richa, A., Touil, S., Fizir, M., & Martinez, V. (2020). Recent advances and perspectives in the treatment of hydroponic wastewater: a review. In Reviews in Environmental Science and Biotechnology, 19(4) 945–966. Springer Science and Business Media B.V. doi:10.100711157-020-09555-9

Rokach, L. Maimon, O. (2014). Data mining with decision trees: theory and applications, (2nd ed.) World Scientific Pub.

Romeo, D., Vea, E., & Thomsen, M. (2018). Environmental impacts of urban hydroponics in Europe: A case study in Lyon. *Procedia CIRP*, *69*, 540–545. doi:10.1016/j.procir.2017.11.048

Safdar, H., Amin, A., Shafiq, Y., Ali, A., & Yasin, R. (2019). Abbas Shoukat, Maqsood Ul Hussan, Muhammad Ishtiaq Sarwar. A review: Impact of salinity on plant growth. *Nature and Science*, *17*(1), 34–40. doi:10.7537/marsnsj170119.06

Sammut, C., & Webb, G. I. (2011). *Encyclopedia of Machine Learning*. Springer US.

Souza, A., Zandonadi, D., Santos, M., Canellas, N., de Paula Soares, C., da Silva Irineu, L., de Rezende, C., Spaccini, R., Piccolo, A., Olivares, F., & Canellas, L. (2021). Acclimation with humic acids enhances maize and tomato tolerance to salinity. *Chemical and Biological Technologies in Agriculture*, *8*(1), 40. doi:10.118640538-021-00239-2

Adafruit. (n.d.). *Temperature Humidity sensor. AM2315C - Encased I2C Temperature/Humidity Sensor*. Adafruit. https://www.adafruit.com/product/5182

Torres-TelloJ.VenkatachalamS.MorenoL.KoS.-B. (2020). Yield measurements in aeroponics for six different crops. Mendeley Data. doi:10.17632/wmyktpx9hv.1

Tudi, M., Ruan, H. D., Wang, L., Lyu, J., Sadler, R., Connell, D., Chu, C., & Phung, D. T. (2021). Agriculture development, pesticide application and its impact on the environment. In International Journal of Environmental Research and Public Health, 18(3), 1–24. MDPI AG. doi:10.3390/ijerph18031112

Tunio, M. H., Gao, J., Shaikh, S. A., Lakhiar, I. A., Qureshi, W. A., Solangi, K. A., & Chandio, F. A. (2020). Potato production in aeroponics: An emerging food growing system in sustainable agriculture for food security. In Chilean Journal of Agricultural Research, 80(1), 118–132. Instituto de Investigaciones Agropecuarias, INIA. doi:10.4067/S0718-58392020000100118

United Nations Department of Economic and Social Affairs, Population Division (2022). *World Population Prospects 2022: Summary of Results*. UN DESA/POP/2022/TR/NO. 3.

United Nations, Department of Economic and Social Affairs, Population Division (2019). *World Urbanization Prospects 2018: Highlights* (ST/ESA/SER.A/421). UN.

Upton, E., & Halfacree, G. (2016). Meet the Raspberry Pi. In E. Upton & G. Halfacree (Eds.), *Raspberry Pi® User Guide*., doi:10.1002/9781119415572

Willmott, C. J., & Matsuura, K. (2005, December 19). Advantages of the mean absolute error (MAE) over the root mean square error (RMSE) in assessing average model performance. *Climate Research*, *30*, 79–82. doi:10.3354/cr030079

Yokoyama, R. (2019). Energy Consumption and Heat Sources in Plant Factories. In Plant Factory Using Artificial Light (pp. 177-184). Elsevier. doi:10.1016/B978-0-12-813973-8.00016-6

Young, M. D., Ros, G. H., & de Vries, W. (2021). Impact of agronomic measures on crops, soil and environmental indicators: A review and synthesis of meta-analysis. In Agriculture, Ecosystems and Environment (Vol. 319). Elsevier B.V. doi:10.1016/j.agee.2021.107551

Zhao, Y., Qin, X.-M., Tian, X.-P., Yang, T., Deng, R., & Huang, J. (2021). Effects of continuous cropping of Pinellia ternata (Thunb.) Breit. on soil physicochemical properties, enzyme activities, microbial communities and functional genes. *Chemical and Biological Technologies in Agriculture*, *8*(1), 43. *Advance online publication.* doi:10.118640538-021-00243-6

KEY TERMS AND DEFINITIONS

Aeroponics: It is a growing soilless technique where the nutrient solution is directly nebulized on the plant's roots.

Atomization: The act or process of splitting into smaller parts, sections, or groups. In this work it refers to the breakdown of water droplets for the irrigation of plants in aeroponics

Controlled Environment Agriculture: CEA refers to the practice of producing crops while adjusting environmental conditions such as lighting, temperature, CO2, humidity, irrigation, fertigation, and other elements that affect plant physiological responses.

Crop Growth procedure: A recipe that specifies the ideal growing environment parameter settings to create a more effective system.

Cross Validation: is an evaluation method used in machine learning to find out how well a machine learning model can predict the outcome of unseen data. In k-fold cross validation the data sample is split into 'k' number of smaller samples.

Hyperplane: In Machine Learning, a hyperplane is a decision boundary that divides the input space into two or more regions, each corresponding to a different class or output label. It is used to separate the dataspace into one less dimension for easier linear classification.

Macronutrients: are nutrients that provide energy to the plants and are required in larger amounts to maintain their development, growth, and photosynthesis. Nitrogen (N), Phosphorus (P), and Potassium (K) are the primary macronutrients for plants.

Micronutrients: are essential plant nutrients that are found in trace amounts in plant tissue, but play an imperative role in plant growth and metabolism

Standardization: is the process of rescaling a dataset to have the properties of a Gaussian distribution (a mean of 0 and a standard deviation of 1).

Vertical farming: is a form of horticulture where plants are grown vertically in layers. Usually, it can be done in a controlled environment with different techniques such as Hydroponics, Aeroponics and Aquaponics.

Chapter 5
Long-Range and Low-Power Automated Soil Irrigation System Using Internet of Things:
An Experimental Study

C. Gnanaprakasam
Panimalar Engineering College, India

Jayavani Vankara
https://orcid.org/0000-0002-3815-6794
GITAM University (Deemed), India

Anitha S. Sastry
Global Academy of Technology, India

V. Prajval
Sampoorna Institute of Technology and Research, India

N. Gireesh
https://orcid.org/0000-0002-1687-1842
Mohan Babu University, India

Sampath Boopathi
https://orcid.org/0000-0002-2065-6539
Muthayammal Engineering College, India

ABSTRACT

In this chapter, the Internet of Things (IoT) system is required for automating irrigation systems and monitoring real-time data from sensors. IoT systems may easily and affordably integrate the long-range wide-area network (LoRaWAN). Four irrigation strategies, including ET (ETc), MP60 (watermark 200SS-5 soil matric potential sensors, (-70 kPa), MP50 (at -50 kPa)), and GesCoN (a decision support system), were developed and put to the test. According to the findings, treatment MP70 had a marketable yield that was greater by 16 percent and 24 percent than that of ET and MP50. Due to improper installation and positioning of the soil moisture sensors, MP40 received relatively little water during irrigation. The GesCoN and ET results were not significantly different from the MP70 results. It has been demonstrated

DOI: 10.4018/978-1-6684-7879-0.ch005

that using sensors and precision irrigation can help farmers conserve water when growing crops. The LoRaWAN-based IoT system nevertheless performed admirably in terms of power usage, connectivity, sensor reading, and valve management.

INTRODUCTION

Ground and surface water are heavily used by agriculture in the United States. Food crop production is anticipated to rise sharply as the world's population continues to rise. Resources for water are becoming scarcer. Consequently, it is crucial to use water well, particularly for crops like vegetables. Irrigation water management can have a significant impact on vegetable yield and quality since water availability can be either excessive or deficient. By administering only, the necessary amount of water to the crop directly, gardeners can minimise plant water stress during key growth stages with the help of precision irrigation, a contemporary irrigation management technique(S. et al., 2022). In order to accurately regulate the time, rate, and distribution of water as necessary, integrated sensing, decision-making, and control systems must be developed for crop production systems before precision irrigation can be used. Depending on the soil, plant, or weather conditions, irrigation may be used(Bagha et al., 2022).

For precision irrigation, many sensor systems and technologies, such as evapotranspiration (ET), plant-based, and soil moisture-based systems, have been researched and tried. For ET-based irrigation, a Class A pan evaporimeter or a comprehensive set of weather characteristics from a nearby weather station are needed to calculate the ET rate. Vegetable fields and protected culture systems have seen extensive testing and deployment of soil moisture sensor-based precision irrigation. Two indications for the amount of water present in the soil that can be used to implement soil moisture-based irrigation systems are soil volumetric water content (VWC) and soil matric potential (MP). MP uses a unit of kPa to measure the amount of energy needed for water to travel through soil, which is always a negative quantity. In this investigation, MP sensors were chosen instead of VWC sensors. According to early research, soil MP sensors can better fit IoT systems and have shown sensitivity to changes in soil moisture for irrigation of vegetables. One of the main technologies for precise and automated irrigation systems is the sensor network, whether it is wired or wireless. For the purpose of planning cotton irrigation, Vellidis et al. (2008) created and assessed a real-time smart sensor array that could measure soil moisture and temperature. In related research, a drip irrigation system's real-time feedback control was handled by a microcontroller(Vellidis et al., 2008).

For automated irrigation systems, a variety of embedded control technologies have been used, including Xbee-PRO technology and GSM Bluetooth-based remote-control systems. Applications for wireless personal digital assistants (PDAs and "tablets") and mobile phones have been created to permit remote sensor data collection and control over actual irrigation systems. Four prototype sites in Brazil, Italy, and Spain were used as part of the SWAMP project, which examined the effectiveness of their IoT-based irrigation system. Sensor data is wirelessly uploaded to a server. The data is then made available for analysis and computation via the internet. To enhance irrigation control, Goap et al. (2018) created an IoT-based smart irrigation management system using a machine learning algorithm(Goap et al., 2018). Crop irrigation management has been examined using a variety of wireless technologies in IoT systems, including Wi-Fi, cellular networks (GPRS and LTE), ZigBee, and LoRaWAN. LoRaWAN technology, which has a maximum range of 10 kilometres and requires little power and money, would make it possible for small farms to finance precision irrigation equipment. Originally released in 2015, this technology was created by the LoRa Alliance, a non-profit organisation that supports the LoRaWAN protocol.

Among other goals, a team of researchers from the University of Michigan set out to create and assess an IoT wireless sensor network system based on LoRaWAN technology for precision irrigation of vegetables. It is speculated that LoRAWAN could be a useful option for the wireless technology in IoT-based irrigation for vegetable fields that are frequently far from the farm centre. Modern irrigation management techniques like precision irrigation only provide the crop with the water it needs. Based on sensor data and scheduling plans in the precision irrigation system, water may be applied precisely. The soil, the plant, or the weather can all be factors in precise irrigation. Automatic irrigation systems have been subject to a variety of control measures. A timer with a predetermined setting is the foundation of the most basic automatic irrigation system. There are also some cutting-edge control ideas that could provide irrigation with more accuracy. IoT-based irrigation is a fresh approach to irrigation that has gained popularity recently. A real-time Internet connection allows users to view field data, which can then be examined and processed on a server. IoT wireless technologies come in a variety of flavours that can be applied to agriculture.

PRECISION IRRIGATION STRATEGIES

For irrigation scheduling, the status of the soil's water supply or its moisture might be employed. Precision irrigation applications frequently use soil VWC, which is the water-to-soil volume ratio. A three-year field experiment was conducted by Zotarelli et al. (2009) to investigate the effects of irrigation timing on tomatoes. According to a study, SUR and SDI used less irrigation water than TIME by 15%–51% and 7%–29%, respectively(Zotarelli et al., 2009). Precision irrigation has also used soil MP sensors in the past. The study found that tomato yield and irrigation water efficiency rose consistently with VWC sensor-based irrigation (iWUE)(Boopathi, Siva Kumar, et al., 2023; Domakonda et al., 2022; Kumara et al., 2023; Samikannu et al., 2022; Vanitha et al., 2023; Vennila et al., 2022).

Soil MP measurements performed better at a threshold of -25 kPa measured at 30 cm soil depth.A reliable model calibration was necessary for a simulation-based strategy to be successful. The soil MP threshold for the best irrigation scheduling varies according to the crop, the type of soil, and the depth of the sensor installation. Higher thresholds are appropriate for crops like rice and some cereals that are more vulnerable to water stress, while lower thresholds are ideal for crops that are tolerant of it. While some researchers have employed a set of procedures to define an appropriate soil MP threshold, others have used simulation models to identify the best soil MP thresholds. According to Cancela et al. (2015), soil-water retention curves and crop coefficient calibration were tested as part of a precision irrigation system for grapes(Cancela et al., 2015). O'Shaughnessy et al. (2012) developed an irrigation plan using the CWSI to express plant water stress in 2012. Results showed that method 2 was superior to technique 1 in terms of yield in the treatments with higher irrigation amounts. But when the irrigation was small, there was little to no difference(O'Shaughnessy et al., 2012).

Based on the cumulative water deficit, which is determined by the ET of the plant-soil system using various equations, ET-based irrigation is used to irrigate the land. To calculate the daily ET rate, ET needs a comprehensive set of weather characteristics from a local weather station. It has been widely used for many different crops. Sensors in the ground or close to the plants are not necessary for ET-based irrigation. For various plants and growth phases, several equations can be used to determine ET0(Babu et al., 2022; Boopathi, Arigela, et al., 2022; Chakravarthi et al., 2022; Jeevanantham et al., 2022). It could produce some accumulated inaccuracies over time because it does not directly measure the water status of a soil or plant. Using the water savings and fruit quality, ET0 was computed using Penman Monteith's equation. The highest iWUE was recorded by RDI, followed by FI, DI, and FAI. The outcomes were revealed.

AUTOMATIC IRRIGATION CONTROL STRATEGIES

Precision irrigation can be applied manually or automatically, depending on the type of sensor used. The simplest and least expensive automatic irrigation systems start and stop an irrigation event based on pre-set intervals using timers. Farm owners have the option of irrigating at regular intervals or at specific times each day. A time-based drip and spray irrigation system. As test crops, onions and cabbage flowers were chosen. Results indicated that applying water at three different times of the day irrigation at each was helpful(S. et al., 2022; Sampath et al., 2022).

Researchers at the University of California, Los Angeles claim that a specific amount of water or a specific amount of time can also be used to determine when to terminate irrigation (UCLan). The static on-off threshold has evolved into a dynamic threshold where the threshold varies depending on the situation. Precision irrigation scheduling with an adaptive algorithm was created (Osroosh et al., 2016). The crop water stress index (CWSI) for apple trees served as the basis for the algorithm. When used in conjunction with an advanced control algorithm, CWSI can be more effective and enhance iWUE. It resolved the issue of excessive irrigation during the crop's first growth phase, which was caused by inaccurate canopy temperature measurement. Automatic irrigation also uses other cutting-edge control theories like PID control, fuzzy control, neural networks, and model predictive control.

APPLICATION OF IOT IN IRRIGATION

Precision irrigation using the Internet of Things (IoT) is based on a wireless sensor network (WSN). Irrigation uses a variety of wireless communication methods with varying data rates, transmission powers, and ranges. among others, cellular networks, Wi-Fi, Bluetooth, LoRaWAN, and others. A low-cost IoT-based irrigation system with a Wi-Fi network has been created by researchers. This system's sensors were used to keep an eye on the air temperature, humidity, and soil VWC. It featured user interaction, one-time setup, decision-making using neural networks, and remote data monitoring(Babu et al., 2022; Boopathi, Arigela, et al., 2022; Jeevanantham et al., 2022; S. et al., 2022; Sampath et al., 2022).

Sensors collected data on soil temperature and VWC, which were then transmitted through the internet to a gateway. In addition to processing sensor data and activating valves, the gateway may also send data to a web application. The results showed that their approach saved 30% more irrigation water than conventional drip irrigation. When compared to applying a fixed amount of water three times each week, the system conserved 90% of the irrigation water volume. Through a website, researchers could programme the irrigation technique and monitor the data. Nodes were placed 150, 300, and 450 metres from the base station, respectively. Each node had an average current usage of 3.13 mA and a battery life of 87 days. The nodes could only operate at a maximum range of 500 m due to the dense foliage. Their research demonstrates that GPRS and WiFi use less power and are more stable over extended distances than older wireless networks.

EXPERIMENTAL SETUP

At the Russell E. Larson Agricultural Research Center's Horticultural Research Farm, open field irrigation management research was carried out. As the test crop, fresh-market Red Deuce F1 tomatoes were employed. The soil had 0.13 percent total nitrogen and 1.13 percent total carbon. The seedlings were sown at the fourth true-leaf stage. The drip tape, which was 16 mm in diameter, had emitters that were 0.30 m apart and flowed at a rate of 1 L/h at 55 kPa. Stake and Florida weave trellising was used for plants. Locations of the sensors and dataloggers on the irrigation plant is shown in Figure 1.

In this study, once crops were transplanted, the entire irrigation system and IoT system were examined. We designed and tested four distinct irrigation schedule treatments. Based on crop evapotranspiration (ETc), Treatment 1 (standard control) involved irrigating the crop with a volume equal to 100% of the ETc calculated daily. After an average cumulative water deficit of 12 mm, irrigation was started. Based on soil matric potential (MP) sensors with differing setpoints of -60 kPa and -40 kPa, treatments 2 and 3 were applied. The introduction of MP sensors suggested the thresholds.

During the experiment, some modifications were done while watching the sensor values to determine the durations. A decision support system (DSS) designed by researchers at the University of Foggia (Italy) to estimate crop N and water requirements served as the foundation for Treatment 4 (dubbed GesCoN). The system would send a proposal for the volume and timing of irrigation when it was necessary. GesCoN uses computers to evaluate the nutrient content of crops and offers advice on irrigation and N fertigation. For the purpose of estimating crop dry matter production, yield, evapotranspiration, and soil moisture, the GesCoN includes sub-models for water balance, plant development, and nitrogen uptake(Vanitha et al., 2023).

The RESTful API method can be used to link the DSS to weather stations in order to automatically get current local climate information (Gallardo et al. 2020). The whole field (planting space) measured 27.74 m x 42.67 m, with 12 raised beds measuring 42.7 m in length. Three beds measuring around 9.14 m in length, with 20 plants each, made up each experimental unit, giving each bed a total of 60 plants. Four replicates of each treatment were used in a randomised complete block design (RCBD). The field was equipped with an irrigation system that included irrigation pipelines, solenoids, pressure sensors, soil MP sensors, and data recorders. The sensors and valves were connected using data recorders. The next sections provide an introduction to sensors, dataloggers, and valves in depth.

Figure 1. Locations of the sensors and dataloggers

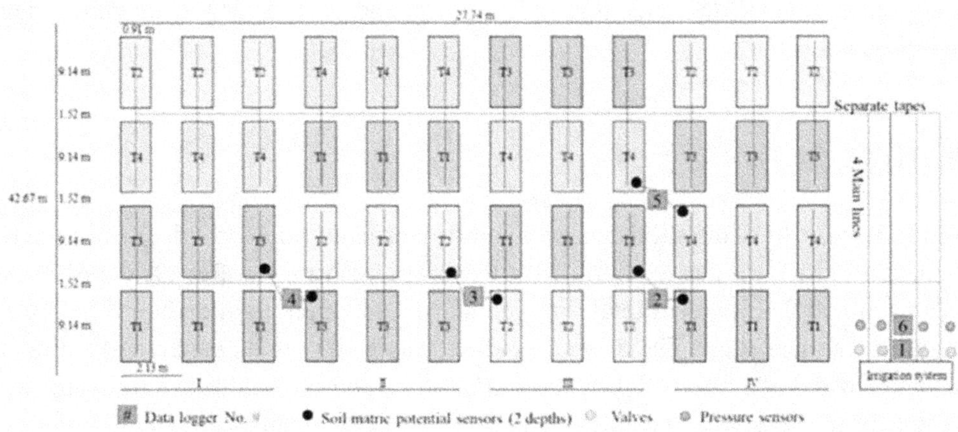

Irrigation System Setup

The irrigation system's structure, including its pipelines, valves, pressure sensors, flow metres, and other components, is depicted in Figure 2. The farm's main water line provided water to the building. Four pipelines, one for each treatment, were split into the main line after a screen filter. During the experiment, the pressure sensors detected the water pressure that reached the following components: When advised by the DSS GesCoN, fertilisers were applied using the fertigation system, which consists of a by-pass with a Venturi injector. The amount of water that could flow to the components below was restricted by pressure regulators (241 kPa) behind the pressure sensors(Boopathi et al., 2012; Boopathi & Sivakumar, 2013; Chakravarthi et al., 2022; Haribalaji et al., 2014).

The pressure gauges were used to calibrate the pressure sensors and gauge the water pressure in real time. The switch in the pipeline was opened when fertigation began, followed by the closing of the injector. For each portion, drip tapes were positioned beneath the mulch. Following that, the drip lines for the same treatment were attached to the pipes along the field walkway. Finally, for each treatment, these pathway pipes were joined to the respective pipelines by 2.54 cm. Pressure regulators at 103 kPa, the recommended pressure for irrigation drip-tape lines, were the last components to be assembled.

Figure 2. The schematic arrangements of one treatment irrigation pipeline

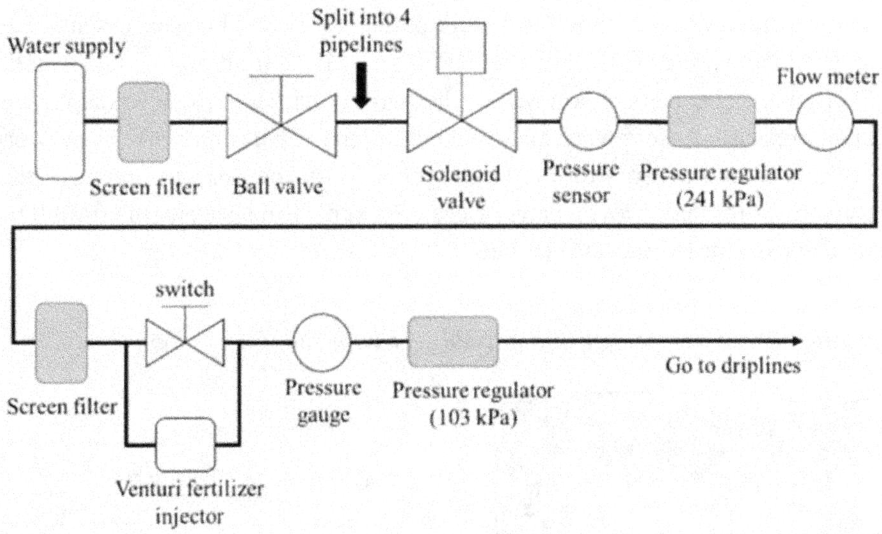

Sensor System Setup

Schematic of soil and pressure sensor with data logger of the real irrigation system are illustrated in Figures 3 and 4 respectively. The creation of the first LoRaWAN (Low Earth Orbit Radiosat) system in the entire world involved the use of six data loggers. The base control board and wireless communication devices with antenna were the main components (LM130-H1, GlobalSat WorldCom Corp., New Taipei City, Taiwan). A solar panel (70 mm 70 mm) was connected to charge the lithium-ion polymer (LiPo) batteries that powered each data logger. Each treatment included the use of four soil MP sensors (Watermark 200SS-5, Irrometer Company, Inc., Riverside, CA)(M et al., 2018).

Figure 3. Schematic of Soil sensor and data logger with real time system

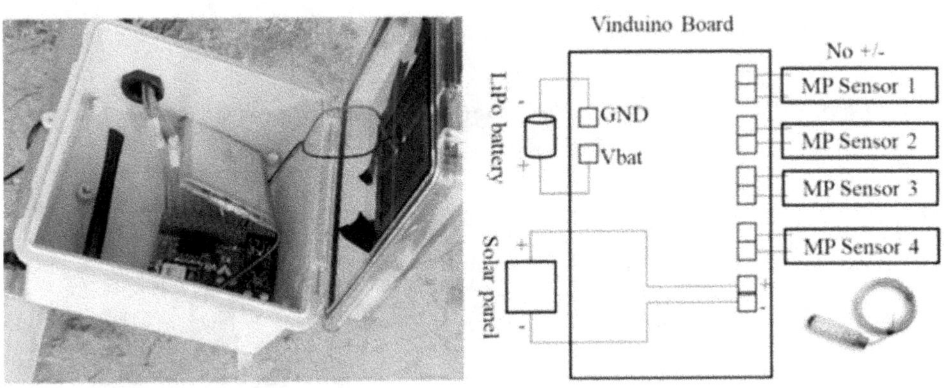

To achieve close contact between the sensors and the soil, soil MP sensors were immersed in water for 24 hours prior to installation in each treatment. A cylinder hole with a depth of 40 cm was dug during installation, and it was then filled with water. Each treatment pipeline has sensors attached to keep track of soil moisture levels and water flow. For the MP60 and MP40 treatments, the irrigation was started at a depth of 20 cm using the average soil MP of the two experiment units. To display the condition of the solenoid valve, pressure sensors were mounted behind the valve (on or off). A network of wires was used to connect the pressure sensors to the data loggers. With the pressure gauges fitted on the same line, the sensors were calibrated. Eleven pairs of pressure sensor measurements from 0 to 207 kPa were used to calibrate each sensor by altering the ball valve's opening.

Figure 4. Schematic of pressure sensor and data logger with real time system

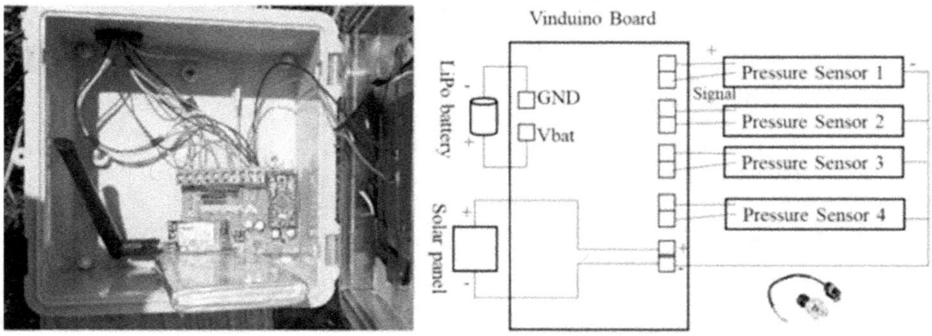

Valve Control

The sensor ports on the Vinduino board were utilised to expand the board's capacity, which originally only supported one valve (Figure 5). Although sensor ports can transmit high- or low-level signals at 3.3 or 0 V, the current is insufficient to operate the valves. For the control of the valve, four 2-way relays were added. The IN1 and IN2 ports receive high/low signals from two sensor ports. The high or low of IN corresponds to Vbat or 0V of OUT. The positive and negative electrodes of the valve are connected to out1 and out2, respectively.

Figure 5. Schematic of different valves and data logger with real time system

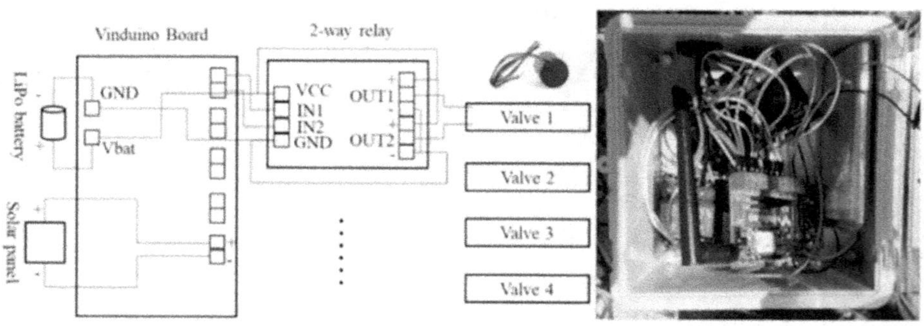

IoT System and Data Collection

A LoRaWAN gateway (SentriusTM RG191, Laird) was positioned 150 metres from the field in an office. The gateway and data recorders were set up on The Things Network, a free IoT server (Figure 6). To store and display the sensor data, AllThingsTalk (AllThingsTalk NV, Mechelen, Belgium) was employed. To match the specifications, the LoRaWAN gateway was initially linked via Wi-Fi to a computer. The recorded gateway ID was used to construct a gateway in the server interface. The intended irrigation system for the tomato field was then represented by an "application" that was constructed in the gateway. The soil MP recording, valve control, and water pressure recording were all monitored by six data loggers. An individual Application Extended Unique Identifier (EUI) was automatically generated for each device by the server. Algorithms have been created.

The AllThingsTalk gateway was linked to the data loggers, which were turned on and ready to post and download sensor data. After processing on the IoT platform, the data was then returned to the server via a web-based interface. In order to monitor and control irrigation, sensor data was uploaded and saved on the IoT platform. One minute was pre-set as the time period for data communication frequency. Both the valve control (downlink) and sensor data (uplink) were transmitted as binary payloads. Through the IoT platform, the valves may be individually turned on or off and show voltages and other data in real-time. By typing a number between 0 and 15 into the textbox, solenoid valves might be controlled. The historical sensor and battery data was archived and exportable as a comma-separated values (CSV) file. A pinboard might show the status of all six devices. A manual control over the IoT platform was employed to initiate and terminate irrigations rather than fully automating the process.

Figure 6. Experimental setup of IoT system

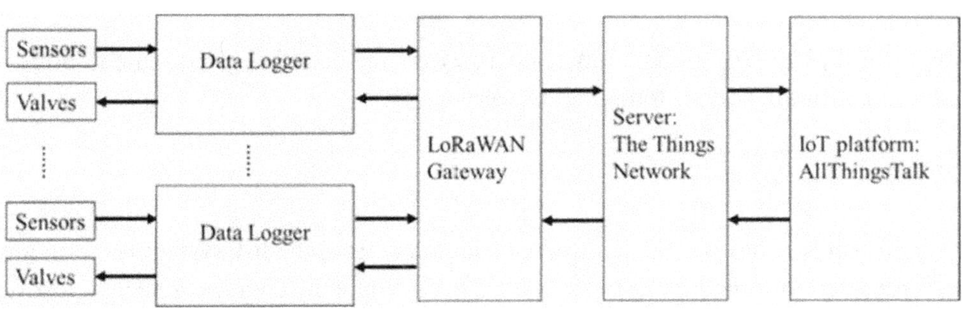

HARVEST, YIELD COMPONENTS, BIOMETRIC ASSESSMENT, AND FRUIT QUALITY YIELD

Harvested tomatoes ranged in colour from mature green to ripe red, with a median output of roughly 1.5 metric tonnes per plant. The final harvest on Day 125 marked the end of the experiment. The U.S. Standards for Grades of Fresh Tomatoes were used to classify tomatoes into fruit that was marketable and fruit that was not.

Fruit Quality

Fruit quality measurements were conducted using subsamples of typical fruit from each experimental unit. From the collected tomatoes, five sample tomatoes per experimental unit were chosen. To calculate the fruit's dry matter content, it was dried in a forced-air oven at 65 °C until the weight remained constant. The percentage of the fresh weight of the sample fruit used to compute the dry matter content. With the aid of a handheld refractometer, fruit TSS was measured (Brix-Stix BX 100 Hs; Techniquip Corporation, Livermore, CA).

Biometric Assessment

Measurements of leaf, stem, fruit, and total plant dry biomass were taken using two representative plants per trial unit. Separated leaves, stems, and fruits were put in paper bags with labels before being dried at 65 °C in a forced-air oven to a constant weight. The combined dry weight of the fruits, stems, and leaves made up the entire plant's dry biomass. In the case of fruit, the total fresh fruit weight of each plant was determined, and the dry matter content was determined by drying a representative subsample in an oven.

Irrigation Water Use Efficiency

The weight of marketable fruit while it is fresh was divided by the total amount of water used to calculate irrigation water use efficiency (iWUE). The four replications' water use was estimated to be the same and measured by treatment.

Weather

To provide an overall picture of global climate change over the next five years, calculations of cumulative solar radiation, rainfall, and air temperature were made.

Statistical Analysis

From the IoT platform AllThingsTalk, 35 days' worth of data from the IoT system was downloaded. The ratio of signals received to signals expected was used to calculate the data loss rate. Microsoft Excel was used to process and display the data logger readings, soil MP sensor readings, and weather information collected during the experiment, such as sun radiation, rainfall, and air temperature. Fruit fresh weight, fruit numbers, fruit total soluble solids (TSS), fruit dry matter, and mean fruit weight were all subjected to analysis of variance (ANOVA). The least significant difference (LSD) test was used to compare all means at $P = 0.05$ with a P value of 0.01 or below.

RESULTS AND DISCUSSIONS

Weather Data

Throughout the experiment, the cumulative precipitation was 300 mm, the mean daily air temperature was 20.3 °C, and the average daily solar radiation was 26.8 MJm-2. Daily air temperatures varied from -1.8 to 21.8 °F (or 16.3 to 34.2 °C) for the minimum and highest, respectively. The entire amount of precipitation fell between 0 and 34 DAT, totalling 167 mm, while the average daytime air temperature was 18.9 °C. For each of the four treatments, the same amount of irrigation water was used, but the irrigation volume was not measured. The cumulative daily precipitation increased from 79 mm to over 200 mm between 35 and 95 DAT. The total amount of precipitation that fell between 96 and 125 DAT was 54 mm, while the average daily air temperature steadily decreased. During this time, irrigation was given less regularly, and frost occurred on 117 DAT. All of the tomatoes in the field were gathered during this time(Boopathi, 2022d, 2022c, 2022b, 2022a; Boopathi et al., 2017; Boopathi, Balasubramani, et al., 2023; Haribalaji et al., 2014; Yupapin et al., 2022).

Data Loss in Communication

In general, the system performed well between 35 and 125 DAT. The IoT system experienced data loss on a few occasions. Due to the gateway's location inside a structure 150 metres away from the field, there may have been some signal loss. If the gateway is installed outside with less obstruction, this can be improved. However, the sensing system was unaffected because data were uploaded at a high frequency every minute and there was no extended period of data loss(Boopathi, Khare, et al., 2023; Palaniappan et al., 2023; Senthil et al., 2023; Trojovský et al., 2023).

Battery

Figure depicts the voltage of these batteries over time. Due to the persistently gloomy and rainy weather, all data loggers experienced voltage drops on DAT 103 and 114. Between 60 and 70 DAT, the data logger 3's solar panel wires were disconnected, which caused a noticeable voltage drop compared to earlier times. Throughout the testing, the battery voltages were consistently higher than 4 V. In an open area, the solar-powered 5000 mAh LiPo battery can power a data logger indefinitely.

Valve Control

The platform notifies the user via email and a mobile application when the soil moisture threshold is met. After a specific number of hours of irrigation, the valves were closed, and the pressure sensors' readings reverted to zero as a result. The pressure sensor measurement remained unchanged as the valve closed, however, and the status of the valve changed to "Green." This may be caused by solenoids that are clogged or by low voltage powering the valves (4 V). The response of the valve can be enhanced in the future by using a higher voltage.

Monitoring soil moisture in tomato fields reveals that using soil moisture sensors in conjunction with GesCoN may have some advantages. Continually dropping between 70 and 80 days, and between 100 and 110 days, were two sensors at MP 4-2. Since just one of the guard rows from which tomatoes were not collected was impacted by this problem, the outcomes were unaffected. The first sensors to respond were those at 40 cm and the initial installation position. In around two hours, the sensor readings stopped rising after increasing for 30 minutes.

At 18:50, when the average measurements fell below -60 kPa, a notification was issued, and the valve was remotely opened. To reach the sensors, irrigation water was dispersed both vertically and horizontally. Usually, water rises to a depth of 20 cm before reaching 40 cm. Sensor values at 20 cm were approximately -60 kPa, and at 40 cm, they were -100 and -140 kPa. Before and after irrigation, sensors at 40 cm were drier than those at 20 cm. Following watering, the MP 2–1 40 cm readings initially fell and subsequently increased. The soil was not evenly moist just after irrigation, and this could be the cause of the water flowing through the area.

Soil Moisture Monitoring

The four treatments' rainfall, irrigation, and soil MP sensor values between 35 and 125 DAT are displayed. The sensor readings rose when irrigation or rainfall occurred. Sometimes after irrigation, the sensors at a depth of 40 cm failed to respond, indicating that the water could not penetrate that depth. The majority of sensor readings for treatment ET were higher than -100 kPa, but because irrigation was based on ETc and did not take sensor readings into account, the sensor readings did not respond consistently.

MP 1-1 values were often lower than MP 2 readings (treatment 1-ET, the second sensor installation place). All of the sensor readings for the MP60 treatment throughout the experiment were higher than -140 kPa. The irrigation was started when the average sensor reading at 20 cm fell below -60 kPa. Due of irrigation, MP 2-1 (20 cm) continually got dryer between 35 and 55 DAT, which was unusual. During the experiment, the volume of each irrigation was changed based on the sensor reaction rather than how much water was being pushed into the plant. The MP 2-2 40 cm sensor values remained low between 65 and 72 DAT. This suggested that there wasn't enough irrigation volume during this time. The irrigation depth was progressively increased to over 10 mm after 85 DAT. The four sensors' responses had improved consistency at this point. Throughout the experiment, all sensor readings for the MP40 treatment (Figure 17) were higher than -120 kPa. The irrigation was started when the average sensor reading at 20 cm hit -40 kPa.

In the range of 35 to 125 DAT, the sensor response remained constant. However, compared to other treatments, the overall irrigation volume was significantly lower. Variations in soil texture may be caused by the soil texture in the experimental field. Only two locations for each treatment had sensors installed, and since the wires for those sensors were only 5 m long, those locations weren't very far apart. The location of the sensors, however, was wetter than the typical treatment condition for MP40.

Fruit Yield Components

At each harvest, the components of the tomato fruit production for each of the four treatments were noted. In general, the MP60 and the ET had the highest marketable yields, whereas the MP40 had the lowest. In comparison to extra-large tomatoes, giant and medium tomatoes only made up a small portion of the first four harvests. At the fourth harvest, MP60 had a higher marketable yield than ET and a higher unmarketable weight than GesCoN. In the second and third harvests, MP40 had fewer extra-large and marketable fresh fruit weights than ET, in comparison. For the majority of yield components, MP60, ET, and GesCoN did not differ considerably. Up until the third harvest, MP40 had less fresh weight for extra-large fruit than ET. Even though MP40 had a higher percentage of unmarketable fruit, the first harvest had a higher overall yield.

MP60 and MP40 are sensor-based irrigation systems that use various thresholds for soil moisture levels. MP40's marketable yield was 12.5 percent less than that of ET and GesCoN but was otherwise comparable to ET. Since MP60 was anticipated to be wetter than MP40, the soil experienced less water stress. Due to incorrect sensor alignment or placing the sensors in an area that was not indicative of the full field, MP40 consumed less water than other treatments. The outcomes of MP60 demonstrate the effectiveness and suitability of sensor-based irrigation with an IoT system for field-scale precision irrigation.

Fruit Numbers

By fruit size categories, the average number of fruits produced per plant was reported. In general, MP60 and ET had the most marketable tomatoes, whereas MP40 had the least. For the majority of categories, there were no appreciable changes between MP60, ET, and GesCoN. In the fourth harvest, MP60 produced more extra-large, marketable, and total fruit than ET. For the majority of fruit size categories, there were no appreciable changes between MP40 and ET. However, during the third harvest, MP40 had fewer extra-large and total tomatoes than ET. Up until the fourth harvest, MP60 had a higher proportion of extra-large, marketable, and total fruit than ET. For the majority of categories, there was no discernible difference between MP60, ET, and GesCoN. The average total number of fruits harvested up until each harvest is shown in Table 5 by category.

MP40 had 3.2 percent less fruit overall than ET, whereas MP60 and GesCoN had 16.8 and 21.9 percent more total marketable fruits, respectively. Fruit number variation is generally less than yield variation. The sensor-based irrigation is appropriate for use in real applications, claims MP60(Boopathi, 2021; Boopathi, Venkatesan, et al., 2023; Saravanan et al., 2022).

FRUIT QUALITY

Mean Fruit Weight

Fruit weight, marketable fruit weight, and mean total fruit weight after each harvest did not significantly differ across the treatments. For marketable or total mean weight, the treatments did not differ significantly until each harvest. Regarding mean fruit weight for each harvest, there was no difference between the treatments. Compared to ET, MP60 and MP40 had lower marketable mean fruit weights of 1.3 and 9.0 percent, respectively. ET, MP60, and GesCoN did not significantly differ from one another. However, the weight of the mean fruit was significantly reduced by the absence of water(Boopathi, Haribalaji, et al., 2022; Gunasekaran et al., 2022).

Total Soluble Solids (TSS) and Dry Matter

Due to the water deficit seen in MP40 fruit, it had a much higher dry matter content than ET and MP60 fruit. The MP40 fruit's TSS and dry matter measurements for the second and third harvests are shown in Table 6. At the third harvest, there was no discernible change in the dry matter between the four treatments. Fruit quality is less affected by irrigation schedule than output and fruit quantity. But the fruit's quality might still be impacted by a lack of water.

Tomatoes Plant Dry Biomass

The dry biomass of tomato plants was measured in relation to the fourth harvest. The treatments for the leaves, stems, fruit, and overall did not differ significantly. For leaves and stems, MP40 had lower dry biomass than ET and GesCoN, but water stress lowered plant dry biomass. To maintain hydroponic plants, the MP60 incorporates sensor-based irrigation scheduling with an IoT system that can be used in conjunction with a cloud-based platform(Boopathi, Jeyakumar, et al., 2022; Haribalaji, Boopathi, et al., 2022; Haribalaji, Venkatesan, et al., 2022; Kavitha et al., 2022).

Irrigation Water Use Efficiency (iWUE)

In comparison to other treatments, MP40 required significantly less irrigation water, which resulted in a low yield. GesCoN, MP60, and MP40 showed iWUEs that were 19.2%, 25.7%, and 27.7% greater than ET. Given this, its lack of MP40 output is not a benefit compared to the other therapies(Boopathi, 2019; Kannan et al., 2022; Myilsamy & Sampath, 2021; Sampath & Myilsamy, 2021).

Future Improvement

Through the IoT technology, the irrigation was manually managed in this trial. In the future, fully automatic irrigation will be accomplished using a battery with a higher voltage. Both the decision support system (DSS) and sensor-based irrigation (MP60 and GesCoN) performed better in general. Field moisture is measured directly using sensors. The number of sensors can lessen the benefits if some of them do not accurately reflect the typical field conditions, as in the case of the MP40. Sensor-based irrigation is successful once the sensors have accurate readings. Sensors are not necessary for the Decision Support System (DSS) to apply irrigation. The DSS-based irrigation can be double-checked using the sensor readings. The accuracy of the model and water treatment can be improved in the future by combining the two methods.

SUMMARY

Both the pressure sensors and the soil matric potential sensors recorded and presented a number of data points. The solar panels that were attached were able to fully charge the batteries for the data loggers. By adjusting the solenoid valves in accordance with the soil moisture content, irrigation was successfully applied. The monitoring of sensor data and valve control showed no discernible harmful impact. The positioning of the interior gateway and the disconnection of each component may have contributed to the IoT system's average data loss rate of 5.5 percent.

Compared to the other treatments, MP40 produces fewer fruits, in lower numbers, and with lower quality. The top three were MP60, GesCoN, and ET. Regarding the production and fruit counts, as well as iWUE, MP60 outperformed the other three treatments. In terms of mean fruit numbers, total soluble solids, and dry matter, MP60 was discovered to be the most effective for growing vegetables. The MP60 treatment did not significantly differ from the other two therapies, though. According to the MP60 results, the created IoT-based system using LoRaWAN technology can potentially be employed for precise and autonomous irrigation applications.

REFERENCES

Babu, B. S., Kamalakannan, J., Meenatchi, N., Karthik, S., & Boopathi, S. (2022). Economic impacts and reliability evaluation of battery by adopting Electric Vehicle. *IEEE Explore*, (pp. 1–6). IEEE.

Bagha, H., Yavari, A., & Georgakopoulos, D. (2022). Hybrid Sensing Platform for IoT-Based Precision Agriculture. *Future Internet*, *14*(8), 233. doi:10.3390/fi14080233

Boopathi, S. (2019). Experimental investigation and parameter analysis of LPG refrigeration system using Taguchi method. *SN Applied Sciences*, *1*(8), 892. doi:10.100742452-019-0925-2

Boopathi, S. (2021). Improving of Green Sand-Mould Quality using Taguchi Technique. *Journal of Engineering Research*, 1–12. doi:10.36909/jer.14079

Boopathi, S. (2022a). An experimental investigation of Quench Polish Quench (QPQ) coating on AISI 4150 steel. *Engineering Research Express*, *4*(4), 045009. doi:10.1088/2631-8695/ac9ddd

Boopathi, S. (2022b). An investigation on gas emission concentration and relative emission rate of the near-dry wire-cut electrical discharge machining process. *Environmental Science and Pollution Research International*, *29*(57), 86237–86246. doi:10.100711356-021-17658-1 PMID:34837614

Boopathi, S. (2022c). (in press). Cryogenically treated and untreated stainless steel grade 317 in sustainable wire electrical discharge machining process: A comparative study. *Environmental Science and Pollution Research International*, 1–10. doi:10.100711356-022-22843-x PMID:36057706

Boopathi, S. (2022d). Performance Improvement of Eco-Friendly Near-Dry Wire-Cut Electrical Discharge Machining Process Using Coconut Oil-Mist Dielectric Fluid. *Journal of Advanced Manufacturing Systems*, 1–20. doi:10.1142/S0219686723500178

Boopathi, S., Arigela, S. H., Raman, R., Indhumathi, C., Kavitha, V., & Bhatt, B. C. (2022). Prominent Rule Control-based Internet of Things: Poultry Farm Management System. *IEEE Explore*, (pp. 1–6). IEEE.

Boopathi, S., Balasubramani, V., & Sanjeev Kumar, R. (2023). Influences of various natural fibers on the mechanical and drilling characteristics of coir-fiber-based hybrid epoxy composites. *Engineering Research Express*, *5*(1), 15002–15020. doi:10.1088/2631-8695/acb132

Boopathi, S., Haribalaji, V., Mageswari, M., & Asif, M. M. (2022). Influences of Boron Carbide Particles on the Wear Rate and Tensile Strength of Aa2014 Surface Composite Fabricated By Friction-Stir Processing. *Materiali in Tehnologije*, *56*(3), 263–270. doi:10.17222/mit.2022.409

Boopathi, S., Jeyakumar, M., Singh, G. R., King, F. L., Pandian, M., Subbiah, R., & Haribalaji, V. (2022). An experimental study on friction stir processing of aluminium alloy (AA-2024) and boron nitride (BNp) surface composite. *Materials Today: Proceedings*, *59*, 1094–1099. doi:10.1016/j.matpr.2022.02.435

Boopathi, S., & Khare, R. KG, J. C., Muni, T. V., & Khare, S. (2023). Additive Manufacturing Developments in the Medical Engineering Field. In Development, Properties, and Industrial Applications of 3D Printed Polymer Composites (pp. 86–106). IGI Global.

Boopathi, S., Kumaresan, A., Manohar, N., & Krishna Moorthi, R. (2017). Review on Effect of Process Parameters-Friction Stir Welding Process. [IRJET]. *International Research Journal of Engineering and Technology*, *4*(7), 272–278.

Boopathi, S., Siva Kumar, P. K., & Meena, R. S. J., S. I., P., S. K., & Sudhakar, M. (2023). Sustainable Developments of Modern Soil-Less Agro-Cultivation Systems. In Human Agro-Energy Optimization for Business and Industry (pp. 69–87). IGI Global. doi:10.4018/978-1-6684-4118-3.ch004

Boopathi, S., & Sivakumar, K. (2013). Experimental investigation and parameter optimization of near-dry wire-cut electrical discharge machining using multi-objective evolutionary algorithm. *International Journal of Advanced Manufacturing Technology*, *67*(9–12), 2639–2655. doi:10.100700170-012-4680-4

Boopathi, S., Sivakumar, K., & Kalidas, R. (2012). Parametric Study of Dry WEDM Using Taguchi Method. *International Journal of Engineering Research and Development*, *2*(4), 1–6.

Boopathi, S., Venkatesan, G., & Anton Savio Lewise, K. (2023). Mechanical Properties Analysis of Kenaf–Grewia–Hair Fiber-Reinforced Composite. In *Lecture Notes in Mechanical Engineering* (pp. 101–110). Springer. doi:10.1007/978-981-16-9057-0_11

Cancela, J. J., Fandiño, M., Rey, B. J., & Martínez, E. M. (2015). Automatic irrigation system based on dual crop coefficient, soil and plant water status for Vitis vinifera (cv Godello and cv Mencía). *Agricultural Water Management*, *151*, 52–63. doi:10.1016/j.agwat.2014.10.020

Chakravarthi, P. K., Yuvaraj, D., & Venkataramanan, V. (2022). IoT-based smart energy meter for smart grids. *ICDCS 2022 - 2022 6th International Conference on Devices, Circuits and Systems*, (pp. 360–363). IEEE. 10.1109/ICDCS54290.2022.9780714

Domakonda, V. K., Farooq, S., Chinthamreddy, S., Puviarasi, R., Sudhakar, M., & Boopathi, S. (2022). Sustainable Developments of Hybrid Floating Solar Power Plants: Photovoltaic System. In Human Agro-Energy Optimization for Business and Industry (pp. 148–167). IGI Global.

Goap, A., Sharma, D., Shukla, A. K., & Rama Krishna, C. (2018). An IoT based smart irrigation management system using Machine learning and open source technologies. *Computers and Electronics in Agriculture*, *155*(May), 41–49. doi:10.1016/j.compag.2018.09.040

Gunasekaran, K., Boopathi, S., & Sureshkumar, M. (2022). Analysis of a Cryogenically Cooled Near-Dry Wedm Process Using Different Dielectrics. *Materiali in Tehnologije*, *56*(2), 179–186. doi:10.17222/mit.2022.397

Haribalaji, V., Boopathi, S., & Balamurugan, S. (2014). Effect of Welding Processes on Mechanical and Metallurgical Properties of High Strength Low Alloy (HSLA) Steel Joints. *International Journal of Innovation and Scientific Research*, *12*(1), 170–179.

Haribalaji, V., Boopathi, S., Mohammed Asif, M., Yuvaraj, T., Velmurugan, D., Anton Savio Lewise, K., Sudhagar, S., & Suresh, P. (2022). Influences of Mg-Cr filler materials in Friction Stir Process of Aluminium-based dissimilar alloys. *Materials Today: Proceedings, 66*, 948–954. doi:10.1016/j.matpr.2022.04.668

Haribalaji, V., Venkatesan, G., Asif, M. M., Pandian, M., Subbiah, R., & Boopathi, S. (2022). Investigation on corrosion and tensile Characteristics: Friction stir welding of AA7075 and AA2014. *Materials Today: Proceedings*, *66*, 743–748. doi:10.1016/j.matpr.2022.04.037

Jeevanantham, Y. A., Saravanan, A., Vanitha, V., Boopathi, S., & Kumar, D. P. (2022). Implementation of Internet-of Things (IoT) in Soil Irrigation System. *IEEE Explore*, (pp. 1–5). IEEE.

Kannan, E., Trabelsi, Y., Boopathi, S., & Alagesan, S. (2022). Influences of cryogenically treated work material on near-dry wire-cut electrical discharge machining process. *Surface Topography : Metrology and Properties*, *10*(1), 015027. doi:10.1088/2051-672X/ac53e1

Kavitha, C., Malini, P. S. G., Charan, V., Manoj, N., Verma, A., & Boopathi, S. (2022). (in press). An experimental study on the hardness and wear rate of carbonitride coated stainless steel. *Materials Today: Proceedings*. doi:10.1016/j.matpr.2022.09.524

Kumara, V., Mohanaprakash, T. A., Fairooz, S., Jamal, K., Babu, T., & B., S. (2023). Experimental Study on a Reliable Smart Hydroponics System. In *Human Agro-Energy Optimization for Business and Industry* (pp. 27–45). IGI Global. doi:10.4018/978-1-6684-4118-3.ch002

M, G., B, R., S, S., & Boopathi, S. (2018). Fabrication of Solar Irrigation System. *International Research Journal of Engineering and Technology (IRJET), 5*(2), 2106–2108.

Myilsamy, S., & Sampath, B. (2021). Experimental comparison of near-dry and cryogenically cooled near-dry machining in wire-cut electrical discharge machining processes. *Surface Topography : Metrology and Properties*, *9*(3), 035015. doi:10.1088/2051-672X/ac15e0

O'Shaughnessy, S. A., Evett, S. R., Colaizzi, P. D., & Howell, T. A. (2012). A crop water stress index and time threshold for automatic irrigation scheduling of grain sorghum. *Agricultural Water Management, 107*, 122–132. doi:10.1016/j.agwat.2012.01.018

Osroosh, Y., Peters, R. T., Campbell, C. S., & Zhang, Q. (2016). Comparison of irrigation automation algorithms for drip-irrigated apple trees. *Computers and Electronics in Agriculture*, *128*, 87–99. doi:10.1016/j.compag.2016.08.013

Palaniappan, M., Tirlangi, S., Mohamed, M. J. S., Moorthy, R. M. S., Valeti, S. V., & Boopathi, S. (2023). Fused Deposition Modelling of Polylactic Acid (PLA)-Based Polymer Composites: A Case Study. In Development, Properties, and Industrial Applications of 3D Printed Polymer Composites (pp. 66–85). IGI Global.

S., P. K., Sampath, B., R., S. K., Babu, B. H., & N., A. (2022). Hydroponics, Aeroponics, and Aquaponics Technologies in Modern Agricultural Cultivation. In *Trends, Paradigms, and Advances in Mechatronics Engineering* (pp. 223–241). IGI Global. doi:10.4018/978-1-6684-5887-7.ch012

Samikannu, R., Koshariya, A. K., Poornima, E., Ramesh, S., Kumar, A., & Boopathi, S. (2022). Sustainable Development in Modern Aquaponics Cultivation Systems Using IoT Technologies. In *Human Agro-Energy Optimization for Business and Industry* (pp. 105–127). IGI Global.

Sampath, B., & Myilsamy, S. (2021). Experimental investigation of a cryogenically cooled oxygen-mist near-dry wire-cut electrical discharge machining process. *Strojniski Vestnik. Jixie Gongcheng Xuebao*, *67*(6), 322–330. doi:10.5545v-jme.2021.7161

Sampath, B. C. S., & Myilsamy, S. (2022). Application of TOPSIS Optimization Technique in the Micro-Machining Process. In Trends, Paradigms, and Advances in Mechatronics Engineering (pp. 162–187). IGI Global. doi:10.4018/978-1-6684-5887-7.ch009

Saravanan, M., Vasanth, M., Boopathi, S., Sureshkumar, M., & Haribalaji, V. (2022). Optimization of Quench Polish Quench (QPQ) Coating Process Using Taguchi Method. *Key Engineering Materials*, *935*, 83–91. doi:10.4028/p-z569vy

Senthil, T. S., Puviyarasan, M., Babu, S. R., Surakasi, R., Sampath, B., & Associates. (2023). Industrial Robot-Integrated Fused Deposition Modelling for the 3D Printing Process. In Development, Properties, and Industrial Applications of 3D Printed Polymer Composites (pp. 188–210). IGI Global.

Trojovský, P., Dhasarathan, V., & Boopathi, S. (2023). Experimental investigations on cryogenic friction-stir welding of similar ZE42 magnesium alloys. *Alexandria Engineering Journal*, *66*(1), 1–14. doi:10.1016/j.aej.2022.12.007

Vanitha, S. K. R., & Boopathi, S. (2023). Artificial Intelligence Techniques in Water Purification and Utilization. In *Human Agro-Energy Optimization for Business and Industry* (pp. 202–218). IGI Global. doi:10.4018/978-1-6684-4118-3.ch010

Vellidis, G., Tucker, M., Perry, C., Kvien, C., & Bednarz, C. (2008). A real-time wireless smart sensor array for scheduling irrigation. *Computers and Electronics in Agriculture*, *61*(1), 44–50. doi:10.1016/j.compag.2007.05.009

Vennila, T., Karuna, M. S., Srivastava, B. K., Venugopal, J., Surakasi, R., & Sampath, B. (2022). New Strategies in Treatment and Enzymatic Processes: Ethanol Production From Sugarcane Bagasse. In Human Agro-Energy Optimization for Business and Industry (pp. 219–240). IGI Global.

Yupapin, P., Trabelsi, Y., Nattappan, A., & Boopathi, S. (2022). Performance Improvement of Wire-Cut Electrical Discharge Machining Process Using Cryogenically Treated Super-Conductive State of Monel-K500 Alloy. *Iranian Journal of Science and Technology - Transactions of Mechanical Engineering*, (pp. 1–17). IEEE. doi:10.1007/s40997-022-00513-0

Zotarelli, L., Scholberg, J. M., Dukes, M. D., Muñoz-Carpena, R., & Icerman, J. (2009). Tomato yield, biomass accumulation, root distribution and irrigation water use efficiency on a sandy soil, as affected by nitrogen rate and irrigation scheduling. *Agricultural Water Management*, *96*(1), 23–34. doi:10.1016/j.agwat.2008.06.007

Chapter 6
Smart Agriculture Irrigation Monitoring System Using Internet of Things

Kondireddy Muni Sankar
Vels Institute of Science, Technology, and Advanced Studies, India

B. Booba
Vels Institute of Science, Technology, and Advanced Studies, India

Sampath Boopathi
https://orcid.org/0000-0002-2065-6539
Muthayammal Engineering College, India

ABSTRACT

In this chapter, smart agriculture irrigation monitoring systems using IoT principles are described. The general information for precision agriculture systems, the integration of the internet of things with irrigation systems, and measuring parameters have been illustrated. Two framework models, such as Agriculture Factor-Based Relevance Vector Analysis (ARV) and advanced irrigation systems for soil moisture prediction systems for smart agriculture irrigation, have been illustrated in this chapter. The environmental factors for the irrigation system have also been monitored and controlled to improve the production rate. The framework model for an advanced irrigation system and the ARV model for prediction and control of water supply to the land based on soil moisture has also been illustrated.

INTRODUCTION

In order to increase output while having the least negative impact on the environment and to better prepare agricultural land for climate change adaptation, the United Nations has established the goal of implementing sustainable food production techniques in agriculture. Precision agriculture is a revolutionary agricultural method that has emerged in agriculture as a means of achieving the UN's goals.

DOI: 10.4018/978-1-6684-7879-0.ch006

This is a method of modern agriculture that makes use of information and communication technologies to boost the efficiency of agritourism. Farmers are utilising sensors more frequently to track the health and productivity of their crops by keeping an eye on factors including plant health, irrigation, crop output, soil health, and the impact of fertiliser and pesticides. Farmers are encouraged to monitor their crops remotely thanks to information and communication technology like wireless sensor networks and internet of things in agriculture.

For real-time monitoring of the agricultural field, an architecture that integrates several systems such as sensor motes, gateways, and handheld devices is presented. The design and creation of a precise, effective, and dependable system with minimal expense and complexity for the farmers is the main objective. The importance of implementing customization in the hardware and firmware design and development is discussed. The sensor mote and gateway process long-range communications to start the data transmission from the agricultural field to the cloud server. Long-range communications and Wireless Fidelity (Wi-Fi) are built into a hand-held IoT device, which gets sensor data from the agricultural field from the cloud server and through long-range communications. Machine learning algorithms are used to anticipate the best harvests using the sensor data collected by the handheld device(Boopathi, Siva Kumar, et al., 2023; Kumara et al., 2023).

The findings point to a promising future for precise, low-cost, and complicated systems that primarily help farmers keep an eye on their fields. Ager, sometimes known as Agri, is a Latin term that means "soil," and "culture" is the act of cultivating soil. The term "agriculture" has evolved in the contemporary era to encompass a variety of activities such as arboriculture, horticulture, domestication, and vegeculture. The 80 percent of the world's population who live in poverty might benefit from increased income, reduced poverty, and improved food security thanks to agriculture. Given that it contributes 4% of the world's GNP in 2018, agriculture is also crucial to economic growth (GDP).

For the purpose of supplying the growing global population with food, conventional agricultural methods are changing. We have developed numerous forms of agriculture during the last few centuries, including 1, 2, and 3.0, as well as 4.0 in the present and coming decades. The first agricultural revolution took place between antiquity and 1920, a time period during which most farming was done manually and relied on methods like tillage and pollination. Agriculture uses 70 percent of the world's freshwater and occupies 23% (11 million KM 2) of the livable space. Thanks to green revolution technologies, agricultural output tripled between 1960 and 2015, and more water, land, and other natural resources were used(S. et al., 2022; Vanitha et al., 2023).

Only a few numbers of species of high yielding wheat and rice variants are being grown in the early stages of the green revolution. The improvement of cultivation uniformity demonstrates that seeds are more prone to illnesses because they lack the capacity to fight off illnesses(Boopathi, 2023; Boopathi, Venkatesan, et al., 2023; Sampath et al., 2022). Due to its considerable contribution to the future potential growth of agriculture, pesticide use has grown to be a serious concern for conserving the environment and biodiversity. The phrase "transition from sustainability" in agriculture refers to the change from one agricultural framework to another that is based on a more comprehensive theory of sustainable agriculture. The third revolution in agriculture, known as precision agriculture, intends to address the sustainability issue as well as the future rise in food demand. According to the World Food Programme, the development and widespread availability of wireless sensors and communication protocols stimulate the use of precision agriculture to increase output (WFP). PA helps farmers monitor the state of their crops by using sensors and communication protocols. IoT, cloud computing, and wireless sensor networks are three of the key technologies that go into PA. PA specifically helps farmers to discover the characteristics required for a healthy crop at a specific point in time. Information gathered from many sources about soil nutrients, plant chlorophyll level, and the presence of pest weeds is combined to increase productivity and efficiency. Predictive analytics can be used to analyse the behaviour and conditions of the crops in terms of disease detection, water management, suitable crop, etc. thanks to data generated by WSN and IoT. With the use of AI, predictive analytics is made possible, and it also offers farmers valuable insights for increasing crop output(Boopathi, Siva Kumar, et al., 2023; Kumara et al., 2023; Vanitha et al., 2023).

Crop rotation, fertilisation, and irrigation are the three main concerns of agriculture. Farmer crop modification has been used since the dawn of civilization. food, raw resources, fibres, and the independent role of agriculture in the economy of the nation. Urban population growth raises the need for water, which may cause a food shortage. Over the last ten years, farmers have employed soil characteristics to comprehend seasonal fluctuations in the cultivation field for crop identification. However, this knowledge usage proved ineffective to target and optimise agricultural inputs, especially in large-scale farming, due to the lack of relevant technologies. To better address today's environmental issues, researchers have improved the farming process with "Precision Agriculture" technology. The main goal of this project is to develop a new agriculture-based decision-making system that would enable high yield production with minimal input from farmers. The series of tasks in the field, such as planting, weeding, watering, fertilising, and finally harvesting, are the responsibility of each farmer. In many wealthy nations, precision agriculture completely relies on the Internet of Things (IoT) rather than using conventional farming techniques(Bagha et al., 2022). The majority of individuals whose livelihoods depend on agriculture live in rural areas with limited access to agricultural technologies. Numerous disciplines, including corporate management, medical advice, weather forecasting, the educational system, and communication systems, have benefited from technological breakthroughs.

An irrigation system that supports decision-making to sustain high-quality crop growth and water stability has been patented by a team from the University of Michigan. It optimises the water level for irrigation. The American Society of Mechanical Engineers will grant the "Agriculture Cultivation Recommender and Smart Irrigation System" patent (ASME). The sensor is a key component of the Internet of Things (IoT), which transforms routine agriculture into precision or intelligent agriculture. Every process, including data sensing and collection, transmission, reception, storage, and decision-making, is the focus of the work mentioned. This technique helps farmers examine crop growth characteristics in relation to soil conditions and provides necessary recommendations for creating precise projections.

PRECISION AGRICULTURE

There is no data on how farmers behave in regards to how much soil is sampled. To ensure the promotion and implementation of effective irrigation systems, the farmer's decisions and insights must be carefully considered. According to equipment, crop consultants, university extensions, news media, and governmental organisations, the study examines southern states in the United States that cultivate cotton. The study advances knowledge of the farmer's perspective on the adoption and application of effective irrigation methods. The adoption of precision farming technologies is impacted in major and varied ways by the many informational sources. For instance, farmers typically combine information from precision farming with data from other sources, such as crop consultation, trade exhibits, advertisements, etc., using the extended version of the information. In order to assist farmers in boosting their profit margins, crop advisors and input suppliers are sharing their knowledge of precision farming.

Leading agricultural practises that use technology include GPS, sensors, variable rate technology (VRT), and yield monitoring (YM). Precision agriculture techniques are attractive to farmers because of their promise to increase agricultural earnings and the environment. Currently, geographic information systems (GIS) and sensors are the most commonly used technologies for this. In 14 southern U.S. states, the adoption of precision farming increased from 63 percent in 2009 to around 73 percent in 2013(Obi et al., 2016).

The soil survey maps, lime, phosphorus, and potassium treatments, and grid and zone soil sampling are the precision agricultural technologies that are most extensively used. According to the International Cotton Growers' Association (ICG), a trade organisation that represents farmers who use cotton as a raw material for animal feed and other products like paper and paperboard, the southern U.S. is the most crucial geographical cotton production region in the world. The majority of southern U.S. states view government organisations, news outlets, crop consultants, university extension agents, other farmers, and marketers of farm equipment as crucial sources of information about precision farming techniques. There is currently no method available to determine irrigation methods over vast areas(Boopathi, Khare, et al., 2023; Palaniappan et al., 2023; Senthil et al., 2023).

USAGE OF IOT IN PRECISION AGRICULTURE

Smart farming is now possible thanks to the Internet of Things. IoT sensors are able to predict climatic change, address irrigation problems, and support farmers in steadily increasing crop output. Agriculture must come up with innovative solutions due to the growing population and increased food demand. According to academics, the main issue with IoT in agriculture is obtaining a high output at a reasonable price. After the Internet of Things (IoT) is used in the farming sector, there will be a 20% boost in the growth of the agricultural industry. The agriculture sector places a strong priority on IoT technology in order to reap benefits quickly and at minimal cost, making farming high-tech and innovative. In irrigation fields, sensors track and save the information for later use. The monitoring cloud network is connected to the sensors. The use of IoT in the irrigation industry has resulted in a significant reduction in manual labour and strenuous physical labour. The Internet of Things can provide farmers a variety of advantages, including improved crop output, reduced water use, crop monitoring, and foresight of environmental effects. Because the sensors collect more precise and accurate data, the IoT cloud environment can make predictions that are more accurate (Obi et al., 2016). It assists the farmer in selecting the appropriate plant for a certain setting, from planting the seeds to choosing crops based on the environment. Water conservation for irrigation can be tracked using an intelligent irrigation planning system. To address the rising demand, machine learning helps improve and modernise the irrigation system. Using IoT and machine learning algorithms, smart decision support systems are becoming a crucial part of agricultural processes(Babu et al., 2022; Boopathi, Arigela, et al., 2022; Jeevanantham et al., 2022).

IOT MEASURES FOR MAJOR AGRICULTURAL ISSUES

Weather Condition

Agriculture is significantly impacted by changes in climate brought on by weather. Utilizing IoT to handle climate change and adapt is an immediate solution to this issue. According to experts, severe climate changes would result in a 30% decline in agricultural production by 2050.

Disease Detection and Diagnosis

IoT-enabled sensors take pictures of harmed plants and provide a suitable and timely decision to save the plant's life. Here, image pre-processing in conjunction with IoT aids plant pathologists in the accurate identification and detection of illness. A lack of effective pesticide management methods and a number of plant diseases have a significant impact on crop productivity.

Usage of Fertilizer

The amount of fertiliser used during farming operations significantly affects the harvest. Based on the requirements of the crop, each farmer should make informed judgments regarding the application of chemicals. The impact of fertiliser consumption destroys the plant's nutrient content and continues to be a major contributor to many acute ailments.

Soil Variations

Based on the needs of the crop, each farmer should choose the right chemicals to use. The use of fertilisers has the negative effect of destroying the plant's nutrient content and continues to be a major contributor to numerous acute illnesses.

Water Quality and Quantity Estimation

The crop loses water through transpiration and evaporation while it is being grown. The farmer should estimate how much water is necessary for agriculture ahead of time. The water depends on a number of variables, including the climate, season, soil type, crop varieties, and stage of growth.

Yield Production Readiness Analysis (YPRA)

Sensors are intended to evaluate how ripe the fruits are. With the aid of this technology, retailers may work with farmers to determine when fruit is ready before sending it to various business sectors. Once farmers are aware of crop price information, they decide to sell the plants when demand is at its highest.

YIELD MONITORING AND FORECAST

The production is maximised and the policy mechanisms are adjusted when the harvest season is correctly predicted. The European Commission, which makes decisions for the EU, is tasked with mapping the production variability of sugar beets(Putri et al., 2012).

The knowledge necessary for a farmer to use precision agricultural technology efficiently is above that of a conventional farmer. However, their expertise is insufficient to convert a technology for uniform crop production into a precise agriculture technology. The potential advantage of a specific precision agriculture information source depends on its ability to close the information gap that could be necessary to employ those technologies efficiently. It depends greatly on how adopting precision agriculture through an information source can increase farm earnings for this indirect advantage to materialise. People decide on information since there may be an advantage to using it. Precision agriculture enables farmers to maximise the use of their inputs by taking advantage of the spatial and temporal variability present in a field. Furthermore, choosing an information source could have a cost attached to it. Endogeneity is a major issue in precision agriculture technology, because it hinders potential end users' acceptance of new advances. The economic agent's human capital endowment and the type of firm operated have a significant impact on the choice and use of information(Comparetti, 2011).

Irrigation uses only 1% of the freshwater used in India, which supplies 4% of the world's population. Developed nations prioritise water conservation efforts and cost-effective smart irrigation systems to guarantee the security and effectiveness of their water supplies. The design of the stated smart irrigation system heavily depends on a few key variables, including rainfall, evaporation, and soil moisture. In this post, we explain how an effective approach for utilising water resources is modelled based on cutting-edge technologies. Wi-Fi sensor integration transforms agriculture in unimaginable ways with smart farming.

FARMING NEEDS FOR PRECISION AGRICULTURE

a. ***Complex technology with illogicality equipment:*** Increased awareness of the benefits of the invention was expected to increase the rate of PA adoption. Farmers should adjust to the hardware and software-related technological challenges. They should also be aware of the parts of the operating PA systems that operate irrationally.
b. ***Lack of products:*** Due to a lack of available goods, the PA combines engineering and technology to facilitate productive farming. This idea integrates the processes of data collection, data processing, and additional monitoring.
c. ***Increase in cost due to maintenance:*** The cost of installing PA equipment has not yet been widely accepted by farmers due to several networking issues.

ADVANCED AGRICULTURAL PRACTICES

By implementing these revolutionary techniques, food quantity and quality were improved. Alternative food supply methods, such as genetically and bioengineered foods, face genetic and DNA issues.The yield can be increased by traditional agricultural practises with little negative impact. Bioorganic food products are a result of similar technology being accepted by society. Numerous studies show that these foods have serious negative impacts on humans, including infertility, changes to human health, ageing, improper insulin regulation, etc. Greenhouse farming is a traditional kind of agricultural production. The 19th century saw a rise in popularity for the concept of plant cultivation(Boopathi, Siva Kumar, et al., 2023; S. et al., 2022; Vanitha et al., 2023). In nations with severe weather, the same was advertised. Moving toward cultural and urban farming necessitates the use of sophisticated and controlled procedures, which are essential. Crop production performance varies based on elements like precision, material, ventilation, decision-support systems, etc. In order to anticipate and adapt to climate change, an IoT-based prototype continuously measures environmental variables including soil temperature, humidity, water level, etc. An accurate and secure link between the participating objects is necessary for timely information transfer and reporting in precision agriculture. In the remote locations where the farmers were found, there were connected energy networks. The agricultural industry needs a robust and large design that takes energy use, dependability, cost, etc. into account(Heideker et al., 2020).

Various ecological metrics are needed to incorporate precision agriculture in the cultivation field in order to indirectly evaluate the environmental impact. Precision farming strives to boost yield and profitability, lessen the effects food production has on the environment, and maximise output at the lowest possible cost. In order to solve the issue of raising yields and profitability, this method calls for a creative strategy. It is necessary to quantify the variability in sugar beet production and quality because it is largely unknown and depends on soil characteristics and microclimate conditions. Several strategies for increasing agricultural income and ensuring efficient resource use are discussed. Here, the highly developed precision agriculture model has been applied to monitor crop growth while accounting for variations in the field, such as air temperature, soil moisture, and soil type. Precision farming reduces the impact of traditional management techniques. The suggested study is focused on identifying the factors that limit crop output. According to the soil types, precision farming has an impact on crop development that is yield-based. The possible methods of precision agriculture could increase the production of the Indian agricultural industry and make use of numerous technologies.

WSN-BASED PRECISION AGRICULTURE

The majority of southern U.S. states view government organisations, news outlets, crop consultants, university extension agents, other farmers, and marketers of farm equipment as crucial sources of information about precision farming techniques. There is currently no method available to determine irrigation methods over vast areas. The Internet of Things (IoT) is a network of internet-connected things that can gather or modify data. It primarily links a computer with an IP address into the global network. Any other device can be connected to it via the global network known as the Internet. In precision agriculture, a vast amount of data is collected and analysed. The sensor node typically runs on battery power with little energy. WSN will aid farmers in increasing crop productivity by utilising contemporary technology. It will also take the place of conventional farming methods(Dhall & Agrawal, 2018). By utilising sensors and machines to monitor and manage water usage, smart irrigation employing the Internet of Things and machine learning techniques attempts to assure optimal water resource management. With the numerous sensing parameters taken into account, this system aids in the prediction of irrigation planning based on irrigation requirements.

METHOD ONE: AGRICULTURE FACTOR-BASED RELEVANCE VECTOR ANALYSIS (ARV) SYSTEM

By 2050, a billion people will be supported by the production of food and fuel thanks to the Internet of Things (IoT). Farmers now gather real-time data from sensors and keep an eye on their crops in order to reap bigger yields and increase revenues. India ranks second in agricultural production due to the variety of fruits and vegetables that are produced there year-round according to different weather conditions. In the majority of developed nations, farmers install sensors in their fields to automate farming operations based on current environmental conditions. It is still difficult for farmers to find answers to problems like preventing plant diseases and reducing crop loss in order to increase productivity(Mekonnen et al., 2020). Recent technology developments like precision-based agriculture increase yields by utilising machine learning and the Internet of Things (IoT). An idea for a computer application to plan irrigation in either a scheduled or automatic manner by analysing favourable climatic parameters like temperature and humidity. Farmers will be able to determine the amount of water needed for cultivation so that healthy plants may be maintained with the least amount of water.

Experimental Setup

The shortage of food is impacted by the growth in the world population. A significant development in the agricultural sector that addresses current environmental problems is the proposed module for precision agriculture, In order to meet the site-specific requirements of the outlined modern agriculture methods, the system recognises differences in environmental elements and suggests their integration in the field. The framework of the proposed ARV recommendation system is shown in Figure 1.

Figure 1. Framework of the proposed ARV recommendation system

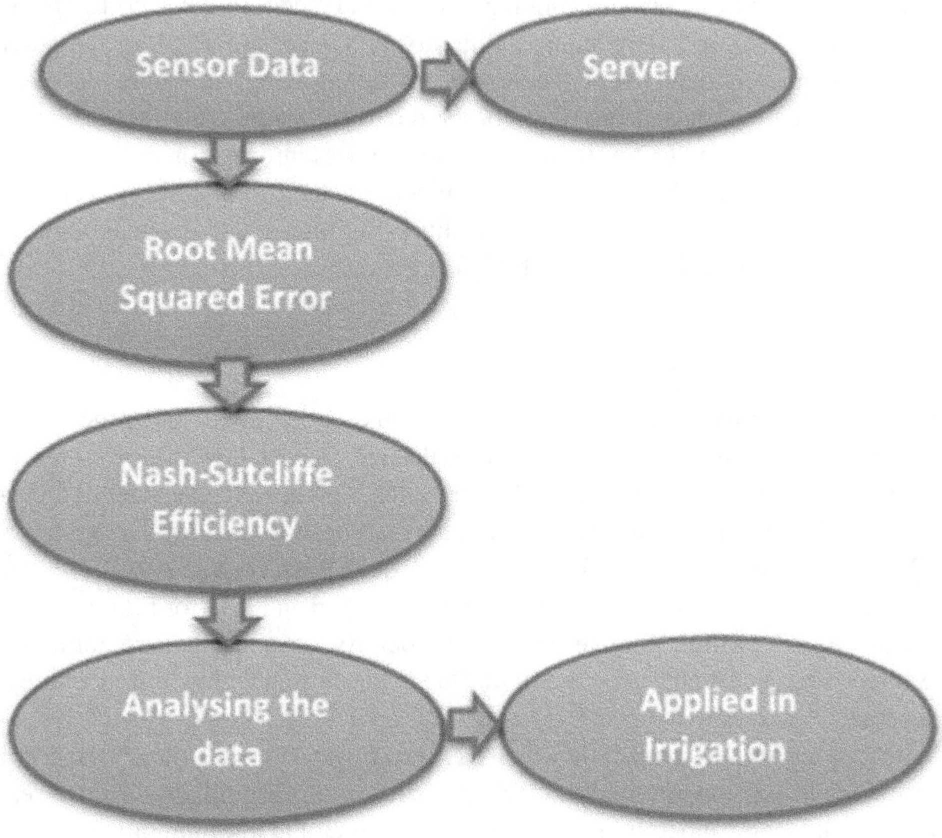

Sensor Data Collection: The ability to improve crop output in real time thanks to data collection is a key component of precision agriculture. Installed sensors in the field continuously collect data and transmit it to a centralised server for archival. Any cultivation must have a seedbed free of weeds in order for the seeds to sprout fast. The best low-cost, low-power sensor design for gathering field data has been developed by several researchers. The best results in the single-tier heterogeneous network with the ZigBee module and the GPRS module were achieved by the soil moisture sensor and soil temperature sensor. In this research, solar-powered irrigation was one of the main points of interest. Indian farmers use ventilation techniques to regulate the temperature and humidity of their crops. The influence of temperature, humidity, and water level on cultivation led to insights into how crucial data on these variables should be considered in this study. Farmers who are attempting to increase their yields receive real-time decision-making help from the data acquired over wireless channels for study by agriculturalists.

Relevant Vector Analysis: Utilizing the cutting-edge ARV algorithm, all data analytics for the ARV Recommendation System were performed. Based on the recorded sensor value, the computer forecasts the important variables and offers suggestions for high-yield farming(Singh et al., 2021). The residuals of errors in the estimation of the data samples used to determine an estimator's chance of success, or its confidence interval, are represented by the RMSE's deviations from the mean.

The real-time sparse dataset from the agricultural field was employed by the ARV recommendation algorithm, which also took into account the training samples and randomly selected 100 plants from 100 different types as testing samples. Temperature, humidity, and water are taken for measurement. Sensors for measuring the microclimate over the course of the full growing season are placed at a distance of 8 metres. Throughout a month, sensors were observed daily at 8 a.m., 10 a.m., 12 p.m., 2 p.m., and 4 p.m. The humidity sensors used to continuously record wetness information do so every 60 minutes. Sensors were utilised to measure the pH value, soil temperature, and soil humidity. The parameter that was sensed has a control system. As shown, an agricultural field is divided into equal tubs, each with a one-unit area and dimensions of 8 metres long by 7 metres wide.

Results and Discussions

The following are the presumptions used for the observation of crop growth under the conditions listed below: A five-minute irrigation period was used. Crops that are grown below the lowest reasonable solar temperature suffer slightly from a lack of water. Under typical lighting and weather circumstances, such as a clear or overcast/drizzly day, crops develop in a normal, healthy manner. The system's effectiveness will be shown following the crop's full growth and yield(Bagha et al., 2022). This example will be performed using the crop growth image that was photographed and kept for future use as proof. The ARV Recommendation algorithm examines the data to identify variables that are not in a satisfactory state and determines whether or not to use the field. Due to the extremely low soil moisture and humidity levels, a necessity.

The "ARV Recommendation" algorithm, which has been suggested, recognises variances in environmental factors and suggests the top farmers for each specific farming discipline. During the experimental period, data for humidity and temperature show high temperatures and low humidity during the daytime, with water evaporation also occurring at a high pace. The "ARV Recommendation" algorithm, which has been suggested, recognises variances in environmental factors and suggests the top farmers for each specific farming discipline. During the experimental period, data for humidity and temperature show high temperatures and low humidity during the daytime, with water evaporation also occurring at a high pace.

METHOD TWO: ADVANCED IRRIGATION SYSTEM FOR SOIL MOISTURE PREDICTION

The minimum water used for cultivation and boost output yield, the Indian government has proposed a water management module for precision agriculture. The "AISM System" proposed module's main goals are to support farmers who practise precision farming and maximise the use of water resources. The created AISM System is useful for keeping an eye on larger crop areas. The management of irrigation in agriculture can be supported by predictions of soil moisture. People today prefer automation over human involvement in all activities. Irrigation is the process of applying water artificially to agricultural fields; it is done by machines rather than people for a variety of reasons, including socioeconomic and environmental considerations. Reduced water use for plants and other agricultural crops is the goal of AISM. How much water is required to develop crops, flowers, and other plant-based organisms depends in part on the moisture content of the soil. A research team has created a machine learning system to forecast soil moisture and schedule irrigation in an area of agriculture. Data from the agricultural sector and web-based weather forecasting platforms like Open Weather and AccuWeather were analysed for the project. The architecture for smart forming is shown in Figure 2.

An infrastructure tool for anticipating soil moisture and intelligent irrigation scheduling in the agricultural sector is the AISM System. The main function predicts soil moisture using a unique SLR (sparse linear regression) technique. The new K SLR (K-Sparse Linear Regimental Algorithm for Smart Irrigation) algorithm is used in the secondary function. The newly created unique SLR (sparse linear regression) algorithm is contrasted with the greatest performance displayed by SVM (SVM). The popular supervised machine learning method SVM is used to solve a variety of classification and regression problems. The kernel function has been altered to accept vectors, pictures, and sequential data. A pair of data points is viewed as a similarity function by the kernel. Based on the provided input information, the AISM System forecasts the soil moisture for the following days. The water pump is connected to a relay switch with Wi-Fi capability for irrigation control. The signal for in-field real-time monitoring is triggered by the online interface.(Heideker et al., 2020).

The farmer plans the irrigation at a predetermined threshold for soil moisture. The R-Pi module notifies the Arduino-Uno and manages the relay to switch the water motor between the ON and OFF states. The mode of irrigation manual or automatic is determined by the Ground Soil Moisture (GSM) anticipated with MSE and its comparison with Tmax. Farmers that rely on the terrestrial tank due to a lack of water can use the AISM System with excellent support. the motor received the signal with the value "1" for "ON." It sends "0" to cease irrigation. The system forecasts how much water will be needed for crops.

Figure 2. Architecture of smart farming

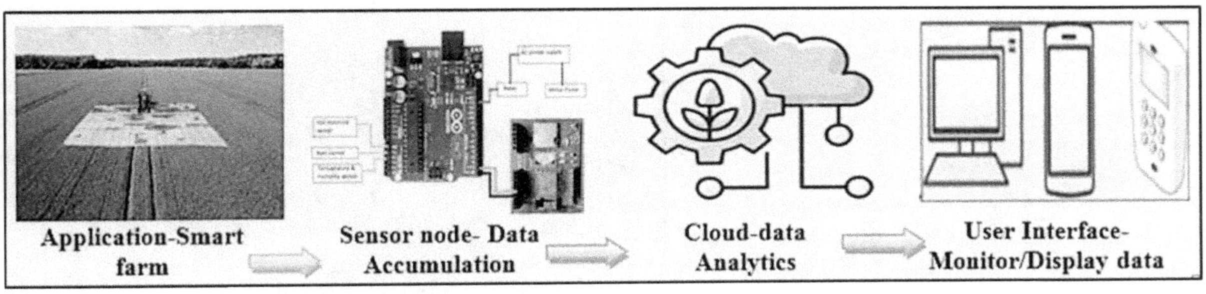

Experimental Setup

The framework of the proposed AISM system is shown in Figure 3. It functions by analysing online data from weather portals like Open Weather and AccuWeather, including temperature, humidity, and other variables related to the weather. **Arduino UNO board:** A microcontroller called Arduino is intended for interactive object-based applications. Six of the fourteen I/O pins on the ATmega328-based UNO board's Arduino are outputs. When linked to USB, the serial setup makes transmission easy. The Atmega328 chip microcontroller is widely used because of the availability of its open-source resources(Chakravarthi et al., 2022; Saravanan et al., 2022). **Moisture Sensor:** The sensor suggests irrigation for plants in the botanical garden and agricultural fields. The operating voltage for ground moisture is 5 V, and the needed current is less than 20 mA. Capacitance is used by the soil moisture sensor to gauge the surrounding soil's dielectric permittivity. The sensor generates a voltage that is proportional to the soil's water content and, consequently, to the dielectric permittivity. The evaporation-related moisture loss is measured by the soil moisture sensor. The humidity sensor software and the Arduino Uno board are mostly used to control the moisture content and its irrigation. The soil moisture is linked to the VCC percent, GND soil moisture, and attached sensor on the Arduino UNO(Jeevanantham et al., 2022).

Figure 3. The framework of the Proposed AISM System

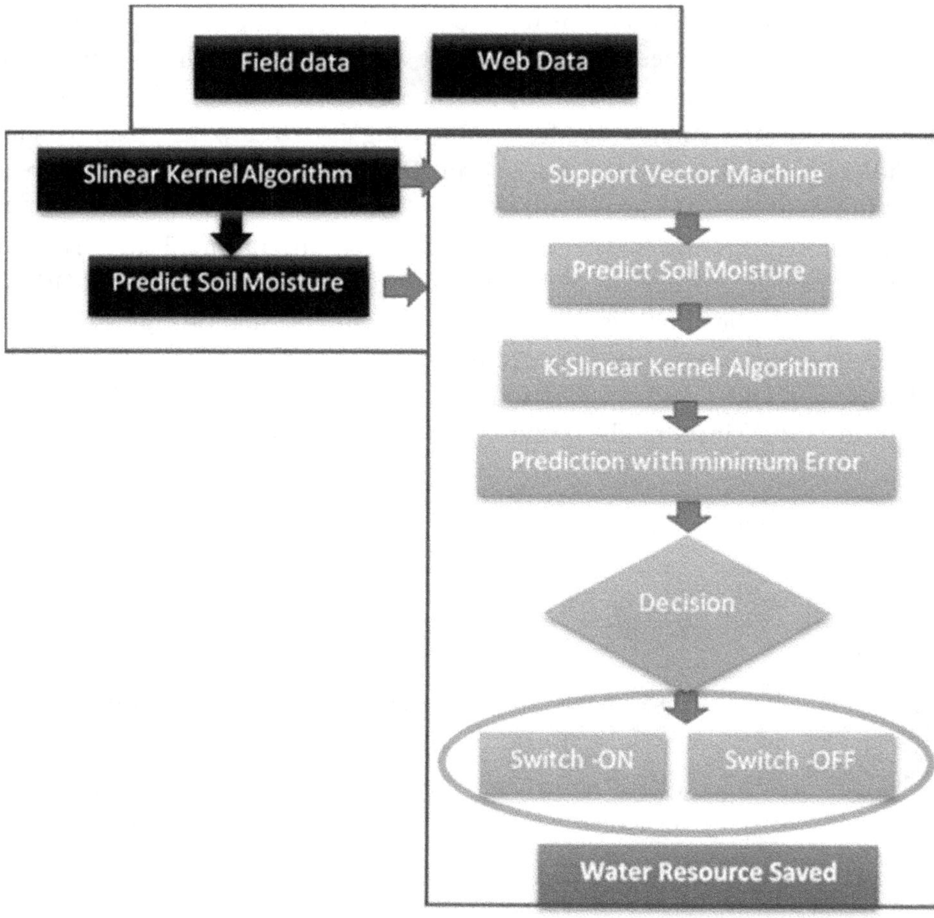

Relay: An essential component of a car's control system is a high-power electric motor relay, which needs to be pre-calibrated to withstand overloads or failures. The common, customary, and normal pins are the three options for connecting the relays. The relay module is very easy to connect to an Arduino. The Arduino digital pin is connected to all of the forwarding pin's pins. The IN1 relay port is connected to a ground, and the relay's GND is connected to the ground.

Water Pump: This is an illustration of a DC motor, a type of motor that is frequently used with DC power distribution systems. The supports are utilised to enable the rotor to continue rotate in the direction of its axis. While some rotors employ supports to hold the conductors in place, some rotors have magnets that the stator uses to grip the conductors.

Results and Discussions

To forecast the needed soil moisture for Indian agriculture, an algorithm has been created. It makes use of sensors to keep an eye on environmental factors including temperature, humidity, and soil moisture. By tracking the irrigation settings on an hourly basis for around three weeks, the programme demonstrated its effectiveness. Algorithm-predicted soil moisture (PSM) measurement instructions are provided by SLR. This method aids in forecasting the soil moisture for the following days. With the Least Mean Square Error and Correlation Coefficient, the K SLR algorithm promotes excellent accuracy. In comparison to the current SVM method, the SLR algorithm exhibits superior accuracy with a lower MSE. For the purpose of predicting the soil moisture difference for the following day, the SLR algorithm is trained using the observed sensor data and the recorded data independently(Ahmed et al., 2018; Bhat et al., 2019).

In addition, Taguchi(Boopathi et al., 2021; Boopathi, Haribalaji, et al., 2022; Boopathi, Jeyakumar, et al., 2022; Boopathi, Thillaivanan, et al., 2022; Boopathi, 2022a; Boopathi, Venkatesan, et al., 2023; Dass james & Boopathi, 2016; Haribalaji et al., 2021, 2022; Kavitha et al., 2022; Sampath & Myilsamy, 2021; Yupapin et al., 2022), and response surface methods(Boopathi & Myilsamy, 2021; Boopathi & Sivakumar, 2013, 2014; Kannan et al., 2022; Myilsamy & Sampath, 2021; Sampath & Myilsamy, 2021) can be applied to the optimization of agricultural systems. Multi-criteria optimization techniques have been applied to improve the overall performance(Boopathi, 2019, 2022b; Boopathi & Sivakumar, 2013; Kavitha et al., 2022; Myilsamy & Sampath, 2021; Yupapin et al., 2022).

The "AISM System" proposed module analyses the sensed data to forecast soil moisture and schedule irrigation effectively. In comparison to the current ANN model, in comparison to manual watering, sensor-based irrigation uses less water and takes less time. The technology uses the soil moisture as the information to prevent both insufficient and excessive irrigation. This module discusses issues with manual irrigation and shows that sensor-based irrigation uses the least amount of water. The most dependable sensor-based irrigation system is the AISM System, which operates automatically without human intervention. The research only considered small-scale agriculture; however, the technology will eventually need to be tested on a much larger scale. The MSE and R analysis demonstrates.

SUMMARY

IoT-based smart agriculture irrigation monitoring systems are discussed in this chapter. The illustration of measuring parameters, internet of things integration with irrigation systems, and general information for precision agriculture systems. In this chapter, two framework models for the irrigation system for smart agriculture have been presented. To increase productivity, the environmental elements affecting the irrigation system have been tracked and managed. With the use of appropriate flow charts, the framework model for an enhanced irrigation system for soil moisture prediction has also been described.

REFERENCES

Ahmed, N., De, D., & Hussain, I. (2018). Internet of Things (IoT) for Smart Precision Agriculture and Farming in Rural Areas. *IEEE Internet of Things Journal*, *5*(6), 4890–4899. doi:10.1109/JIOT.2018.2879579

Babu, B. S., Kamalakannan, J., Meenatchi, N., Karthik, S., & Boopathi, S. (2022). Economic impacts and reliability evaluation of battery by adopting Electric Vehicle. *IEEE Explore*, (pp. 1–6). IEEE.

Bagha, H., Yavari, A., & Georgakopoulos, D. (2022). Hybrid Sensing Platform for IoT-Based Precision Agriculture. *Future Internet, 14*(8), 233. doi:10.3390/fi14080233

Bhat, M. A., Adil, A. W., Sikander, B. M., Lone, Y., & Malik, J. A. (2019). Waste Management Technology for Sustainable Agriculture. In *Waste Management Technology for Sustainable Agriculture* (pp. 156–176). IGI Global. doi:10.4018/978-1-7998-0031-6.ch009

Boopathi, S. (2019). Experimental investigation and parameter analysis of LPG refrigeration system using Taguchi method. *SN Applied Sciences, 1*(8), 892. doi:10.100742452-019-0925-2

Boopathi, S. (2022a). (in press). Cryogenically treated and untreated stainless steel grade 317 in sustainable wire electrical discharge machining process: A comparative study. *Environmental Science and Pollution Research International*, 1–10. doi:10.100711356-022-22843-x PMID:36057706

Boopathi, S. (2022b). Experimental investigation and multi-objective optimization of cryogenic Friction-stir-welding of AA2014 and AZ31B alloys using MOORA technique. *Materials Today. Communications*, *33*, 104937. doi:10.1016/j.mtcomm.2022.104937

Boopathi, S. (2023). An Investigation on Friction Stir Processing of Aluminum Alloy-Boron Carbide Surface Composite. In *Materials Horizons: From Nature to Nanomaterials* (pp. 249–257). Springer. doi:10.1007/978-981-19-7146-4_14

Boopathi, S., Arigela, S. H., Raman, R., Indhumathi, C., Kavitha, V., & Bhatt, B. C. (2022). Prominent Rule Control-based Internet of Things: Poultry Farm Management System. *IEEE Explore*, (pp. 1–6). IEEE.

Boopathi, S., Haribalaji, V., Mageswari, M., & Asif, M. M. (2022). Influences of Boron Carbide Particles on the Wear Rate and Tensile Strength of Aa2014 Surface Composite Fabricated By Friction-Stir Processing. *Materiali in Tehnologije, 56*(3), 263–270. doi:10.17222/mit.2022.409

Boopathi, S., Jeyakumar, M., Singh, G. R., King, F. L., Pandian, M., Subbiah, R., & Haribalaji, V. (2022). An experimental study on friction stir processing of aluminium alloy (AA-2024) and boron nitride (BNp) surface composite. *Materials Today: Proceedings*, *59*, 1094–1099. doi:10.1016/j.matpr.2022.02.435

Boopathi, S., & Khare, R. KG, J. C., Muni, T. V., & Khare, S. (2023). Additive Manufacturing Developments in the Medical Engineering Field. In Development, Properties, and Industrial Applications of 3D Printed Polymer Composites (pp. 86–106). IGI Global.

Boopathi, S., Lewise, K. A. S., Subbiah, R., & Sivaraman, G. (2021). Near-dry wire-cut electrical discharge machining process using water-air-mist dielectric fluid: An experimental study. *Materials Today: Proceedings*, *49*(5), 1885–1890. doi:10.1016/j.matpr.2021.08.077

Boopathi, S., & Myilsamy, S. (2021). Material removal rate and surface roughness study on Near-dry wire electrical discharge Machining process. *Materials Today: Proceedings, 45*(xxxx), 8149–8156. doi:10.1016/j.matpr.2021.02.267

Boopathi, S., Siva Kumar, P. K., & Meena, R. S. J., S. I., P., S. K., & Sudhakar, M. (2023). Sustainable Developments of Modern Soil-Less Agro-Cultivation Systems. In Human Agro-Energy Optimization for Business and Industry (pp. 69–87). IGI Global. doi:10.4018/978-1-6684-4118-3.ch004

Boopathi, S., & Sivakumar, K. (2013). Experimental investigation and parameter optimization of near-dry wire-cut electrical discharge machining using multi-objective evolutionary algorithm. *International Journal of Advanced Manufacturing Technology*, *67*(9–12), 2639–2655. doi:10.100700170-012-4680-4

Boopathi, S., & Sivakumar, K. (2014). Study of water assisted dry wire-cut electrical discharge machining. *Indian Journal of Engineering and Materials Sciences*, *21*(1), 75–82. https://doi.org/http://nopr.niscair.res.in/handle/123456789/27453

Boopathi, S., Thillaivanan, A., Azeem, M. A., Shanmugam, P., & Pramod, V. R. (2022). Experimental investigation on abrasive water jet machining of neem wood plastic composite. *Functional Composites and Structures*, *4*(2), 025001. doi:10.1088/2631-6331/ac6152

Boopathi, S., Venkatesan, G., & Anton Savio Lewise, K. (2023). Mechanical Properties Analysis of Kenaf–Grewia–Hair Fiber-Reinforced Composite. In *Lecture Notes in Mechanical Engineering* (pp. 101–110). Springer. doi:10.1007/978-981-16-9057-0_11

Chakravarthi, P. K., Yuvaraj, D., & Venkataramanan, V. (2022). IoT-based smart energy meter for smart grids. *ICDCS 2022 - 2022 6th International Conference on Devices, Circuits and Systems*, (pp. 360–363). IEEE. 10.1109/ICDCS54290.2022.9780714

Comparetti, A. (2011). Precision Agriculture: Past, Present and Future. *Agricultural Engineering and Environment*, *16*, 17.

Dass james, A., & Boopathi, S. (2016). Experimental Study of Eco-friendly Wire-Cut Electrical Discharge Machining Processes. *International Journal of Innovative Research in Science, Engineering and Technology*, *5*(1), 666–675.

Dhall, R., & Agrawal, H. (2018). An improved energy efficient duty cycling algorithm for IoT based precision agriculture. *Procedia Computer Science*, *141*, 135–142. doi:10.1016/j.procs.2018.10.159

Haribalaji, V., Boopathi, S., & Asif, M. M. (2021). Optimization of friction stir welding process to join dissimilar AA2014 and AA7075 aluminum alloys. *Materials Today: Proceedings*, *50*, 2227–2234. doi:10.1016/j.matpr.2021.09.499

Haribalaji, V., Boopathi, S., Mohammed Asif, M., Yuvaraj, T., Velmurugan, D., Anton Savio Lewise, K., Sudhagar, S., & Suresh, P. (2022). Influences of Mg-Cr filler materials in Friction Stir Process of Aluminium-based dissimilar alloys. *Materials Today: Proceedings*, *66*(xxxx), 948–954. doi:10.1016/j.matpr.2022.04.668

Heideker, A., Ottolini, D., Zyrianoff, I., Neto, A. T., Salmon Cinotti, T., & Kamienski, C. (2020). IoT-based Measurement for Smart Agriculture. *2020 IEEE International Workshop on Metrology for Agriculture and Forestry, MetroAgriFor 2020 - Proceedings*, (pp. 68–72). IEEE. 10.1109/MetroAgriFor50201.2020.9277546

Jeevanantham, Y. A., Saravanan, A., Vanitha, V., Boopathi, S., & Kumar, D. P. (2022). Implementation of Internet-of Things (IoT) in Soil Irrigation System. *IEEE Explore*, (pp. 1–5). IEEE.

Kannan, E., Trabelsi, Y., Boopathi, S., & Alagesan, S. (2022). Influences of cryogenically treated work material on near-dry wire-cut electrical discharge machining process. *Surface Topography : Metrology and Properties*, *10*(1), 015027. doi:10.1088/2051-672X/ac53e1

Kavitha, C., Malini, P. S. G., Charan, V., Manoj, N., Verma, A., & Boopathi, S. (2022). (in press). An experimental study on the hardness and wear rate of carbonitride coated stainless steel. *Materials Today: Proceedings*. doi:10.1016/j.matpr.2022.09.524

Kumara, V., Mohanaprakash, T. A., Fairooz, S., Jamal, K., Babu, T., & B., S. (2023). Experimental Study on a Reliable Smart Hydroponics System. In *Human Agro-Energy Optimization for Business and Industry* (pp. 27–45). IGI Global. doi:10.4018/978-1-6684-4118-3.ch002

Mekonnen, Y., Namuduri, S., Burton, L., Sarwat, A., & Bhansali, S. (2020). Review—Machine Learning Techniques in Wireless Sensor Network Based Precision Agriculture. *Journal of the Electrochemical Society*, *167*(3), 037522. doi:10.1149/2.0222003JES

Myilsamy, S., & Sampath, B. (2021). Experimental comparison of near-dry and cryogenically cooled near-dry machining in wire-cut electrical discharge machining processes. *Surface Topography : Metrology and Properties*, *9*(3), 035015. doi:10.1088/2051-672X/ac15e0

Obi, F., Ugwuishiwu, B., & Nwakaire, J. (2016). Agricultural Waste Concept, Generation, Utilization and Management. *Nigerian Journal of Technology*, *35*(4), 957. doi:10.4314/njt.v35i4.34

Palaniappan, M., Tirlangi, S., Mohamed, M. J. S., Moorthy, R. M. S., Valeti, S. V., & Boopathi, S. (2023). Fused Deposition Modelling of Polylactic Acid (PLA)-Based Polymer Composites: A Case Study. In Development, Properties, and Industrial Applications of 3D Printed Polymer Composites (pp. 66–85). IGI Global.

Putri, D. A., Saputro, R. R., & Budiyono, B. (2012). Biogas production from cow manure. *International Journal of Renewable Energy Development*, *1*(2), 61–64. doi:10.14710/ijred.1.2.61-64

S., P. K., Sampath, B., R., S. K., Babu, B. H., & N., A. (2022). Hydroponics, Aeroponics, and Aquaponics Technologies in Modern Agricultural Cultivation. In *Trends, Paradigms, and Advances in Mechatronics Engineering* (pp. 223–241). IGI Global. doi:10.4018/978-1-6684-5887-7.ch012

Sampath, B., & Myilsamy, S. (2021). Experimental investigation of a cryogenically cooled oxygen-mist near-dry wire-cut electrical discharge machining process. *Strojniski Vestnik. Jixie Gongcheng Xuebao*, *67*(6), 322–330. doi:10.5545v-jme.2021.7161

Sampath, B. C. S., & Myilsamy, S. (2022). Application of TOPSIS Optimization Technique in the Micro-Machining Process. In Trends, Paradigms, and Advances in Mechatronics Engineering (pp. 162–187). IGI Global. doi:10.4018/978-1-6684-5887-7.ch009

Saravanan, A., Venkatasubramanian, R., Khare, R., Surakasi, R., Boopathi, S., Ray, S., & Sudhakar, M. (2022). POLICY TRENDS OF RENEWABLE ENERGY AND NON. *Renewable Energy*.

Senthil, T. S., Puviyarasan, M., Babu, S. R., Surakasi, R., Sampath, B., & Associates. (2023). Industrial Robot-Integrated Fused Deposition Modelling for the 3D Printing Process. In Development, Properties, and Industrial Applications of 3D Printed Polymer Composites (pp. 188–210). IGI Global.

Singh, R. K., Berkvens, R., & Weyn, M. (2021). AgriFusion: An Architecture for IoT and Emerging Technologies Based on a Precision Agriculture Survey. *IEEE Access : Practical Innovations, Open Solutions*, *9*, 136253–136283. doi:10.1109/ACCESS.2021.3116814

Vanitha, S. K. R., & Boopathi, S. (2023). Artificial Intelligence Techniques in Water Purification and Utilization. In *Human Agro-Energy Optimization for Business and Industry* (pp. 202–218). IGI Global., doi:10.4018/978-1-6684-4118-3.ch010

Yupapin, P., Trabelsi, Y., Nattappan, A., & Boopathi, S. (2022). Performance Improvement of Wire-Cut Electrical Discharge Machining Process Using Cryogenically Treated Super-Conductive State of Monel-K500 Alloy. *Iranian Journal of Science and Technology - Transactions of Mechanical Engineering*, (pp. 1–17). Springer. doi:10.1007/s40997-022-00513-0

Chapter 7
Futuristic Adoption of Blockchain Technology in the Smart Agriculture and Food Industry:
Basic Concepts, Structure, Characteristics, Applicabilities

S. K. Geetha
KPR Institute of Engineering and Technology, India

J. K. Kiruthika
KPR Institute of Engineering and Technology, India

Rajasekaran Thangaraj
KPR Institute of Engineering and Technology, India

G. H. S. A. Devipriya
SRM Institute of Science and Technology, India

M. Naveenkumar
KPR Institute of Engineering and Technology, India

ABSTRACT

Agriculture is the backbone of human activity, and it is playing a vital role in human survival. Smart agriculture has a favorable impact on society as a result of technological advancements. The major goal is to increase production quality and quantity. Many industries, including healthcare, law, real estate, banking, and other applications are being transformed by using blockchain technology. On the other hand, agriculture is a known application that uses blockchain technology to totally transform the lives of human being in the world. With such a growing population it is obvious that the food demand will increase significantly. To fulfill the food demand of the world population, the agriculture industry is making use of Artificial Intelligence (AI), blockchain, and other food automation techniques and technologies to meet the demands through innovative production. This chapter presents an overview of various blockchain technology use cases, benefits and obstacles of food industry, and deployment of blockchain technology in agriculture.

DOI: 10.4018/978-1-6684-7879-0.ch007

INTRODUCTION OF BLOCKCHAIN

Blockchain is a decentralized digital ledger that may be used to store different types of data. As a result, it is a database that stores encrypted data blocks and connects them to form a single trustworthy source for the data. The structure of the blockchain is shown in Figure 1

Figure 1. Structure of blockchain

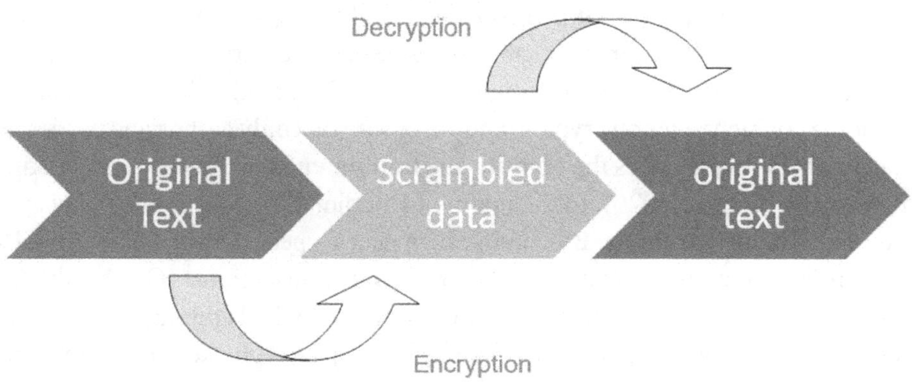

A simple analogy is that Google Documents is a perfect model for thoughtful blockchain technology. When document need to disseminate with the group of users, instead of sending duplicated copies, the document is shared among the group without any duplication (Nguyen.,2020). Anything store on blockchain cannot be changed by its structure, despite the fact that blockchain technology got to prominence as a result of its engagement in the financial sector, it has a wide range of applications that extend beyond cryptocurrencies. Technology is projected to radically disrupt numerous industries, including healthcare, law, real estate, banking, and many more. The agro sector can benefit from blockchain technology in a various use case such as demand pricing, seed quality, harvest and yield etc. as shown in Figure 2.

Figure 2. Application of blockchain in agriculture

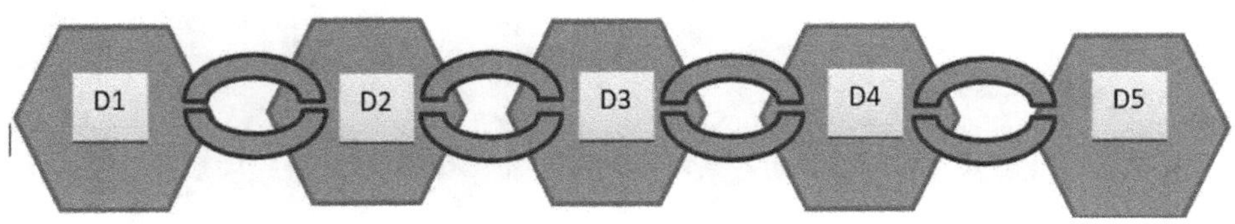

Working of Blockchain

Blockchain is a revolutionary technology that integrates three key components: cryptography for secure key management, a decentralized peer-to-peer network with a distributed ledger, and implied computing capabilities for storing transactional data and network records. This powerful combination enables a tamper-proof and transparent system for conducting and recording transactions, without relying on a central authority. By utilizing blockchain, individuals and organizations can engage in secure and efficient transactions, with increased trust and transparency (Geetha sk., 2011). Cryptography is a highly effective technique used to protect sensitive information and communications from unauthorized access by encoding them with complex codes. It involves the use of two keys, namely the private key and the public key, which enable secure information exchange between two parties. By maintaining these keys, individuals can establish a trusted and secure digital identity. Secure identity is a crucial component of Blockchain technology, which relies on cryptography to ensure that only authorized parties have access to the data on the network. This enhances the security and transparency of transactions, making them more reliable and trustworthy (Gupta., 2020). In the realm of Bitcoin, digital signatures play a crucial role in authorizing and regulating transactions. By combining a peer-to-peer network with digital signatures, a vast network of experts can come to an agreement on various transactions. (Geetha sk., 2022). Once a transaction is agreed upon, it is verified through complex mathematical computations, ensuring a secure transfer of assets between parties. Blockchain users rely on cryptography keys to conduct different types of digital transactions through a decentralized network. Figure 3 and Figure 4 demonstrate the process of public key encryption, where the original text is encrypted using the public key before being sent between two parties. The information is then decrypted using the private key on the receiving end. The public key is available to all network participants, while the private key remains confidential, ensuring that only authorized parties can access the information.

Figure 3. Public key encryption

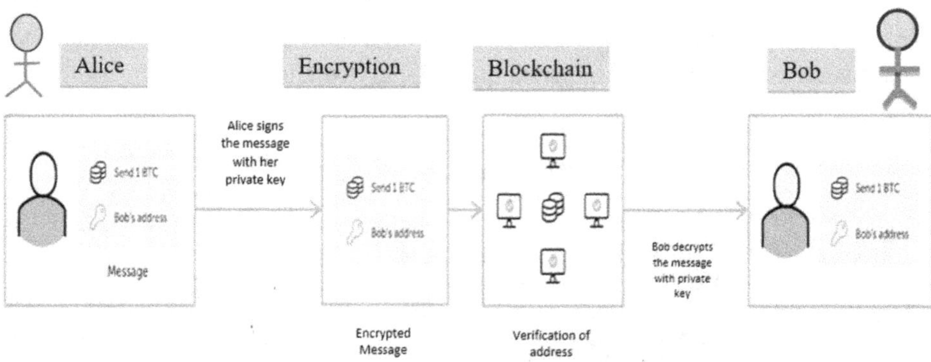

Figure 4. Cryptographic implementations

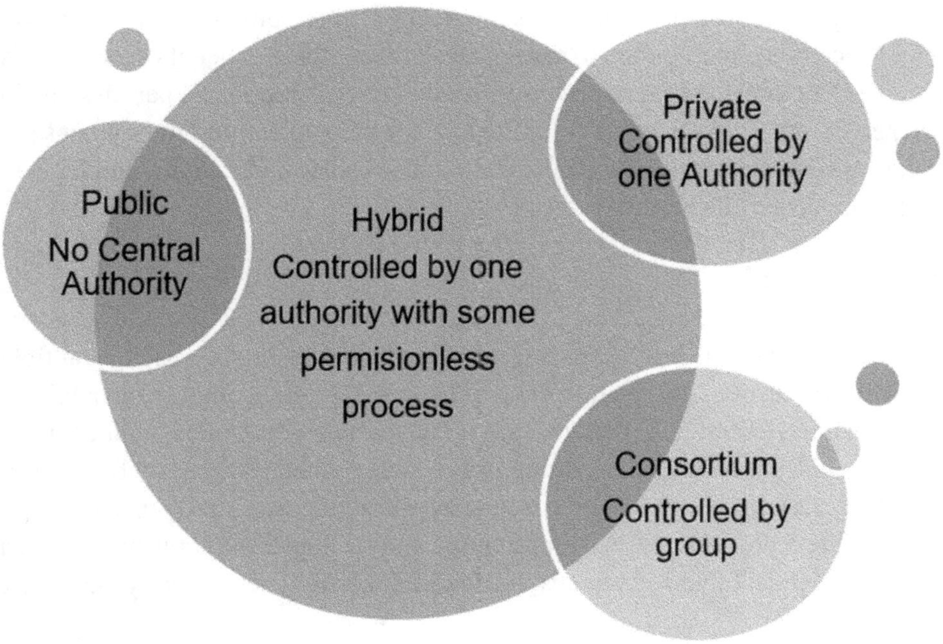

BLOCKCHAIN STRUCTURES

Blockchain technology can be classified into three distinct categories: public blockchain, private blockchain, and consortium blockchain. Each of these categories represents a different type of network configuration that is tailored to specific use cases and requirements. Understanding the differences between these categories is crucial in selecting the appropriate blockchain solution for a given application.

Public Blockchain

Public blockchain refers to a decentralized system that provides open access to information and resources to all participants in the network, with examples including Bitcoin, Ethereum, and Litecoin. (Viriyasitavat., 2019). A key advantage of a public blockchain is its inherent security, as transaction records are immutable and cannot be altered. Transactions are only approved and validated once a majority of network participants verify them, making it difficult for hackers to compromise the network. Public blockchains also offer transparency, as they are built on open-source codes, and all transactions are publicly visible and verifiable (Andreev., 2018). Another benefit of public blockchains is anonymity, as the distributed nature of the ledger makes it challenging to determine the initial address of the user involved in a transaction.

Private Blockchain

The private blockchain framework is managed, as it were, by clients from a certain company or approved clients who have been granted permission to participate. Because the number of users on a private blockchain is less, it takes less time to reach a consensus to validate a transaction. A permission blockchain is a private blockchain. Since, the network is controlled by one or more entities, transactions must be done through third parties. Full anonymity, high efficiency and speedier transactions, and higher scalability are all advantages of private blockchain.

Consortium Blockchain

Federated blockchain, or consortium blockchain, is a type of blockchain governed by multiple organizations and is considered a permissioned platform rather than a public one. Compared to public blockchains, consortium blockchains offer faster transaction speeds due to a more controlled number of users. Access to the consortium platform is restricted, resulting in a more regulated and predictable network environment. Additionally, transaction costs remain constant as everything is centrally managed in the consortium blockchain. Figure 5 provides a visual representation of the logical relationships between different blockchain structures, while Table 1 compares the unique characteristics of each type of blockchain.

Figure 5. Relationship between types of blockchain

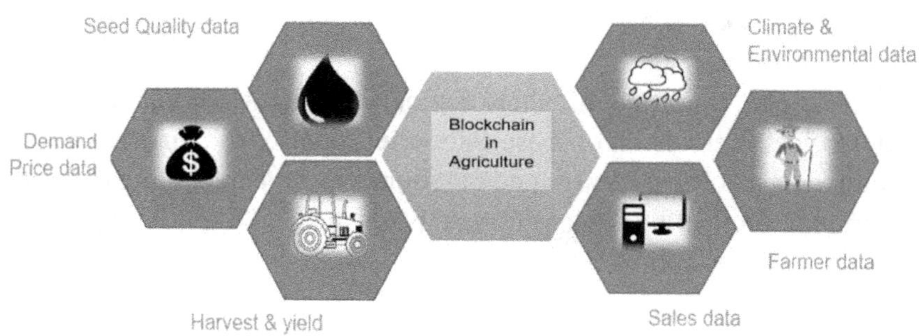

Table 1. A comparison of blockchain types

Property	Public blockchain	Consortium blockchain	Private blockchain
Choosing an agreement	All miner nodes	selected set of nodes	Restricted inside the company
Reading authorization	Public	Public or restricted	Public or restricted
Level of Immutability	Tampering is impossible	Could be Tampered	Could be tampered
Resource utilization	Low	High	High
Centralization	No	Partial	Yes
Consensus process	Permissionless	Needs permission	Needs permission

BLOCKCHAIN CHARACTERISTICS

The Figure [6] shows the vital characteristics of the blockchain characteristics as mentioned below

Increased Capacity

The collective power of thousands of computers working together as a group is one of the most notable advantages of blockchain technology, as it can greatly enhance the capacity of an entire network. This distributed approach offers more computing power than relying on a few centralized servers, making blockchain technology a powerful tool for a wide range of applications.

Better Security

The robust security measures of blockchain architecture make it virtually impossible to shut down any of the frameworks (Viriyasitavat., 2019). Despite being a high-value financial system, blockchain technology remains largely immune to hacking attempts. Bitcoin, for instance, has never been successfully hacked, thanks to the distributed network of computers called nodes that verify transactions and protect the system. This decentralized approach adds an extra layer of security to the blockchain architecture, making it an extremely reliable and secure technology for various use cases.

Immutability

Immutability is a key characteristic of the blockchain ledger, which ensures that the information stored within it remains unchanged and permanent. This means that the data in the blockchain cannot be modified, providing a high degree of integrity and transparency. Even over time, the original records are likely to be preserved in their original form, further enhancing the immutability and reliability of the blockchain technology.

Faster Settlement

One of the key issues with traditional money-keeping systems is the amount of time it takes to settle transactions, which can often take several days (Viriyasitavat., 2019). To address this, financial institutions have been exploring alternative accounting architectures, such as blockchain technology. By leveraging the speed and efficiency of blockchain, cash exchanges can be settled much faster, providing a more streamlined and efficient financial system.

Decentralized System

The concept of decentralization involves shifting control and decision-making authority from a centralized entity, such as an individual, organization, or group, to a distributed network. This network can manage various resources, ranging from contracts to data. By using a key associated with their account, the owner has complete control over their assets and can exchange them with any other account. Blockchain technology has the potential to significantly transform industries and decentralize the internet.

Minting

The process of adding transaction records to the Bitcoin blockchain is commonly known as "blockchain mining." It is the way in which Bitcoin transactions are securely processed and recorded on the blockchain.

Figure 6. Key characteristics of blockchain

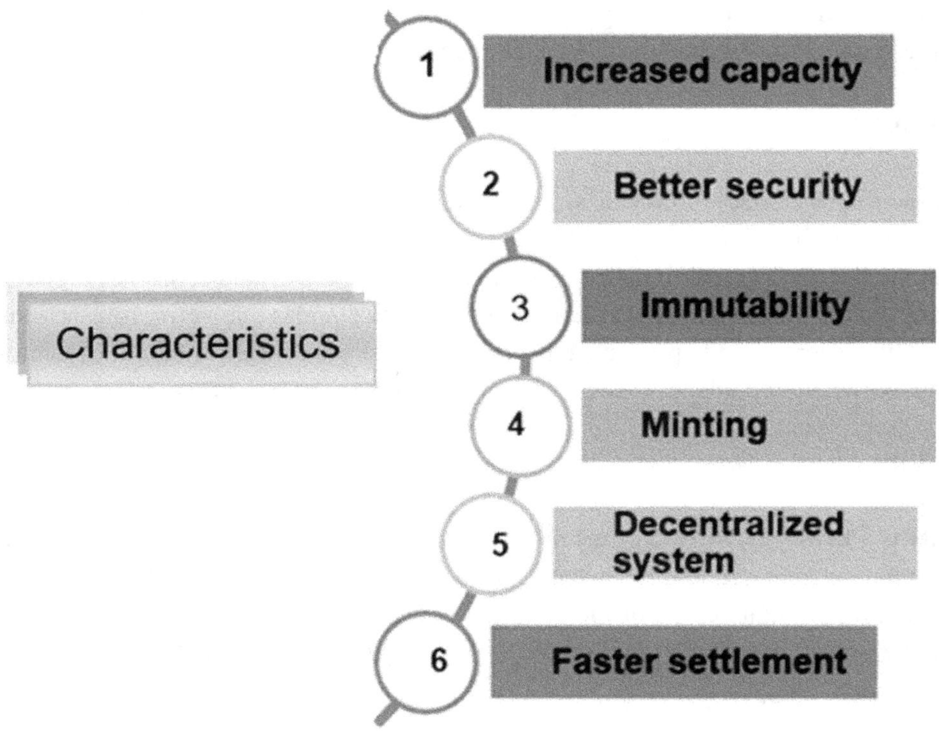

APPLICABILITY OF BLOCKCHAIN IN AGRICULTURE SECTOR

Blockchain has the ability to completely change agriculture. Blockchain technology can assist agriculture in a variety of ways.

Blockchain in Agriculture: Use Cases

There are various use cases in the agro-industry that are currently being developed. Some of the most common blockchain applications are as follows (Sajja, 2021):

- Food Supply Chain Management
- Registration of Agricultural Land.
- Agricultural Insurance

Food Supply Chain Management

Agriculture products always have a large potential market since it is the need of everyone. Nowadays, every consumer is highly keen on knowing from where their food is originated. Consumers can learn about the origins of agricultural products known as "Traceability". Traceability ensures the quality of the product, safety and provenance. Trusted traceability aids in minimizing food contamination concerns and helps to maintain the reputation for regulatory compliance and inventory management. From "farm to fork," the agri-food chain refers to the interrelated processes that occur throughout the production of food.

The supply chain in the food business is depicted in Figure 7. The food supply chain process includes farmers, processing engineers, wholesalers, retailers, distributors, and consumers. Food production, processing, packaging, transportation, distribution, and finally consumption are the six phases involved in the food supply chain system.

Figure 7. Food supply chain cycle

Producer

The early stages of the food chain are the cultivation of crops and animals, as well as the harvesting of seafood from wild fisheries or aquaculture. Transport crops, animals, and by-products to the next level of the food chain after they have been carefully selected.

Processing

There are numerous types of processing in the food chain, which are covered here.

Primary Processing: Primary food processing transforms agricultural products, such as uncooked rice or animals, into foods that may be consumed. This product category has been manufactured for thousands of years through drying, threshing, winnowing, and grinding grain, as well as slaughtering animals for meat.

Secondary Processing: Secondary processing includes physical procedures like pressing, grinding, and dehydration, as well as chemical processes like hydrolysis, hydrogenation, and enzyme usage.(Kamilaris., 2019). Secondary food items that have been refined, purified, extracted, or modified from minimally processed foods include cheese, flours, edible oils, sugars / sweeteners, and starches.

Ultra-Processed Foods

Ultra-processing food is the commercial production of processed food that is ready to eat or heat and serve food. Such as Cakes, sweets, jams, soft drinks, and ready meals are examples of items made by combining primary and secondary food ingredients to create a ready-to-eat food.

Packaging

Food packaging is an essential component of the food industry and helps to keep food and beverages in a sanitary manner; Packaging protects food from physical, chemical, and biological degradation throughout the food supply chain.

Logistics

The process of transferring and arranging commodities, food, and equipment from one location to another is referred to as logistics. To be successful, the food industry relies heavily on the timely and smooth transportation of raw ingredients.

Distribution

The food is transported and delivered to the proper store or supplier when it is ready to eat. Distributors deliver the commodities, manage inventories, minimize costs, and do a variety of other things to increase the value of the food item.

Consumers

The consumer is the final link in the product flow chain. Customers are individuals or organizations who consume services or products to suit their requirements. Customers are people who will only utilize these things and will not resale them to third parties. Customer purchasing decisions can impact the agri-food chain.

Figure 8. Food supply chain system based on blockchain

Blockchain in Food Supply Chain

Blockchain, a type of Distributed Ledger Technology (DLT), is gaining interest in supply chains, for example, in proving product provenance and implementing product track-and-trace throughout the supply chain. The blockchain has the capability to solve the traceability. It can link and facilitate efficiency, openness, and responsibility among participating parties when implemented properly. Better and more trustworthy data can aid in the optimization of corporate decisions and the attainment of greater production, efficiency, and sustainability standards.

Blockchain technology provides unique digital identifiers for food commodities, which are linked to information about their growth conditions, batch numbers, and expiry dates. This enables the tracking of food products through supply networks, reducing food waste and allowing customers to evaluate the environmental impact of their meals. Additionally, the immutable nature of the blockchain ledger helps prevent fraud and facilitates the identification of the source of foodborne illnesses. The digital nature of these technologies also allows for on-farm data sharing, potentially enhancing supply chain management (Caro., 2018). Figure 8 illustrates the application food Supply Chain System Based on Blockchain.

Food Supply Chain Using Blockchain

The blockchain food supply chain can avoid food fraud by adopting the following steps:

Step 1: Data generation by IoT sensors or data is stored by farmers
Step 2: Distribution of cultivated crops to food processing firms
Step 3: Processed Food Distribution to Wholesalers and Retailers
Step 4: Backward tracking of supply chain by the Consumers

Step 1: Data Generation by IoT Sensors or Data is Stored by Farmers

"Smart farming" is a contemporary process that alludes to farm management that makes use of the Internet of Things (IoT) (Vijayaganth., 2021). robotics, drones, and artificial intelligence (AI) Naveenkumar M ., 2021). to raise the volume and quality of products produced while reducing the amount of human labour required. Smart farming sensors give critical information on the crops that have been sown in the fields. Suppose the farmer is not employing any technology-assisted techniques. People can then use their mobile app to save important information such as crop quality, seed kind, and the meteorological circumstances under which a crop should be sown. Data collected by IoT sensors or manually by farmers is saved in IPFS, a distributed storage network that uses blockchain addresses to store data.

Step 2: Distribution of Cultivated Crops to Food Processing Firms

Once the crops have been planted, the food processing firms begin bidding on the bidding site.IoT-enabled trucks may transport crops to refineries while also recording the temperature conditions in which they are stored and delivered. When the bid is approved via smart contracts, the crops are processed, and firms save data recorded at each stage of the process on the blockchain. The information gathered from refineries can assist distributors and merchants in determining whether the food delivered is of high quality. Data stored on the blockchain can also be used to ensure that the food supply chain is always compliant.

Step 3: Processed Food Distribution to Wholesalers and Retailers

During the processing of food or crops, wholesalers and merchants can utilize the auction platform to bid on the products they desire. (Ganesh Kumar., 2017). Food is delivered to wholesalers and merchants by IoT-enabled vehicles in the same way as crops are sent to refineries. Traceability is enabled by the blockchain supply chain, which enables food companies to execute product shortages or inquiries quickly and easily.

Step 4: Consumers Can Track the Supply Chain Backwards

Customers can explore anything by tracking the supply chain because details are digitally linked to the food goods within the blockchain, such as:

- Specifics about agriculture origin
- Information about transportation
- Batch codes
- Processing of food

- Data from the factory
- Expiry date
- Temperature of storage

Various stakeholders can access the information on the food's quality at each stage of the process using a blockchain-based food supply chain. Table [2] represents the relationship between food item and intention of tractability for each food item.

Table 2. Commercial applications in the agri-food supply chain

FOOD ITEM	INTENTION OF TRACTABILITY	REFERENCE
Fish	To trace harvested fish with certified social sustainability claims from the sea to tuna cans	Cook, Bubba, and W. N. Zealand. "Blockchain: Transforming the seafood supply chain." World Wide Fund for Nature (2018).
San Domenico coffee production chain	To provide dependable, permanent coffee documentation as well as complete transparency from the plantation to the cup.	Kamilaris, Andreas, Ian R. Cole, and Francesc X. Prenafeta-Boldú. "Blockchain in agriculture." In Food Technology Disruptions, pp. 247-284. Academic Press, 2021.
Milk	Data collection and registration automation in the dairy supply chain to combat food fraud	Shingh, Shuvam, V. Kamalvanshi, Sarthak Ghimire, and Sudarshan Basyal. "Dairy supply chain system based on blockchain technology." Asian J. Econ. Bus. Account 14 (2020): 13-19.
Pasta	Identification of the complete supplier chain (i.e., the producer, the products and flours used, the drying method, and the mode of transportation).	Maouchi, Mourad El, Oğuzhan Ersoy, and Zekeriya Erkin. "DECOUPLES: a decentralized, unlinkable and privacy-preserving traceability system for the supply chain." In Proceedings of the 34th ACM/SIGAPP Symposium on Applied Computing, pp. 364-373. 2019.
Beef	To tackle food fraud, they must develop and integrate blockchain-based technology into their supply chains.	Cao, Shoufeng, Warwick Powell, Marcus Foth, Valeri Natanelov, Thomas Miller, and Uwe Dulleck. "Strengthening consumer trust in beef supply chain traceability with a blockchain-based human-machine reconcile mechanism." Computers and electronics in agriculture 180 (2021): 105886.

Barriers of Blockchain Technology

The hurdles to supply chain management are divided into four groups, as shown in Figure 9

- Barriers inside the organization
- Barriers between organizations
  -Barriers to the system
- External obstacles

Figure 9. Barriers of blockchain technology

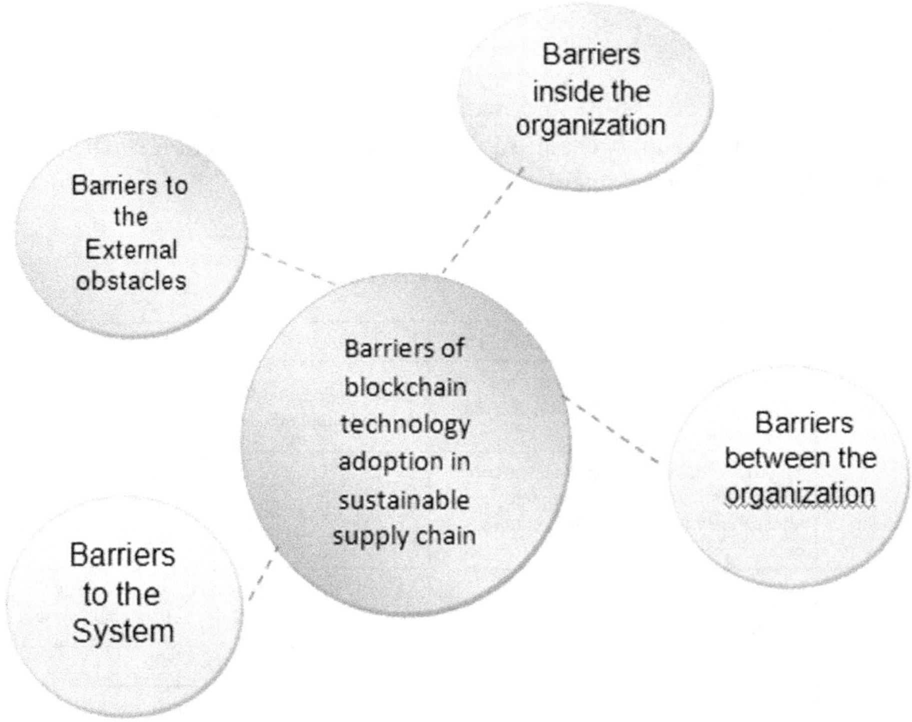

One of the major obstacles to integrating blockchain technology is the continual expansion of the blockchain ledger. Throughout the supply chain, data is generated and transmitted to the blockchain as transactions. Since each transaction is stored in every node of the blockchain, the storage capacity needed can become significant, and the volume of storage required is linked to the rate at which transactions are processed. Each node in the blockchain network keeps its own copy of the ledger, and if the application is large, data storage capacity may become an issue(Ganesh Kumar, C.,2017).

Two things influence blockchain storage requirements:

i) The number of transactions per second increases as the number of users grows.
ii) As transactions are added to the blockchain, the size of the ledger increases, resulting in a corresponding increase in memory usage.

Agricultural Insurance

The agricultural sector plays a vital role in the economic growth of many countries, both directly and indirectly. However, natural disasters such as floods, earthquakes, and other calamities can have a significant impact on farmers, leading to devastating losses. To mitigate these risks, governments often implement farm insurance plans. However, traditional insurance systems are centralized and come with complex rules and conditions, leaving many farmers uninsured. This is particularly true for farmers with limited education or knowledge of insurance procedures. In cases of crop failure, farmers often face insurmountable debt and may even resort to suicide or government loan forgiveness programs. Fortunately, crop insurance offers a more viable solution to this problem, providing a safety net for farmers in times of crisis (Jha, Nishant.,2021).

Crop insurance and government subsidies are important tools for farmers to safeguard their investments. By purchasing crop insurance, farmers can protect themselves from financial losses resulting from crop failures. (Iyer Vinayak., 2021). This helps alleviate the economic difficulties faced by marginal and small farms and increases farmers' risk tolerance by providing greater economic security. Despite its benefits, crop insurance has historically been challenging for farmers to understand and, in some cases, unprofitable. Consequently, farmers may be hesitant to purchase crop insurance due to a lack of confidence in insurance organizations and concerns over potential delays or denial of claims. The various stages of the insurance life cycle are illustrated in Figure 10.

Figure 10. Insurance life cycle

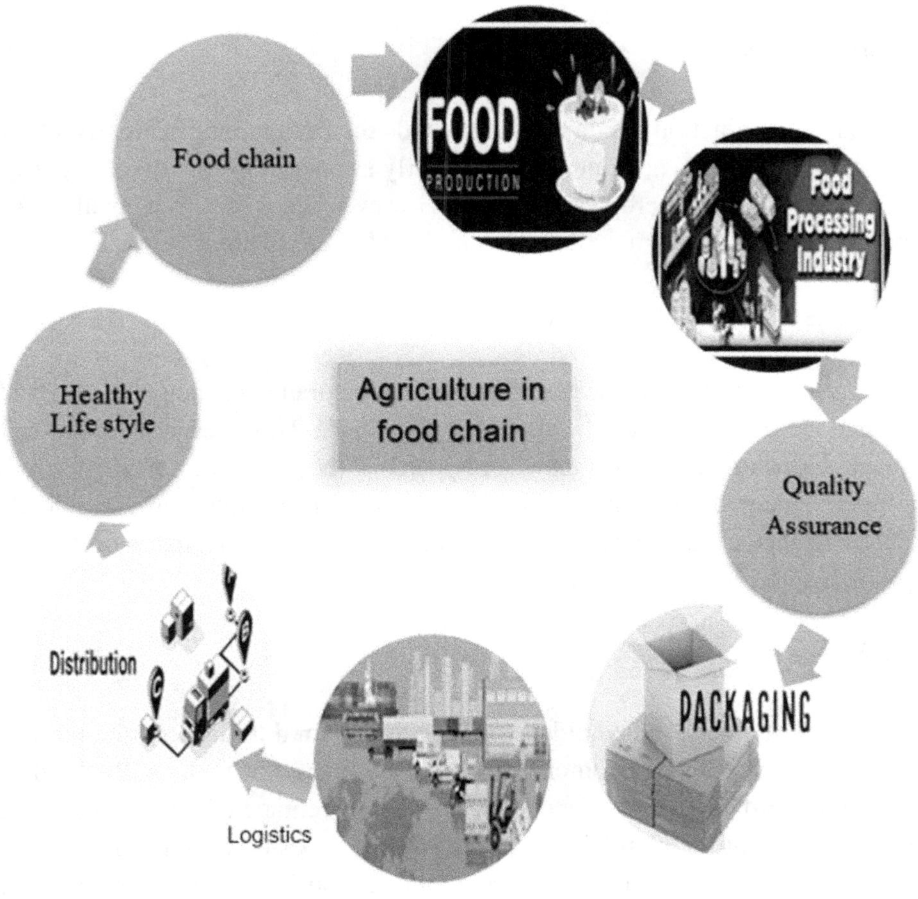

Publish Policy Information

The Risk Management Agency initially publishes the insurance plan, which includes coverage levels, guaranteed prices, premium rates, and subsidy amounts, for specified commodities and farming techniques that may be insured in a country.

Policy Closing Date

Farmers are subject to strict deadlines when applying for or modifying their crop insurance coverage. Failure to meet these deadlines may result in the rejection of their application by the approved insurance provider (AIP). However, policyholders are permitted to make changes to their previous year's insurance coverage before the deadline, if necessary.

Notification of Affirmation

Following the policy end date, the insurance provider promptly reviews all submitted applications to determine eligibility based on the qualifying criteria. Eligible applicants receive a coverage summary and the necessary policy documents. It is important to note that a policyholder is not permitted to cancel their insurance during the first year of coverage after their application has been accepted.

Informing About Production

The insurer maintains a production document that records their output and serves as a historical reference for insurance purposes. This document is particularly important in calculating the Actual Production History (APH) for the current crop year's yields. However, there is a strict deadline for submitting production reports to ensure their timely inclusion in the APH calculation.

Attachments to Insurance

Once the annual agricultural planting season commences, insurance coverage becomes effective and applies to crops that are planted within the specified time frame. However, if there are delays or if the crops are not planted, certain restrictions may be imposed on the insurance coverage. It is important to note that crops must be planted before the planting deadline and the insurance provider should be notified accordingly. The insurance coverage may only be applicable to the areas that are affected. In general, perennial crops are insured on an annual basis.

Reporting on Acreage

When obtaining crop insurance, policyholders are usually required to disclose the number of insured and non-insurable acres, as well as the completion date of the plantation to the AIPs. To ensure adequate coverage, farmers must provide further details such as the percentage of crops planted, farming techniques employed, location of the plantation, and the specific types or varieties of crops grown before the deadline. AIPs use this information to determine the appropriate level of coverage and premiums that the farmer will require. However, if the application is submitted after the deadline, the AIPs hold the authority to reject it at their discretion.

Coverage Overview

AIPs are responsible for reviewing and processing crop insurance applications that are submitted on time. After completion, the AIPs provide a summary report to the insurer which contains information regarding the insured acres, the insured amount for the covered land unit, and crop details. Once the AIPs and the policyholder have reached an agreement, the policyholder may modify the full acreage report if desired.

Premium Payment

Upon the commencement of insurance coverage, policyholders are required to pay the premiums which are considered "earned" at that point. AIPs will generate invoices for the premiums which must be paid on or before the deadline provided in the actuarial documentation related to the cornfield. The amount billed will be based on the information provided in the acreage reports. In addition to the premium amount, administrative fees may also be included in the invoice. In the event that the policyholder fails to pay the premium amount or administrative fees before the deadline specified in the policy, AIPs may add interest charges to the outstanding balance.

Notice of Production Loss or Damage

When loss or damage occurs, policyholders are required to notify AIPs in writing within 72 hours. (Shetty Ankitha., 2022). Depending on the crop, there may be additional conditions to fulfil. This written notification facilitates AIPs in assessing the quality and condition of the affected crop. A detailed report will then be generated based on the damage, which will inform subsequent decisions.

Inspection

Crop damage assessment and timely identification of crop pests pose significant challenges. When policyholders report crop damage or loss on their farms, an agro-inspector will conduct a field examination. Based on the results of the inspection, the insurance organization will initiate the payment process.

Payment of Indemnity

When a claim is approved, AIPs will issue an indemnity check to the policyholder. The AIPs maintain a "summary of indemnity payment," which outlines any deductions made to the indemnity amount owing to unpaid premiums, interest, or administrative costs.

Challenges in Current Insurance Applications

Agricultural insurance presents pricing difficulties due to regional associations, unpredictable losses from one season to another, and insufficient data. (Niranjana Murthy., 2019). The early days of agricultural insurance were plagued with problems that restricted the types and availability of insurance products. Fig 11 illustrates the existing challenges in agricultural insurance applications.

Figure 11. Challenges in insurance sector

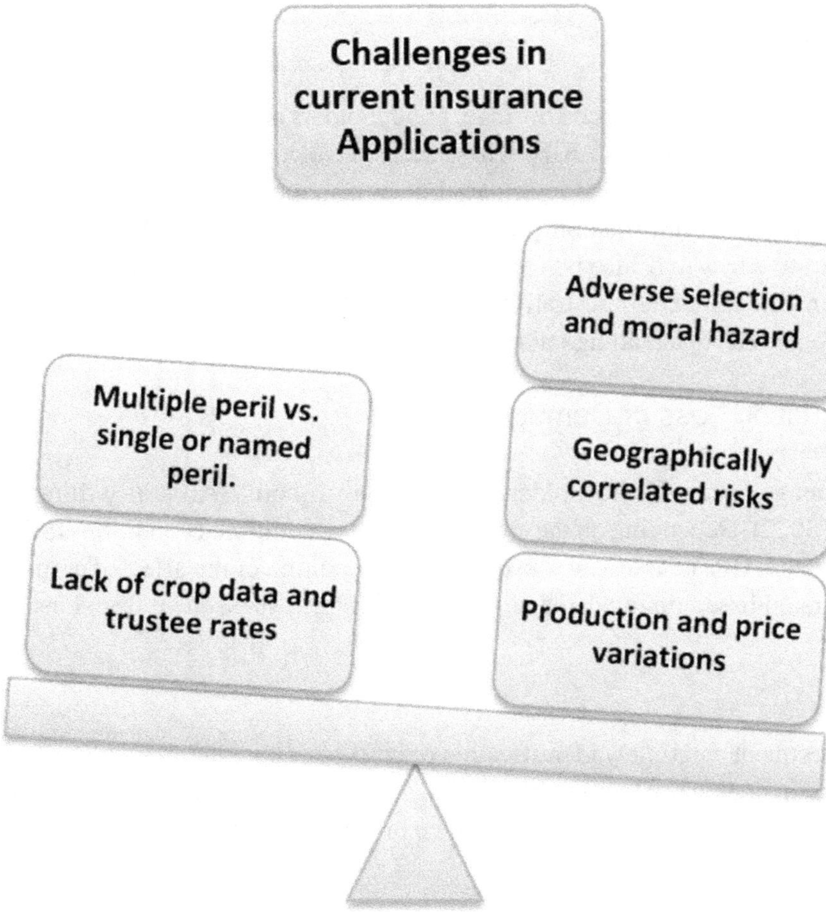

Lack of Crop Data and Trustee Rates

During the late 1980s and early 1990s, addressing this concern was of paramount importance. Despite advancements in data availability and grading systems, several critical issues persist, primarily due to uncertainties in yield reporting and increasing concerns related to climate change.

Production and Price Variations

Farm-based incomes are generally unstable due to a range of factors, including commodity price fluctuations. Farming activities are susceptible to the influence of weather, natural disasters, and other unforeseeable events. Commodity prices are primarily determined by market demand and supply chain factors. Non-farming commodities and services typically exhibit higher volatility in prices from year to year and even day to day.

Single or Named Peril vs. Many Perils

Private crop insurance plans that provided coverage for various risks and revenue streams were unsuccessful before the introduction of the federal crop insurance policy. Unlike named-peril insurance, multiple-peril plans provide coverage for all risks under broad categories. However, losses sometimes exceeded the annual premium, which posed a significant challenge for insurance providers from smaller geographical regions. The introduction of revenue insurance has made it more challenging to sustain losses from significant price declines.

Geographically Correlated Risks

Spatially correlated risks, such as droughts, floods, and other natural disasters, affect large geographic areas simultaneously, leading to widespread losses from a single event. Compared to uncorrelated risks, correlated risks are more likely to cause significant losses across an insurer's portfolio. Agricultural production is vulnerable to a range of spatially associated and unrelated risks. While the private market has been successful in insuring certain unrelated risks like hail and fire, it has faced challenges in covering spatially correlated risks that have a significant impact on agriculture.

Adverse Selection and Moral Hazard

Adverse selection and moral hazard are commonly identified as significant issues in the private insurance market for agriculture. Adverse selection occurs when high-risk farmers are more likely to purchase insurance at the same premium rate as low-risk farmers. This is because the insured individual has more knowledge about the potential for loss than the insurer. Moral hazard refers to the tendency of an insured individual to take on more risk after obtaining insurance. These issues can make it challenging for private insurers to accurately price and manage risk in the agriculture industry

SWOT Analysis of the Adoption of Blockchain

A SWOT analysis (Niranjana Murthy., 2019). highlighting the benefits and drawbacks of this technology is shown in Figure 12. And the table 3 listed the various proposed blockchain based crop insurance for the agro industry.

Figure 12. Swot analysis

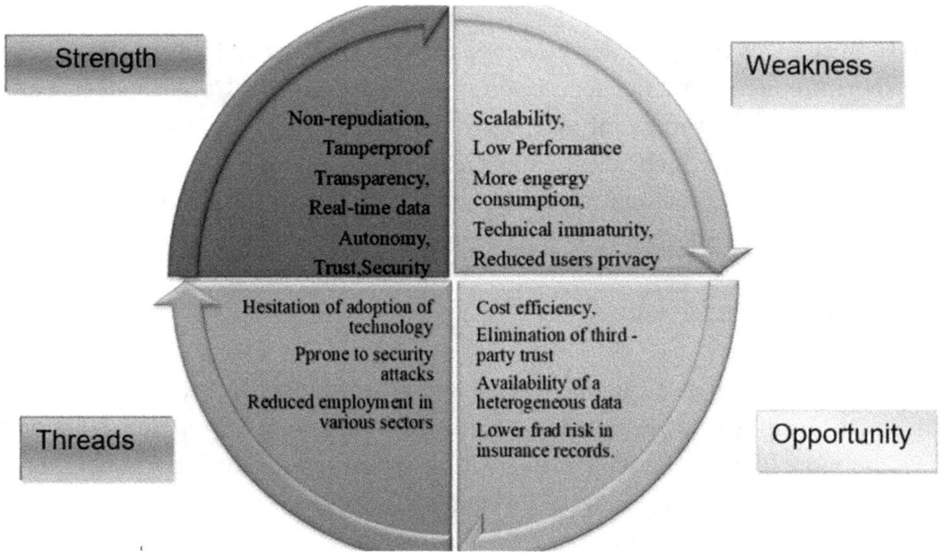

Table 3. Various proposed blockchain based crop insurance

REFERENCE	PROPOSED	FINDING
Nguyen, T. Q., A. K. Das, and L. T. Tran. "NEO smart contract for drought-based insurance." In 2019 IEEE Canadian Conference of Electrical and Computer Engineering (CCECE), pp. 1-4. IEEE, 2019.	Drought-based Insurance using NEO Smart Contracts	The proposed solution for crop insurance is based on a blockchain architecture utilizing the NEO blockchain. The framework is built around an Oracle server, which serves as the core component of the system.
Iyer, Vinayak, Kshitij Shah, Saurabh Rane, and Radha Shankarmani. "Decentralised Peer-to-Peer Crop Insurance." In Proceedings of the 3rd ACM International Symposium on Blockchain and Secure Critical Infrastructure, pp. 3-12. 2021.	Crop Insurance on a Peer-to-Peer Basis	An end-to-end prototype of a decentralized insurance system for severe rainfall has been developed, utilizing blockchain technology to eliminate the need for third-party involvement and creating a more trustworthy environment. This approach is particularly useful in situations where trust is traditionally lacking.
Kevorchian, Cristian, Camelia Gavrilescu, and Gheorghe Hurduzeu. "A Peer-to-Peer (p2p) Agricultural Insurance Approach Based on Smart Contracts in Blockchain Ethereum." Agric. Econ. Rural Dev. New Ser 1 (2020): 29-45.	Blockchain based agircultural insurance using smart contracts	The implementation of a peer-to-peer insurance system supported by blockchain technology has the potential to decentralize the agricultural insurance system and reduce bureaucratic hurdles. The use of smart contracts on the Ethereum platform ensures a more transparent payment process, addressing farmers' concerns about the reliability of the insurance product. The 'Etherisc' system leverages smart contracts to manage supply and demand chains in a decentralized manner.
Sharifinejad, Maedeh, Ali Dorri, and Javad Rezazadeh. "BIS-A blockchain-based solution for the insurance industry in smart cities." arXiv preprint arXiv:2001.05273 (2020).	blockchain-based solution for the insurance industry in smart cities	The proposed programme is a smart city insurance system powered by blockchain technology (BIS). This system involves various entities such as smart city managers, insurance companies, consumers, sensors, and other devices. Users are identified through their public keys, which can be changed to maintain anonymity. By utilizing blockchain technology, the BIS aims to create a more efficient and secure insurance system for smart cities.
Gera, Jaideep, Anitha Rani Palakayala, Venkata Kishore Kumar Rejeti, and Tenali Anusha. "Blockchain technology for fraudulent practices in insurance claim process." In 2020 5th International Conference on Communication and Electronics Systems (ICCES), pp. 1068-1075. IEEE, 2020.	Insurance based on blockchain to protect against fraudulent activities	This article suggests that a consensus algorithm-based insurance system can help overcome the problem of fraudulent claims and claim manipulation. The system involves farmers, supply firms, and marketplaces, and is designed to promote decentralization and build trust. By leveraging the power of consensus algorithms, the system is able to maintain its integrity and ensure that all parties involved are accountable.

India's Land Registration Process

Land registration is necessary to confirm that the property is legally owned. Land records and registration are time-consuming, and there are a few ambiguous techniques for keeping land records in the country. Fig.13 displays the various stages of land registration process in India.

Figure 13. Land registration process

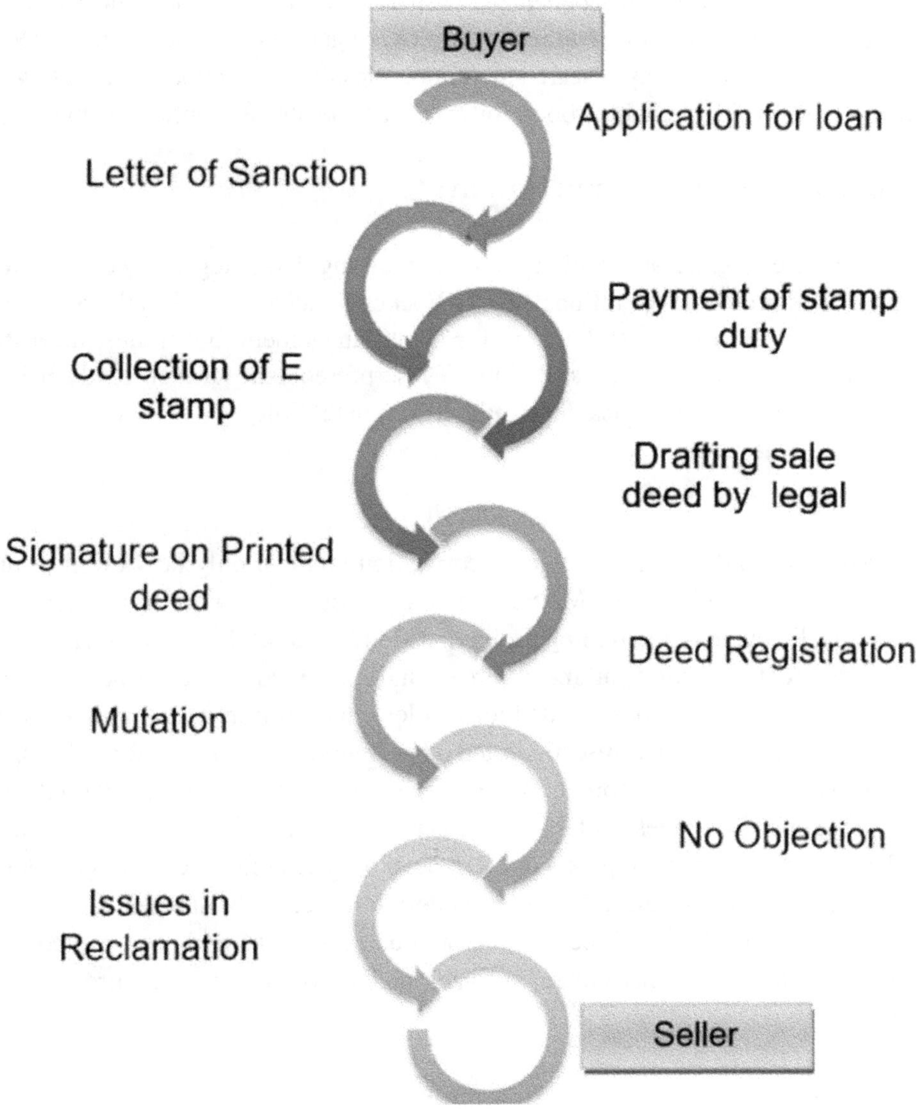

Issues In Existing Land Record Management

Due to the practical difficulty, the government is facing several issues, which include the maintenance and availability of old records. Currently, the Indian legal framework lacks the ability to provide any assurance regarding ownership (Imran., 2013). Various departments are working together at district or village level. The information is stored by various departments. Another issue is that data maintained by the departments may sometimes not be synchronized. The resultant record may contain inconsistent information. The main reason for the difficulty in land record management is the poor handling of the records. Based on the World Bank's survey, India is in 154th place in terms of land registration. The index is based on various factors, including quality, transparency, time, and cost. The following are some issues in the current system: i) the involvement of multiple agencies and their lack of cooperation. ii) Expensive outdated/traditional survey/research methodologies iii) Corruption and fraud; iv) insufficient IT support(Thakur., 2020). Blockchain is boon for resolving the above-mentioned challenges.

Need of Blockchain in Land Registration

Land administration faces significant challenges in maintaining data integrity, security, and validation while minimizing installation and operational costs. Blockchain technology has the potential to address these challenges and be a disruptive technology for land management, but it must meet three explicit prerequisites (Data Integrity, Security, Validation). These prerequisites are currently being addressed and show promise for the future of blockchain in land administration.

Data Integrity

The guarantee that digital information is free of corruption and can only be viewed or updated with authorization is known as data integrity. Maintaining the consistency, correctness, and trustworthiness of data throughout its lifecycle is an ongoing problem. (Mendi., 2020). From the beginning, blockchain handles data integrity. All transaction parameters are set and controlled by users in advance and as needed in the blockchain. A new link cannot be added to the blockchain if transaction protocols and expectations are not satisfied. Unauthorized transactions and other fraudulent manipulations of property data are not permitted to be carried out and recorded in the immutable shared ledger. The solid data integrity of blockchain has numerous advantages. Notably, there is no longer any need to double-check and validate individual work or reconcile conflicting data that can develop when multiple offices are engaged on the same property transaction. Instead, all stakeholders benefit from real-time data and the assurance that the property information displayed on the blockchain is actionable, current, and uncorrupted. Data exposed by a blockchain system has the required integrity, ensuring data veracity from a third-party perspective.

Security

Security necessitates the protection of digital data and other IT assets from both internal and external, deliberate and unintentional dangers. And it's rightfully become a top priority for nearly all government entities in recent years. Fortunately, blockchain technology provides a powerful choice for decision-makers concerned about the security of sensitive government property data. Because the data is scattered throughout a large network, hackers would have to infect the entire network, which is an impossible task.

This also implies that no single person, even if permitted, has the authority to alter any property data. To participate, all end-users must be approved, and they must agree on transaction regulations that are embedded into the blockchain acceptance protocols.

Validation

Validation ensures that activities are based on clean, correct, and valuable data. Various validation criteria and restrictions ensure that the data entered into the system is correct, meaningful, and secure. Third-party users (notaries, cadastres, bank clerks, and so on) can utilise a blockchain distributed network to query information and entities that have the necessary permissions, as well as to confirm and verify the data recorded in the blockchain itself.

Examples of Blockchain Implementations in Land Registry Management

A Blockchain-based land registration process is already implemented by many countries. The Georgian government is the first to be tasked with implementing blockchain technology in land record management. The idea behind the system is that all the transactions related to land are stored on blockchain, mainly to minimize the inconsistency in registration and in the process of property exchange. Transaction history is automatically entered in the distributed digital ledger, with no need for notary verification (Mendi., 2020). Based on the nature of blockchain, the records are immutable, and the information is generally stored in the cloud securely instead of storing the information on paper or in a database. Even in India, the state of Andhra Pradesh successfully implemented blockchain-based land record management. Transaction details are stored on numerous nodes of the Ethereum network. The developed system allowed for ownership transfer between buyer and seller, with approval given by the land department and other government bodies involved in the process (Shuaib., 2022).

Blockchain Technology Implementation Challenges in Developing Countries (Agri/Food)

Technology allows for a faster, easier, and less time-consuming process of property registration while maintaining data security. Like any other technology, blockchain has limitations. Fig 14 list the various limitations of blockchain implementation. Because blockchain is a new concept, there are only a few proofs of concept: A few case studies and implementations in India's various government sectors have been completed. Some state governments are eager to adapt their applications to use this technology. Particularly for supply chain sectors. Several states, including Andhra Pradesh, Madhya Pradesh, Punjab, and Haryana, have announced that they will launch bids for land records and registration systems based on blockchain technology (Rana., 2021).

i) A scarcity of specialists because this technology is still in its early stages, there is a scarcity of development experience, and those who are already familiar with it find it difficult to deal with different blockchain platforms such as private, public, and hybrid. Only a few companies have a proven track record of implementation success.

ii) Difficulty in updation: Blockchain's fundamental feature is immutability, yet this property might make consumers fearful of the technology. It is extremely difficult to change data that has already been saved on the network. One of the reasons many apps continue to use the old methods rather than adopting new technologies is because it is easy to alter.

iii) Key management would be a problem, especially in private Blockchain: In the case of public Blockchain implementations, the Government PKI infrastructure and eSign mechanism (http://cca.gov.in) is not yet linked to key management. In the case of private Blockchain, the Indian government has yet to issue any recommendations on the use of keys or key management. Cyber-attacks and security flaws have resulted in the closure of existing PKI infrastructures.

iv) There are no regulatory norms that govern cross-jurisdictional application: There are no rules regarding the format of the transactions. In India, no regulatory agency has established the minimal set of features required for transaction recording. As a result, the implementor State should identify a minimal set of attributes for each sector in order to implement Blockchain Technology. A certifying or recommending organization must make decisions like the amount of Blockchains that can be maintained. Blockchains can be maintained on a topic basis. Maintain a single Blockchain for all transactions in the current land records management ecosystem. If a significant number of Blockchains based on topics are required, the infrastructure needs increase.

v) There are no cross-jurisdictional regulatory standards: There are no rules governing the format of the transactions. No regulatory agency in India has established a minimum set of qualities for registering a transaction. As a result, for each sector, the implementor State should identify a minimum set of properties. An approving or recommending organisation must make decisions such as the maximum number of Blockchains that can be maintained. Blockchains can be kept up to date on a theme basis. We maintain a single Blockchain for all transactions in the current land records management ecosystem. If there is a need to maintain a large number of Blockchains depending on themes, the infrastructure needs increase.

vi) The cost of setting up a large network for the application of Blockchain is expensive at first: **It's** a significant job to set up a network with enough bandwidth to keep the nodes running. The availability of networks in India is still quite low. Furthermore, in several states, land records databases are still kept at the district/taluka level. The updating of land records data is still done at the Taluka level, where the Tehsildar is in charge of the basic registers.

vii) A successful Blockchain network with a large number of ledger records may experience latency issues: For complete knowledge of the transactions and the many entities linked with the transactions, a network and interactions of the nodes with traditional database servers are required. High-end servers are necessary not just to handle the current demand, but also to handle the additional connections required for transaction replication to all participating nodes. Furthermore, the layout nodes and their efficient implementations have yet to be studied, making it problematic in the event that implementations are unavailable. To suit business objectives, for example, in private Blockchain implementations, several channels can be created to facilitate a group of nodes and separate them from the rest. Multiple smart contracts must be used to integrate the various channels. The proper application of business rules and smart contracts is critical for a successful Blockchain implementation.

To address these difficulties, the Indian government has established specific technical excellence centres, known as Centres of Excellence (COEs), with the necessary manpower and resources.Furthermore, these centres work with a variety of technical institutes, including Indian Institutes of Technology (IITs) and National Institutes of Technologies (NITs).The Indian government has begun supporting Blockchain projects for several start-ups.With the aid of industry, the government of India has begun to improve the PKI infrastructure and public network.

Figure 14. Implementation challenges

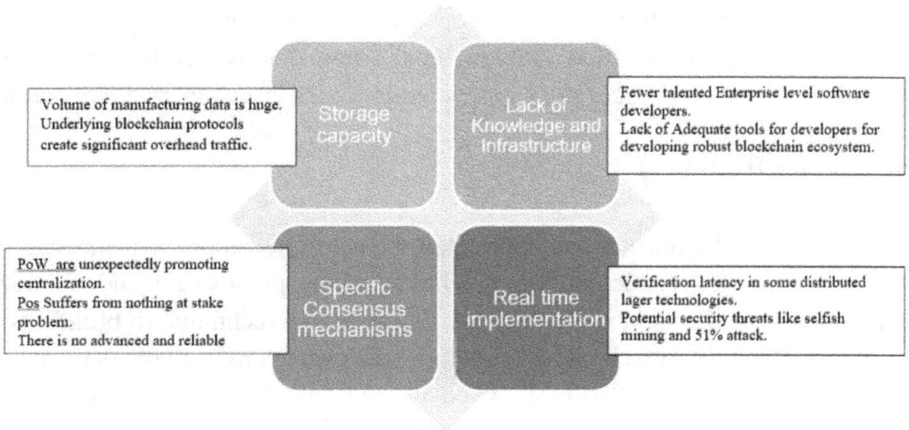

BLOCKCHAIN BENEFITS

Blockchain is an upward technology and it has begun to transfigure more and more industries and there will be a plenty of opportunities for blockchain technology in the agricultural field. Now, the agricultural field has worth of 2.4 trillion dollars and 1 billion people involved. Agriculture symbolizes a main source of an employment & contribution towards a mixture of countries around the earth. In 2030, the food demand is expected to grow by 35% because of the more and more production systems that require an increasing population (Rana., 2021).The Blockchain technology permits an end –to-end communications without the intervention of third party or intermediary such as bank / middleman in an agricultural field. The technology can change the way that the trust is provided by removing the necessity of the middle authority. Cryptographic techniques can be utilized for restoring the trust in between the producers and consumers in which that can compress the transaction costs in the agriculture food industry. The blockchain technology proposes the feasible and consistent techniques for monitoring the transactions in between unknown applicants. Blockchain technology can promotes transparency among the involved participants and facilitates the collected data. The collection of transparency, integrity, visibility and immutability provides an enormous benefit to increase the trust in the food supply chain (Fran Casino., 2019). Major benefits of the blockchain technology in agriculture are:

1. Increased food safety and its quality
2. Improved farmers efficiency
3. Food supply chain traceability

Increased Food Safety and Its Quality

In the globe, crop failure is considered as a frequent problem that is faced by farmers because of the climatic and weather conditions, like unequally distributed rainfall and unpredictable weather. The industries like JBS, Aero farm systems, Cargill, Nestle and COSMOSID are harvesting the new technologies for solving the similar issue. For example, creating an IoT devices that will permit farmers to track the factors that can affect the crops like pests, soil quality and irrigation. Blockchain ledger is used for connecting these devices. The outcome can be increasingly upgraded and monitored in real time by using the application through mobile. This will permit the farmers to ensure the factors of all that they should be. If anything goes wrong, it will be indicated to the person immediately to do some alternate arrangements. In addition, an issue like food safety outbreaks immediately identified and monitored with this ledger. This can save time and cost of the supply chain process even can save lives of the farmers.

Improved Farmers Efficiency

Software companies are developing various applications recently to record and manage the farmers data and information effectively. Particularly, the techniques like spreadsheets and visualization used for monitoring the processes. Though, it requires a lot of work and technique to broadcast the data to other service providers. Blockchain technology will allow the farmers to keep the data and monitor the information in a place easily and effectively. Blockchain ledger is able to monitor the things such as, business goals, plan to achieve the business goals, number of animals and its health issues, eating habit of animals, pattern of animals, crop varieties, farmer schedule, payment of the farmers, working hour of the farmers, income and expenses etc., Tracking of everything on a single application instead of having multiple techniques that clarifies the process and reduces the chance of valuable information getting lost.

Food Supply Chain Traceability

An expectation of the consumer will focuses on food standards and that are increasing day by day. Notably, different consumers want to know the food that they actually can come from (V. Vijayaganth., 2021).

CONCLUSION

The book chapter presents a comprehensive analysis of the challenges encountered in food supply chain management, registration of agricultural land, and agricultural insurance. It highlights the potential of blockchain technology to address these challenges by providing a secure, transparent, and decentralized platform for data management and transactions. The studies and proposals presented in this chapter demonstrate how blockchain can improve the efficiency, accountability, and trustworthiness of food supply chain management, land registration, and agricultural insurance. As blockchain technology continues to evolve and mature, it holds immense promise for transforming the agricultural industry and promoting sustainable food production and distribution.

REFERENCES

Andreev, R. A., Andreeva, P. A., Krotov, L. N., & Krotova, E. L. (2018). *Review of blockchain technology: types of blockchain and their application.* Intellekt. Sist. Proizv.

Caro, Miguel Pincheira, Muhammad Salek Ali, Massimo Vecchio, and Raffaele Giaffreda. (2018). Blockchain-based traceability in Agri-Food supply chain management: A practical implementation. In 2018 *IoT Vertical and Topical Summit on Agriculture-Tuscany*, p.1-4.

Casino, F., Kanakaris, V., Dasaklis, T. K., Moschuris, S., & Rachaniotis, N. P. (2019). Modeling food supply chain traceability based on blockchain technology. *IFAC-Papers On Line,* Volume 52, Issue 13, ISSN 2405-8963, https://doi.org/, p.2728-2733. doi:10.1016/j.ifacol.2019.11.620.0

Ganesh Kumar, C., Pachayappan, M., & Madanmohan, G. (2017). Agri-food supply chain management: Literature review. *Intelligent Information Management*, *9*(2), 68–96. doi:10.4236/iim.2017.92004

Geetha, S. K., & Kalpana, R., M. SK, K. S and P. PS.(2022). Health Record Management Using Blockchain Technology. *8th International Conference on Smart Structures and Systems (ICSSS)*, Chennai, India.

Geetha, S. K., S. G. R. Preethi, and R. Sreenithi. (2021). Educational certificate verification using blockchain based framework. *Turkish Journal of Physiotherapy and Rehabilitation*, 1833-1843.

Gupta, S., Sinha, S., & Bhushan, B. (2020). Emergence of blockchain technology: Fundamentals, working and its various implementations. *In Proceedings of the International Conference on Innovative Computing & Communications.* 10.2139srn.3569577

Hans, Ronny, Hendrik Zuber, Amr Rizk, and Ralf Steinmetz. (2017). *Blockchain and smart contracts: Disruptive technologies for the insurance market.*

Imran, S., Ferdous, T., Adeel, M., & Faheem, M. (2013). Existing land record management system of Pakistan and proposed Arc GIS parcel data model. *In 2013 International Conference on Aerospace Science & Engineering (ICASE)*, pp. 1-3 . IEEE. 10.1109/ICASE.2013.6785567

Iyer, V., Shah, K., Rane, S., & Shankarmani, R. (2021). Decentralized Peer-to-Peer Crop Insurance. *In Proceedings of the 3rd ACM International Symposium on Blockchain and Secure Critical Infrastructure*, p.3-12 10.1145/3457337.3457837

Jha, N., Prashar, D., Khalaf, O. I., Alotaibi, Y., Alsufyani, A., & Alghamdi, S. (2021). Blockchain Based Crop Insurance: A Decentralized Insurance System for Modernization of Indian Farmers. *Sustainability (Basel)*, *13*(16), 8921. doi:10.3390u13168921

Kamilaris, A., Fonts, A., & Prenafeta-Boldύ, F. X. (2019). The rise of blockchain technology in agriculture and food supply chains. *Trends in Food Science & Technology*, *91*, 640–652. doi:10.1016/j.tifs.2019.07.034

Mendi, A. F., Sakaklı, K. K., & Çabuk, A. (2020). A blockchain based land registration system proposal for Turkey. *In 2020 4th International Symposium on Multidisciplinary Studies and Innovative Technologies (ISMSIT),* IEEE, p.1-6

Naveenkumar, M., Srithar, S., Rajesh Kumar, B., Alagumuthukrishnan, S., & Baskaran, P. (2021). *InceptionResNetV2 for Plant Leaf Disease Classification*. Fifth International Conference on I-SMAC (IoT in Social, Mobile, Analytics and Cloud) (I-SMAC), Palladam, India. 10.1109/I-SMAC52330.2021.9641025

Nguyen, D.-H., Tuong, N. H., & Pham, H.-A. (2020). Blockchain-based Farming Activities Tracker for Enhancing Trust in the Community Supported Agriculture Model. *International Conference on Information and Communication Technology,* (pp.737-740). IEEE. 10.1109/ICTC49870.2020.9289297

Niranjana Murthy, M., Nithya, B. N., & Jagannatha, S. J. C. C. (2019). Analysis of Blockchain technology: Pros, cons and SWOT. *Cluster Computing*, *22*(6), 14743–14757. doi:10.100710586-018-2387-5

Rana, R. L., Tricase, C., & De Cesare, L. (2021). Blockchain technology for a sustainable agri-food supply chain. *British Food Journal*, *123*(11), 3471–3485. doi:10.1108/BFJ-09-2020-0832

Sajja, G. S., Rane, K. P., Phasinam, K., Kassanuk, T., Okoronkwo, E., & Prabhu, P. (2021). Towards applicability of blockchain in agriculture sector. *Materials Today: Proceedings*.

Shetty, A., Shetty, A. D., Pai, R. Y., Rao, R. R., Bhandary, R., Shetty, J., Nayak, S., Dinesh, T. K., & Dsouza, K. J. (2022). Block Chain Application in Insurance Services: A Systematic Review of the Evidence. *SAGE Open*, *12*(1), 21582440221079877. doi:10.1177/21582440221079877

Shuaib, M., Hassan, N. H., Usman, S., Alam, S., Bhatia, S., Koundal, D., Mashat, A., & Belay, A. (2022). Identity model for blockchain-based land registry system: A comparison. *Wireless Communications and Mobile Computing*, *2022*, 1–17. doi:10.1155/2022/5670714

Thakur, V., Doja, M. N., Dwivedi, Y. K., Ahmad, T., & Khadanga, G. (2020). Land records on blockchain for implementation of land titling in India. *International Journal of Information Management*, *52*, 101940. doi:10.1016/j.ijinfomgt.2019.04.013

Vijayaganth, V., & Naveenkumar, M. (2021). *4 Smart Sensor Based Prognostication of Cardiac Disease Prediction Using Machine Learning Techniques. Applications of Machine Learning in Big-Data Analytics and Cloud Computing*. River Publishers.

Viriyasitavat, W., & Hoonsopon, D. (2019). Blockchain characteristics and consensus in modern business processes. *Journal of Industrial Information Integration*, *13*, 32–39. doi:10.1016/j.jii.2018.07.004

Chapter 8
Integration of Technology and Nanoscience in Precision Agriculture and Farming

Rakshi Anuja Dinesh
Sathyabama Institute of Science and Technology, Chennai, India

Jayashree Shanmugam
Sathyabama Institute of Science and Technology, Chennai, India

Kunal Biswas
Sathyabama Institute of Science and Technology, Chennai, India

ABSTRACT

In addition to other cutting-edge computer technologies, the internet of things (IoT), artificial intelligence (AI), and others have been used in agriculture. However, there is currently increased focus on the use of such technologies. It is now possible to monitor the agricultural environment and ensure high-quality production through cutting-edge IoT technologies in amalgamation with nanoscience. Nanotechnology is the branch of science that deals with manipulating materials on a nanoscale. By encountering problems that cannot be solved conventionally, working with the smallest particles imaginable inspires hopes for enhancing agricultural productivity. By releasing fixed nutrients, efforts are made to restore soil fertility in the management aspects. Nano biosensors and satellite systems are used to assess the input needs for crops in controlled conditions for agriculture and precision farming, and the necessary amounts are then delivered at the correct moment. These disadvantages can be addressed by various cyber-physical systems, that is briefly explained in this chapter.

DOI: 10.4018/978-1-6684-7879-0.ch008

INTRODUCTION

Given that 60% of the world's population depends on agriculture for subsistence, it is still a crucial component of the global economy. Agriculture has advanced as a consequence of this interdependence, originally via the labor-intensive techniques of traditional agriculture (Ang, 2019). Due to the recent rapid expansion in the world's population, it is now urgently important to balance supply and demand by adopting modern technologies to increase food production (Poonia et al., 2018) (which is expected to reach 8.9 billion by 2050). This development strains natural resources since agriculture presently requires 70% of the freshwater available on Earth for irrigation. Therefore, it will be very difficult to produce enough nutritious food to feed the population due to limited resources and the effects of climate change (Senthil Kumar & Janet Joshiba, 2019). A worldwide project called Smart Agricultural aims to protect resources and uphold sustainable agriculture (Bennett et al., 2014). Smart agricultural methods are being adopted across the nation and are quickly emerging as a key trend. They sparked an era of change in agriculture and have lately received increased attention. There are various limitations associated with the adoption and marketing of such automation technologies, even though combining these three smart technologies may increase agricultural efficiency (Talaviya et al., 2020). The goal of this study is to update the many promising uses of AI, IoT, and nanotechnology in different phases of agriculture. It also intends to highlight the key obstacles to advanced farming's adoption as well as provide suggestions for the future.

PRECISION FARMING AND AGRICULTURE

A farming management concept called precision agriculture (PA) or precision farming (PF) is focused on watching, measuring, and reacting to inter- and intra-field variability in farming. Numerous writers (Fahad et al., 2021; Lee et al., 2021; Lencsés et al., 2014; Mulla, n.d.) discuss revolutions during the time of industrial agriculture. The first revolution was sparked by the invention of tractors, which combined harvesters with mechanization. The emergence of biotechnology, notably the much-discussed practice of genetic modification, catalyzed the second revolution. Later, computer technology was used in agriculture to improve existing practices and develop new ones. Precision agriculture sometimes referred to as "Agriculture 4.0," has been popular recently. Precision agriculture, also known as Agriculture 4.0, is where the phrase "Industry 4.0" originated. It alludes to cutting-edge methods and equipment used in agriculture to improve productivity, lower costs, boost efficiency, and take use of sophisticated processing, data analysis, and other features that will modernize the industry. To improve crop performance and environmental quality, precision agriculture uses technology and concepts to manage the spatial and temporal variability linked with every aspect of agricultural production. All farms may presently conduct precision agriculture using automated machine control, a large tracked tractor or a tiny compact tractor with a centrifugal spreader. Agriculture 4.0 also enables considerable natural and monetary resource savings through the development of autonomous section control systems and the use of locally variable nutrient treatments. Due to the increase in global land area and farm size, this information could not be obtained effectively. It was more common to view a location as a homogeneous area with unrealized potential. Only the availability of technology and the necessary technical tools has changed this situation. It provides the entire spectrum of information needed from various sources together with their in-depth analysis. Decisions on agronomic activities, such as modifying the variable application of fertilizers and pesticides in the right quantity and place or predicting the status and characteristics of the soil or crop, are aided by the findings. (Balafoutis et al., 2020; Bhatnagar et al., 2019; Gagliardi et al., 2021; Mendes et al., 2020). Precision farming technology has advanced fairly quickly as well. Precision Livestock Farming (PLF) is the term used in other countries to describe precision agriculture in animal feeding and breeding. The goal is to increase farm operation accuracy and support the farmer's quick decision-making. The selection of fodder crop types, their cultivation, harvest, nutritional content, silage, or storage are the new directions in contemporary farming, but animal nutrition and product quality are still priorities.

Nano-Agriculture

Agricultural production of feedstock for industrial processes will be improved through nanobiotechnology. In the meantime, tropical agricultural products like rubber, cocoa, coffee, and cotton, as well as the small-scale farmers who grow them, will become antiquated and irrelevant in a new nano economy of flexible matter where the characteristics of industrial nanoparticles can be altered to create more accessible, intelligent replacements. According to the hypothesis, agribusiness will have more influence over all facets of world agriculture and food production as a result of private nanotechnology. This is done under the pretense of giving food to those in need, protecting the environment, and giving customers more options. From seed to stomach, genome to the gullet, proprietary nanotechnology will be deployed, just as genetically modified (GM) agriculture led to unprecedented levels of corporate concentration across the food chain. During the past two generations, scientists have been molecularly altering food and agriculture. The industrial food chain is further broken down and connected through agro-nano. GM plants may now be transformed into atomically modified plants using novel nanoscale methods for combining and harnessing genes. Pesticides may be manufactured more precisely to kill undesired pests and include synthetic flavors and aesthetically pleasing natural nutrients. Molecular sensors, molecular delivery systems, and inexpensive labor may now be used to realize visions of an automated, centralized industrial farm. The European Union (EU) plans to use biotechnology's potential for the benefit of the economy, society, and environment of the EU to realize its vision of a knowledge-based economy. This industry is facing new difficulties, such as an increase in the demand for wholesome, safe food, a rise in disease risk, and dangers posed by changing weather patterns to agricultural and fisheries output. The development of a bio-economy, however, is a difficult and intricate process requiring the fusion of several scientific fields. Nanotechnology has the potential to change the agriculture and food sectors by providing new tools for the molecular treatment of diseases, rapid disease detection, boosting the ability of plants to absorb nutrients, etc.

Methods for Smart Delivery Based on Nanoparticles

Long before symptoms become visible at the macroscale, nanoscale devices have the potential to identify and cure an infection, vitamin shortage, or another health issue. The effects of medications, nutraceuticals, minerals, food supplements, bioactive substances, probiotics, pesticides, insecticides, fungicides, and immunizations may all be monitored using smart delivery systems. Delivery systems are crucial in agriculture for applying fertilizers and pesticides as well as for genetically modified plant enhancement. Pesticide application techniques must concentrate on improving effectiveness and controlling spray drift, while fertilizer bioavailability issues are brought on by soil chelation, overapplication, and run-offs. Controlled delivery systems for the application of pesticides and fertilizers provide a workable solution to these issues. To maximize biological activity and reduce negative effects, controlled delivery techniques strive for metered release of essential and appropriate quantities of agrochemicals throughout time. To achieve this, the use of micronic and sub-micronic particles as pesticide delivery systems were investigated. Nanoparticles (100 nm) have the benefit of effective loading over micronic particles (1000 nm) because of their higher surface area, ease of attachment, and quick mass transfer. The delayed release properties of the nanomaterial, the chemicals' attachment to the material, and the ambient factors all contribute to the controlled release of the active component. The transfer of genetic material faces challenges including a limited host range, cell membrane transit, and shuttling to the nucleus. However, researchers are looking at the use of NPs to aid in the transfer of genetic material to create plant kinds that are resistant to insects. To blast plant cells and tissues with genes, DNA-coated gold NPs are used as bullets in a gene gun device, for example (Vijayakumar et al., 2010).

Pesticides Delivery

Nanotechnology can efficiently distribute chemical and biological pesticides by using agrochemical formulations based on nanomaterials or nanosized preparations. Nanomaterial-based formulations have several benefits over conventional pesticides and their formulations, including greater efficacy because of increased surface area, greater solubility, induction of systemic activity because of the smaller particle size and higher mobility, and reduced toxic effects because organic solvents are not present (An et al., 2022). When using NPs in the field to deliver pesticides and biopesticides, there are several obstacles to overcome, including various environmental disturbances, sizable regions that must be covered by the spray, and ultimate cost-effectiveness. Because it is easier to apply the chemical when using a large volume, low-value preparation, the whole crop is sprayed with it in the standard spraying regimen. A low-volume, high-value use is anticipated for nanomaterial-based preparations. These regulated nanoparticulate delivery devices will need a targeted distribution strategy that is based on understanding the life cycle and behavior of the virus or pest.

Application of Fertilizers

According to (Wilson et al., 2008), the use of excessive fertilizer in the form of urea, nitrate, or phosphate compounds, as well as ammonium salts, is hazardous. Furthermore, since they are lost to run-off and leaching, which causes pollution, a large portion of fertilizers are inaccessible to plants. The gradual release of fertilizers may benefit from the use of nanomaterials. Due to their higher surface tension than ordinary surfaces, nanocoatings, or lacquers of nanomaterials, on fertilizer particles, more effectively retain the plant material. Additionally, nanocoatings shield bigger particles' surfaces from damage. As the sulfur content is advantageous, particularly for sulfur-deficient soils, fertilizers containing sulfur nanocoating are helpful slow-release fertilizers. Since the coating was stable, the pace at which the fertilizer dissolved was lessened, and it was possible to release the sulfur-coated fertilizer over a long period. To address the needs of the soil and crops, sulfur nanocoatings as well as the encapsulation and release of urea and phosphate would be helpful. Kaolin and polymeric biocompatible nanoparticles utilized for the sustained release of NPK fertilizer sources such as urea, calcium phosphate, and potassium chloride are further nanomaterials with potential applications (Corradini et al., 2010).

Assessment of Crop Nutrition With Nanotechnology

Fertilizers considerably enhanced India's production of food grains with the introduction of high-yielding and fertilizer-responsive crops kinds during the "green revolution" era. The yields of many crops have begun to plateau as a consequence of uneven fertilization and a decline in the organic matter content of soils, notwithstanding the enormous success in grain production. In addition to harming aquatic habitats, excessive nitrogen fertilizer usage also has an impact on groundwater. The fact that fertilizer usage efficiency is 20–50% for nitrogen and 10–25% for phosphorus is alarming. The nutrient buildup in soils, which causes eutrophication and taints drinking water, may be avoided with the development of nano-fertilizers as alternatives to traditional fertilizers. In reality, advances in nanotechnology have created new possibilities for enhancing the effectiveness of nutrient utilization while lowering environmental protection expenses. Recent research showing that microorganisms and plant roots may directly extract nutritional ions from the solid phase of minerals has been made public. Exceptional substitutes for soluble fertilizers include nano-fertilizers and nanocomposites that release slowly. Throughout the development of the crop, nutrients are released more slowly, and plants can absorb the majority of them without losing any of them. Using zeolites, a class of naturally occurring minerals with a layered crystal structure resembling a honeycomb, it is possible to create a slow release of nutrients into the environment. In addition to other slowly dissolving components comprising phosphorus, calcium, and a full range of minor and trace nutrients, it is possible to load nitrogen and potassium into its network of interconnecting tunnels and cages. Nutrients are held in zeolite and gently released as needed. Nano-membranes that enable the gradual release of nutrients may be applied to fertilizer particles. According to (Tarafdar, Agrawal, et al., 2012), nano-fertilizer technology is quite cutting-edge and has a wide range of applications. Foliar fertilization using nanoparticles has been shown to significantly enhance yields. It was shown that cluster beans and pearl millet yielded 80 kg ha^{-1} phosphorus equivalent yield with 640 mg ha^{-1} foliar spray (40 ppm concentration) of nano phosphors in a dry environment. The development of nano-composites to give all the necessary critical nutrients in the right proportions using intelligent delivery systems is currently being researched. Nanotechnology has the potential to provide balanced fertilization (Tarafdar, Raliya, et al., 2012). Nano-composites that are being considered to provide all the nutrients in the proper amounts using "Smart" delivery methods also need careful examination. The loss of 50–70% of the nitrogen provided by traditional fertilizers results in poor nutrient usage efficiency. By boosting plant absorption, new nutrient delivery techniques that take use of plants' porosity nanoscale structures might decrease nitrogen loss. The absorption of nutrients will be increased by fertilizers included in nanoparticles (Tarafdar, Xiong, et al., 2012). Nutrients may be delivered either in response to an environmental factor or at a predetermined period. By coating and cementing nano- and sub-nano-composites, it is possible to regulate how quickly nutrients are released from the fertilizer capsule. It has been shown that a proprietary nano-composite including N, P, K, micronutrients, mannose, and amino acids enhances grain crops' ability to absorb and use nutrients (Hatti et al., 2020). The primary potential use of the technique, according to (Tapan et al., 2010), is nano fertilizers for plants' effective use of nutrients and water that release slowly. One of these brand-new facilities incorporates nutrients within a nanoparticle and operates in one of three ways. The nutrition may be administered as nanoscale-sized particles or emulsions, coated with a thin polymer film, or enclosed in nanoporous materials (Rai et al., 2012). Nano fertilizers will also use nanodevices to time fertilizer-N and -P release with crop absorption, preventing unintentional nutrient losses to the soil, water, and air by direct ingestion by crops and eradicating nutrient contact with soil, microorganisms, water, and air (DeRosa et al., 2010).

Smart Sustainable Agriculture

Precision farming, smart irrigation, and smart greenhouse are just a few of the terminology used to describe recent extensive research into sustainable smart agriculture. Precision farming is a practice that uses IoT and Information and communication technologies (ICT) to manage crops and conserve resources. It gathers real-time information on the state of farm components, such as crops, soil, and air, to save the environment while assuring profitability and sustainability (Mohanraj et al., 2016). Using IoT-based smart irrigation systems, smart irrigation is a technique for increasing the effectiveness of irrigation operations, lowering water losses, and conserving already existing water resources (Patrício & Rieder, 2018). Drones are employed in several agricultural applications, such as monitoring field crops and animals and screening large areas, while sensors on the ground collect a broad range of data (Crosson, 2016). Smart greenhouses encourage crop development with the least amount of human interaction possible by continuously monitoring the climatic conditions, automating activities based on the assessed changes, and deploying corrective action to preserve the most favorable conditions for growth (Bibri, 2018). Farm Management Systems (FMS) control and run a large number of sensors and tracking devices to give farmers a variety of information that has been gathered. The obtained data is evaluated to perform complex decision-making tasks before being saved on a medium. This enables the application of the best agricultural data analysis methodologiess (Arjun Kumar et al., 2012). By keeping an eye on the physical, chemical, and biological characteristics of the soil, soil monitoring systems aid in tracking and enhancing its quality. To improve animal welfare and production, livestock monitoring systems offer real-time assessments of these factors (Van Erp-van der Kooij & Rutter, 2020). To help farmers make decisions, reduce operational expenses, and increase productivity, the IoT/AI SSA platform Cloud offers real-time information. It is described as the process of establishing, monitoring, managing, processing, and analyzing data gathered from different agricultural resources, such as fields, crops, livestock, and others, to ensure the longevity and quality of agricultural products and further improve stakeholder decision-making. **Figure 1** is an example of smart and modern technologies.

Figure 1. Smart and advanced computing technologies

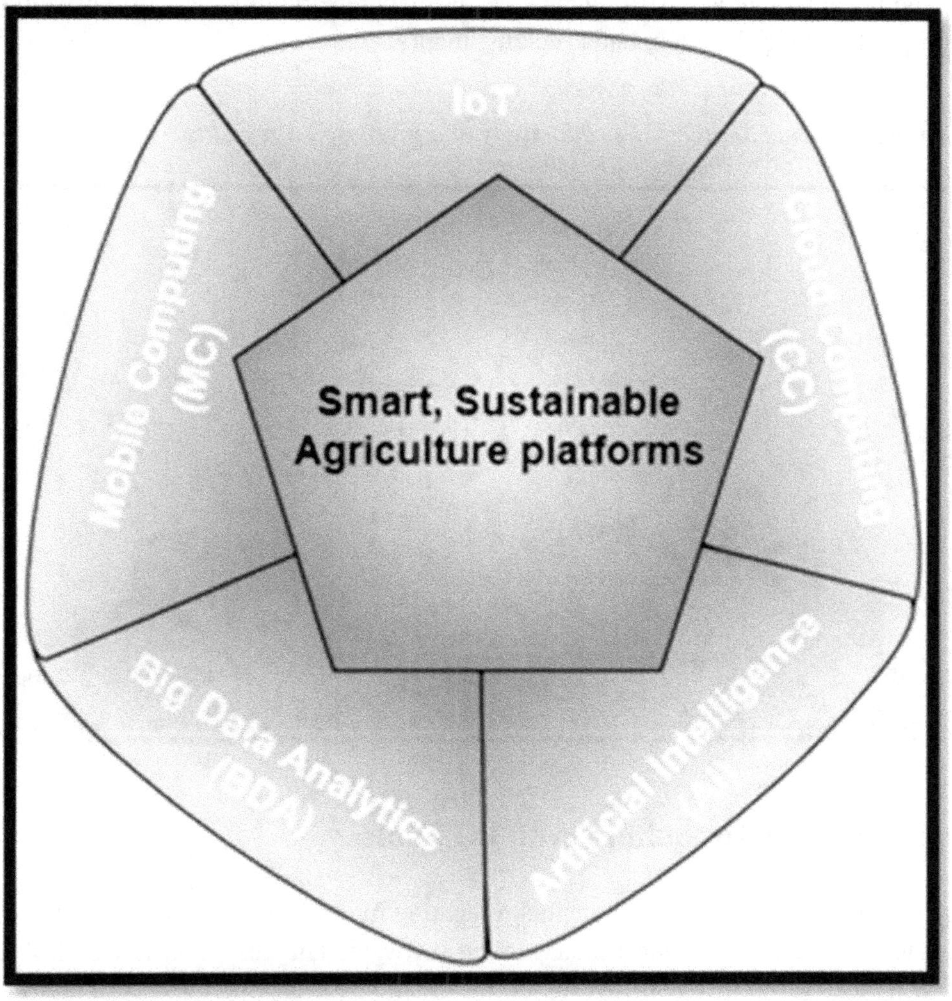

COMPUTER SCIENCE IN AGRICULTURE

Internet of Things and Precision Agriculture

As we go toward more cultivated and urban farming, the use of cutting-edge scientific innovations like IoT-based technology is becoming more and more essential in a variety of agricultural systems for a wide range of uses. They assist in enhancing several agricultural techniques to boost productivity while conserving or reducing the influence on the product's uniqueness. The advancements in the underlying IoT technology drive ongoing changes in precision agriculture. To increase productivity and agricultural yields, this evolution strives to achieve several essential characteristics. These characteristics are listed in **Figure 2**. Precision agriculture's core characteristics, such as the creation and comparison of Key Performance Indicators (KPIs), rely on technical aspects including sensors, mobility, real-time monitoring, connection, and simplicity of application implementation. The agriculture industry is already participating in the next industrial revolution, and farmers are using cutting-edge technology to fulfill the need for food (Zambon et al., 2019). This gives farmers the ability to support agricultural choices with the appropriate timing, location, and crucially, inputs.

Figure 2. Critical elements to increase productivity and increase crop yields

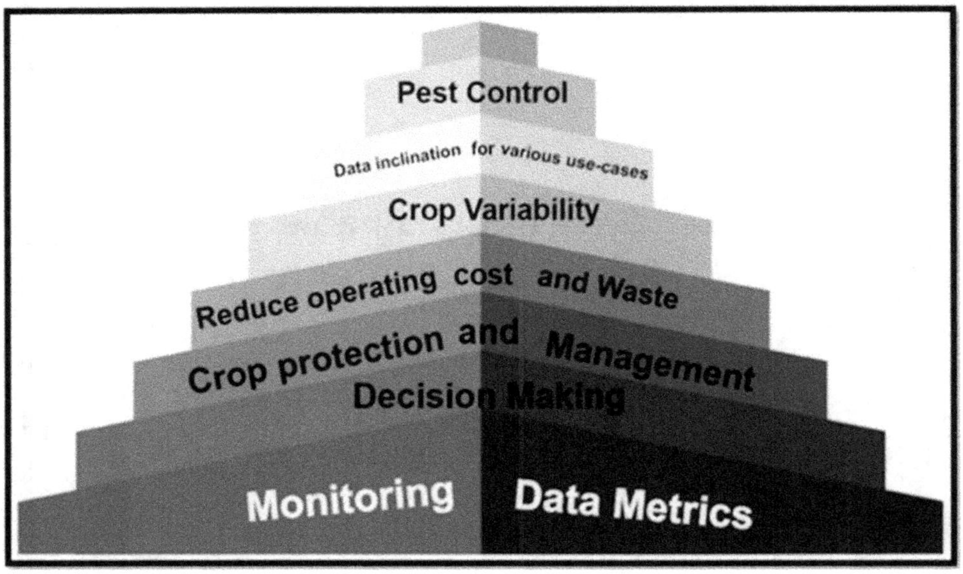

Advancing Precision Agriculture With IoT Tools

The primary facilitators in data collecting and aggregation for precision agriculture (PA) are emerging IoT technologies like sensors and farm management software. They use a variety of IoT solutions to monitor geographical variability and investigate information for an improved agricultural operation. Remote sensors, VRT, soil sampling, and other equipment are some of them. Deep analyses of soil resistivity and information about prior crops are made possible by other technologies, such as geolocating a field. Information about environmental constants may be obtained by long-term observation of the environment. These point indicators support the identification of crop factors such as disease, water stress, and others. The Internet of Things (IoT) provides a prediction and control technique to find and manage static indicators before planting altogether, such as soil and crop history. Furthermore, most decisions made throughout the crop cycle are based on forecasts developed using big data, ML, and artificial neural networks. Farmers may manage their fields more successfully thanks to the information system and networking technologies offered by the IoT. IoT devices, drones, and robots are often employed in combination with a distinct set of PA applications. By using attachments like multispectral and hyperspectral sensors for managing the stress and wellness of the plant, unmanned aerial vehicles also aid in risk reduction and catastrophe management. Due to the shrinking of sensors, there has been a significant advancement in their capacity and aptitude to detect various crop health metrics, including biomolecular, electrical, chemical, etc. Precision farming software uses this data to manage tools and other management duties. Continually moving from short-range (like ZigBee or Wi-Fi) to long-range (like cellular connection, Low-Power Wide-Area Networks (LPWAN), etc.) connectivity protocols are required to develop an IoT solution for intelligent farming. To monitor things like crop biomass, water level, etc., other location monitoring technologies like satellites are also employed. The growth of equipment guidance and Controlled Traffic Farming (CTF) systems has been made possible by the Global Navigation Satellite System's (GNSS) greater precision. These IoT-based solutions are beneficial for the PA, which not only helps the farmers but also allows governments, policymakers, and other stakeholders to exploit the data produced by the farms.

Using IoT to Monitor Yield

The FarmTRX mobile app connects to a yield monitor that can be installed on any linked harvester. This yield monitor shows real-time produced data and instantly uploads it to the platform of the manufacturers (Ayaz et al., 2019). With the help of other farm management tools, the farmer may export the high-quality yield maps produced by this software for further study. To accurately assess the quality and productivity, a fruit growth assessment might be very helpful. Monitoring the productivity of large-scale crops may be done with the use of satellite images. Sentinel-1A interferometric images were utilized to record rice crop productivity in Myanmar using this technique (Torbick et al., 2017). It is noticed extensively that different sensors were employed for detection of different altered conditions in the agricultural fruits and vegetables like different stages of mango ripening stages were analysed using RGB depth photo analysis (Wang et al., 2017). In the similar form, papaya shrinkage have been monitored at the drying condition employing characteristic optical sensors (Udomkun et al., 2016).

Enhancing Privacy and Security in Precision Agriculture

Together, on-field sensors and machinery provide a productive agricultural solution. Vulnerabilities and cyberattacks have been made public by the introduction of heterogeneous equipment and connectivity in the agriculture sector. Unhealthy agricultural conditions are caused by these attacks, which vary from utilising tractors, unmanned aerial vehicles (UAVs), and more to flooding fields and liberally applying poisons. Large-scale, well-organized strikes in the agriculture industry can completely collapse a nation's economy. Nevertheless, the usage of IoT and smart communication technologies exposes smart farming ecosystems to a wide range of cybersecurity risks and vulnerabilities. Such cyberattacks have the potential to destabilize the economies of nations with significant agricultural dependence. It is understood that in this dynamic and dispersed cyber-physical environment, multi-layered architecture seems to be pertinent in the field of precision agriculture involving different security concerns. The security challenges are enhanced using a multi-layer smart farming architecture displaying many elements relevant to real-time use cases supported by edge and cloud settings. A report (Mutschler, 2018) describes the many dangers associated with PA and the need for more research in this area. Although IoT technologies are increasingly widely used, it is essential to establish and follow compliance and regulation to protect the solutions. In Pennsylvania, security is of the utmost importance; even tiny security breaches can disrupt the whole food supply chain. National regulations or adherence to cybersecurity operating standards set out by an individual or municipal authority may be used to enforce the obligations. Additionally, even though the food supply chain may have cyber insurance coverage, the inclusion of AI and smart applications makes it much harder to evaluate and forecast the associated cyber risk (Gupta et al., 2020). The design and optimization of cyber-physical systems for agriculture have received numerous contributions. A continuing trend is the employment of a variety of technologies to improve cyber security. In a similar line, solar photovoltaic water systems are used by (Selmani et al., 2019) to guarantee bidirectional communication in the greenhouse.

Transportation and Food Safety Using IoT

Given the current hunger issue brought on by population expansion, there is great potential to reduce food waste and increase the supply of food by simply implementing a temperature-controlled transportation infrastructure. Food safety could be increased by employing self-contained system, which involves in-built wireless electronical system for auto-detection of differential temperatures changes of food storage. In this manner, the continuous and updated temperature profile of the food storage could be achieved, which would be entirely based upon data and quantitative values (Bouzembrak et al., 2019). Owing to the recent development in the cloud computing, the data retrieved from such analysis could be stored in cloud and could be accesed by any device using conventional internet connections (Bharati & Mondal, 2021; Podder et al., 2021). It is due to the advancements of Internet-of –Things (I-o-T), the monitoring and sensing of such data is regulated and ascertained in a real-time mode, whereby any alterations in the temperature sensing and malfunctioning of the system would be notified to the user promptly, thereby controlling and managing the system without any delay in reparing and corrections on time (Popa et al., 2019).

Soil Moisture Sensors

The moisture sensor is one of the crucial sensors in Precision Agriculture (PA). The level of water in the soil would be provided via a moisture sensor. An agronomist determines the appropriate water level for a certain crop. The amount of water required to irrigate the crops would be known based on the water level feedback mechanism. The use of nitrogen by the roots might be impacted by too much water in the soil. At the same time, there will not be enough oxygen at the roots. Different amounts of water would be needed for different crops. As a result, depending on the type of crop, moisture will suggest a different amount of water. Soil moisture may be determined by two different methods. Both contact-based and contact-free-based are methods. The sensor's detecting region must be immediately contacted by the dirt being detected in the contact-based mode of detection. Capacitance sensors (Brillante et al., 2015; Z. Zhang et al., 2015), heat pulse sensors (B. Zhang et al., 2015), and fiber optic sensors (Ciocca et al., 2012) are only a few examples of contact-based sensors that use different approaches dependent on detecting characteristics. Contact-free based is another method of gauging soil moisture. The sensor's detecting area does not have to come into direct contact with the dirt being detected in a contact-free system. Passive microwave radiometers and synthetic aperture radars, as well as thermal techniques, are examples of contact-free based sensors.

Artificial Intelligence's Role in Advanced Farming

By enhancing the productivity of conventional farming and removing the obstacles and disadvantages that traditional farmers encounter, artificial intelligence-based technologies help agriculture. Artificial intelligence (AI) is the process by which people create artificial computers that are comparable to the human brain but can handle bigger quantities of data than the human brain. Although artificial intelligence (AI) is closely related to computer science, it should go beyond this to advance agricultural technology (Jha et al., 2019). AI has been used to produce a variety of technological tools and equipment that have been tested on agricultural fields and improved. Several parameters like storage management, post-harvest activities, treating sick crops, inadequate irrigation for meeting crop demands etc. have been taken care by such advanced technology at the agricultural interface. The quality and the quantity of the produce have been enhanced effectively by the farmers (Talaviya et al., 2020) It could be further understood that, different environmental concerns like unregulated utilisation of irrigation measures causing water loss, contamination of water with fertilisers, robust utilisation of pesticides etc. have bee seem to be reduced by the applications of AI technologies and interventions. Different scientists in the recent past have developed numerous AI –based systems for plant based applications (Bannerjee et al., 2018). The effective dependency of labour towards different services have been reduced using AI based technologies. Due to the automation in the procedure of collecting and analysing data from the different Government websites and to the real life field data, AI-system based technology enabled for effective large problem solution unlike met by human dependency. In order to obtain desired and effective agricultural related outputs mediated by the AI-based technologies, farmers and the persons associated with the different AI-based farm related activities need to be awared and taught in different dimensions of AI-mediated agricultural practices and its widespread applications.

Agriculture-Related AI Technology Foundations

Machine learning forms the foundation stone in incorporating effective AI based technology in different field of applications. In order to make the AI based information understandable to humans at field level, the informations associated with the AI based procedures must be in machine readable form and the corresponding responses must be in human understable format. Defined database should be utilised in the AI mediated process, as the different current problems are analysed using the entered datasets in the system. In order to interpret real-time parameters in the AI-mediated process, real-time data is of paramount importance for drawing a positive and rational conclusion and analysis. It is understood further that determining the weather forecast is of prime importance for analysing the decision of different cropping seasons aspects. Chatbots are machines that converse with farmers who have less experience interacting with technology to help them remotely. Government and institutional agricultural officials often use unmanned aerial vehicles (UAVs) to geolocate possible threats to the fields, such as the development of forest fires, insect invasions, disease assaults, and many others (Talaviya et al., 2020). Some of the approaches include fuzzy logic, expert systems, neuro-fuzzy logic and the artificial neural networks (ANNs) (Jha et al., 2019). ANNs seems to be of optimal utilisation when employing AI-based technological interventions. Similar to human brain-ANN works in the brain replication system. It is clearly understood that analogous to human brain, AI based system works, which is mediated by different algoristhms and programmes like Silva, Delta-bar-delta, Almeida etc. which works in the similar way brain impulses passes signals through synapses and nerve endings in brain. Thus ANN forms the basis of an effective AI-based technology. In this context, in the domain of agricultural practices, farmers and ground-level technical persons should develop some software-hardware embedded system for designing an effective user-friendly AI-based system for agricultural applications for running smoothly in future.

AI for Crop Management Techniques

Sensors and embedded systems have been developed to support growth chamber parameters including intensity of light, humidity, and O_2 and CO_2 levels. They may alter crop conditions in real-time to meet optimized settings based on the most recent crop growth data (Lakhiar et al., 2018). A technical company called Trace Genomics harvests DNA from soil samples taken from agricultural regions and counts the bacteria that live there. Utilizing machine learning techniques, the soil data is analyzed. To offer remedies to the farmer that would be supported by evidence based on historical occurrence data, this biological data is coupled with the chemical characteristics of the soil sample. This would make it easier to choose the best crop for the land or vice versa. Crop production requires a steady supply of water that is also easily accessible. It is strongly encouraged not to use more water resources than the crop needs since freshwater is so scarce. To alter the quantity of water and automate the start and finish of the water supply, AI technologies that collect real-time data on the level of soil moisture and the weather are needed (Talaviya et al., 2020). A couple of these AI-powered automated irrigation techniques were enumerated by (R. Kumar et al., 2015). According to (Mahmood et al., 2016), excessive pesticide use poses dangers to human health, biodiversity, and the ecosystem by posing eutrophication concerns to rivers and posing threats to human health. The quantity of pesticide typically sprayed may be reduced by up to 55% thanks to Rover, a tiny vehicle-like equipment invented by (Facchinetti et al., 2021), which also improves crop coverage.

The Use of AI in Pest and Weed Control

The introduction of modern weed-specific robots which could easily recognise the weeds in the agricultural fields as introduced by (Partel et al., 2019), minimises the requirement of any unwanted exposure of agro-chemicals towards the agricultural fileds and thereby reducing the wastage of weedicides. Examples of pest control techniques are Pasqual and Mansfield, SMARTSOY, and CORAC (Bannerjee et al., 2018).

AI for Product Marketing and Storage

Before reaching the customer, agricultural goods must be kept in appropriate storage environments to retain quality. For these items to last longer after harvest, several sensors have been designed for storage chambers. Traditional farmers only comprehend the traditional marketplaces for their goods, but current market developments, pricing choices, and consumer consumption information are accurately analyzed by market data and may recommend the farmers' next crop (Talaviya et al., 2020).

Aerial Mapping to Detect Pathogen Affected Crops

Any disease that affects agricultural plants sufficiently enough to result in morphological and physiological changes may be detected via remote sensing. Cotton root rot, a soilborne disease brought on by the fungus Phymatotrichopsis omnivorous, washed away cotton fields in the late 1920s according to aerial images taken from an airplane. In-depth research into the spectral response traits of healthy and stressed crops as well as the ideal film and camera settings for diagnosing and discriminating certain cereal crop illnesses led to an increase in the use of aerial photography for crop disease detection. Airborne imaging systems with multispectral and hyperspectral sensors have been used to map agricultural diseases for a few decades now. Both root-knot nematodes and Phymatotrichum root rot in cotton were found utilizing aerial color-infrared (CIR) videography. For Phytophthora foot rot in citrus orchards, mapping late blight in tomato fields (Yang, 2020), and mapping cotton root rot in cotton fields, airborne digital multispectral photography was tested. To detect tomato yellow leaf curl (Lu et al., 2018), grapevine leafroll virus (MacDonald et al., 2016), and yellow rust in wheat, airborne hyperspectral photography was examined. For the detection of citrus greasy spot, cotton root rot (Yang et al., 2009), and huanglongbing or citrus greening (Arun Kumar et al., 2012; Li et al., 2014), aerial multispectral and hyperspectral imaging methods were applied. Mapping agricultural diseases has also been investigated using satellite photography. Even at a spatial resolution of 30 m, Landsat imagery was able to map severe take-all disease infestations in wheat (Yang, 2020). Satellite sensor improvements have substantially increased the spatial resolution of images. High accuracies were attained with severe infections at late development stages when powdery mildew and leaf rust in winter wheat were detected using QuickBird satellite photography (Mandal & Sinha, 2021). Early diagnosis of agricultural diseases is still difficult, even though both airborne and satellite imagery have been utilized to effectively identify and map numerous crop diseases. In most instances, crop damage may already have occurred by the time remote sensing images can identify any disease indications. This delayed finding could be too late for certain crops to stop the infection during the current growth period, while it might be early enough for other crops to prevent further damage with particular actions.

One of the key instruments for the growth of agriculture is the assessment of agricultural land suitability. To collect and process agricultural information, several new technologies and innovations are being used in agriculture. The Internet of Things (IoT) has been regarded as a viable tool for automation and decision **Nanosensor-enabled AI in agriculture land suitability** -making in the field of agriculture, which has sparked the construction of small, inexpensive nanosensor devices. For the assessment of agricultural land suitability, we here offer an expert system that combines nanosensor networks with Artificial Intelligence systems like neural networks and Multi-Layer Perceptron (MLP) (Mehra et al., 2018). This system's four decision classes—more appropriate suitable, moderately suitable, and unsuitable—will assist farmers in classifying agricultural land for production. This evaluation is made using data gathered from several nanosensor devices that are utilized to train the system. When compared to the other current models, the results produced utilizing MLP with four hidden layers are shown to be successful for the multiclass classification system. This trained model will be employed to analyze subsequent evaluations and categorize the land following each crop (Vincent et al., 2019).

PRECISION AGRICULTURE USING A WIRELESS NETWORK OF NANOSENSORS

Underground Wireless Sensor Networks

Wireless Subterranean Sensor Networks, a different kind of WSNs, are its underground cousin (WUSNs) (Lin & Hao, 2020; Raut & Ghare, 2020; Sadeghi et al., 2022). The wireless sensors are buried deep into the ground in this variant. It is used for environmental monitoring of harmful compounds as well as the examination of soil qualities. Higher frequencies are severely attenuated in this environment, although relatively lower frequencies may pass through the earth (Banaseka et al., 2021; Singh et al., 2022). Thus, the network needs more nodes to cover a larger region as the communication radius becomes constrained.

By needing a comparatively fewer number of sensors, the use of wired sensors improves network coverage. The sensors and cables in this arrangement, however, may be exposed to agricultural operations. Below is an illustration of a typical subsurface sensor network-based agricultural application. In contrast to the TWSN-based applications shown in **Figure 3**, the sensor nodes in this illustration are buried under the ground. To send the data gathered by the subsurface sensor nodes to the surface sink positioned above the ground, one gateway node is also installed. After then, the data may be sent via the Internet to be stored in distant databases and utilized to warn a user of a mobile phone. However, more nodes must be installed for usage in WUSNs because of the relatively shorter connection distance.

Figure 3. Underground sensor networks: a quick overview

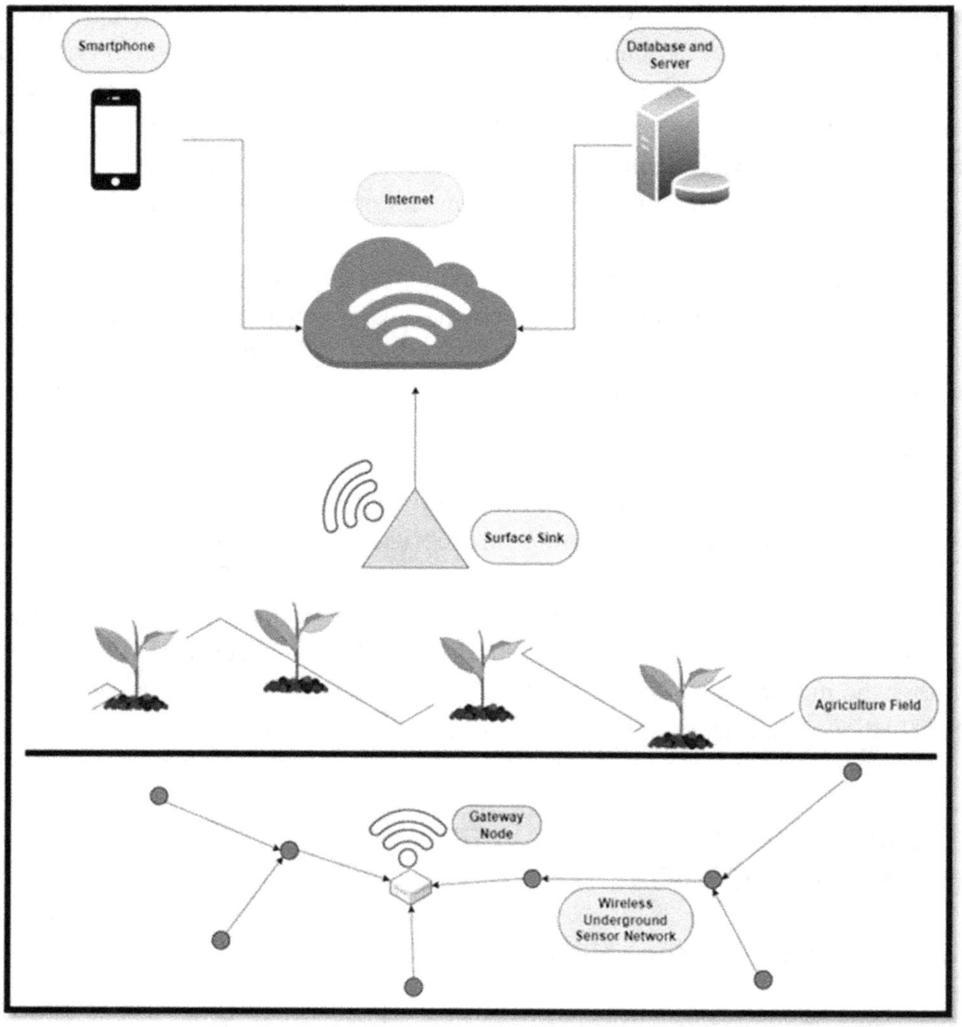

Terrestrial Wireless Sensor Networks

To build a smart network in the agricultural sector utilizing tiny, inexpensive sensors, sensor nodes are positioned above the ground. WSNs, which are often deployed to serve a particular application purpose, are networks of wirelessly linked, battery-powered sensors (Akyildiz & Kasimoglu, 2004). These robust sensors enable a sensor node or mote to precisely gather environmental data. These nodes subsequently network among themselves to carry out the application needs based on the detected data. Consider, for instance, a precision agricultural setting where WSNs are installed all over the field to automate the watering system. The soil moisture content is determined by all of these sensors, and they also work together to schedule the timing and duration of irrigation in that area. The choice is then sent over the same network to the sensor node connected to a water pump. One such automated irrigation system utilizing a WSN and GPRS module was presented by (Gutiérrez et al., 2013).

An illustration of a typical wireless sensor network used in the field for agricultural applications can be seen below. Sensor nodes with application-specific onboard sensors power the field. The industrial, scientific, and medical (ISM) radio bands (such as 902-928 MHz and 2.4-2.5 GHz) used for radio-frequency (RF) communications are used by the nodes in the on-field sensor network to interact with one another. To establish a link between the sensor network and the outside world, a gateway node is often put alongside the sensor nodes. As a result, the Global System for Mobile Communications (GSM) or GPRS powers the gateway node. The field's sensors and actuators may be managed remotely, along with the field's overall condition. For instance, when the water level applied to the field reaches a certain threshold amount, a user may turn on or off a pump or valve. Mobile phone users may remotely monitor and manage the on-field sensors as well. The mobile user is linked through GPRS or even a Short Message Service (SMS). For both types of users, it is also feasible to build recurring information updates out from sensors and provide on-demand system administration. The flowchart of the process is been visualized in **Figure 4.**

Figure 4. Terrestrial wireless sensor networks: a quick overview

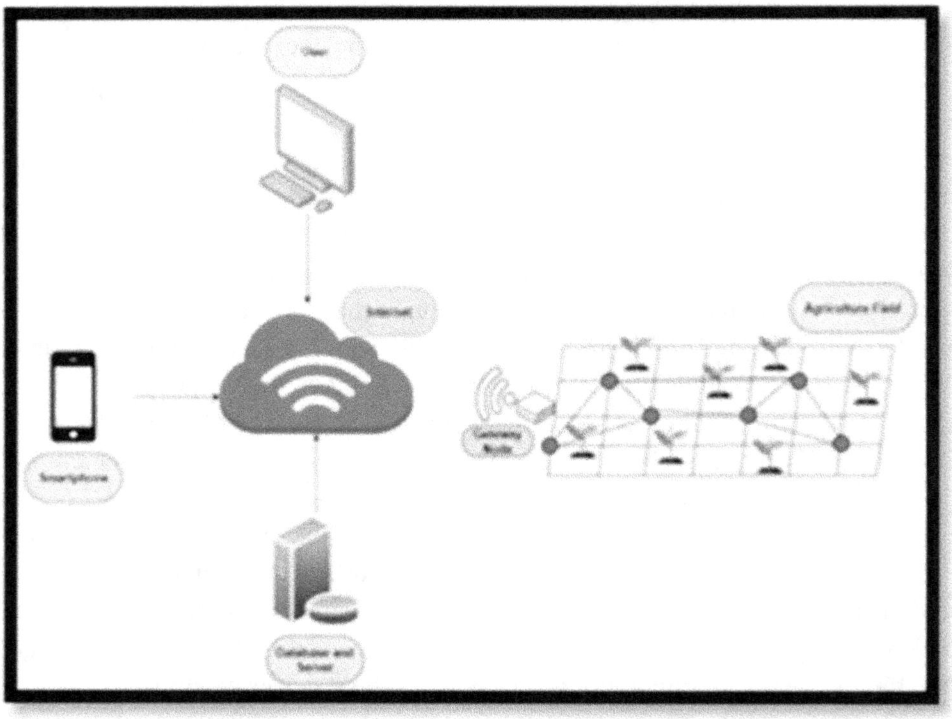

CHALLENGES

In the foreseeable future, a greater supply of high-quality food will be required owing to the population's fast growth (Ayaz et al., 2019). Therefore, the use of technology in the field should boost both the quantity and quality of agricultural produce. A greater yield is elusive for the other 98% of the agricultural population, even though 2% of them perform better in terms of quantity and quality due to access to contemporary equipment. This is obvious given that industrialized nations have been adopting modern technology and equipment for the previous 50 years and have achieved exponentially larger yields, including Australia and the majority of the European Union. It follows that modern equipment and technology both help farms become more productive and environmentally friendly while also improving yields (Zha, 2020). Given this situation, it is expected that agriculture will transform into a high-tech sector in the future, with networked systems utilizing artificial intelligence and big data capabilities. The resultant solutions will integrate into a unified system that links farm management and equipment from sowing to production projections. It is anticipated that owing to the introduction of cloud-based Artificial Intellegince, big data, agricultural robotics and machine learnings, the agricultural production seems to be augmented in the days to come. The main difficulty in implementing technology in advanced farming is that the farmers who use it in the field lack the necessary skills. To maximize the devices in the future, it is thus strongly advised that farmers be concurrently educated about the benefits of modern equipment and that an accurate knowledge base be created from each particular farmed area. Farmers in the agricultural industry have hesitation toward using these technologies due to a fear that they would replace the necessary labor force. Since field management and disease control tasks that were formerly carried out by seasoned farmers are now handled by machines, there is a good probability that these technologies won't be used. Therefore, it is evident from practical experience that younger farmers who have more hands-on experience with the technologies are more willing to adopt them into their fields than older farmers. Consequently, they need to be gradually allowed and encouraged in this domain of science (Jha et al., 2019). Agriculture includes both the domestication of species and their cultivation (Harris & Fuller, 2014). Scientists and agro-technical experts have mostly been in charge of domestication since the majority of farmers nowadays are only concerned with farming. If AI were to be employed in the field of domestication, it would therefore be easier for farmers to put their understanding of wild kinds to use and test them for domestication. Different sub-disciplines like the mixed-crop livestock farming, horticultural practices also should be parallel encouraged and compliance with AI-enabled technologies besides conventional agricultural practices, for obtaining optimum products and services from such important practices. Only those aspects of agriculture where profit is the primary goal have seen AI development and engagement. Except for greenhouses, where they are subjected to harsh circumstances, IoT devices are used in open settings in the majority of agricultural sectors. Because external factors like rain, high temperatures, humidity, and strong winds may have an impact on their operation, safety devices are necessary for IoT hardware (Farooq et al., 2020).

The use of IoT in agriculture should reduce network and communication outages, and hacking of host assets, farm data, and agricultural data. Due to the scattered network of sensor nodes used by the Internet of Things, which makes a single security protocol inappropriate, it is imperative to plan for data loss (Paul et al., 2020). The most typical inquiries we receive from our clients are listed below. But, it is essential to have a system that utilizes and integrates these technologies into every facet of farm management. Governments should thus be willing to fund agricultural research for the benefit of their inhabitants, and it is important to encourage research and development in this particular area.

CONCLUSION

To meet the world's food needs, precision agriculture is becoming very essential. IoT gathers crucial data from the farms with the aid of nano-sensors and transmits it over a secure network to the cloud application. IoT, AI, and ML are used to analyze the acquired data to seek any discrepancies and utilize them as a basis for decision-making. Different applications, including a disease prediction alarm system, aid in reducing farmers' uncertainty. Agriculture 5.0, a new set of frameworks for smart agriculture, is built on top of it. This study has shown the significance of using cutting-edge computer technologies, in particular nanotechnology, AI, and IoT, in the agriculture industry. The effective technological interruptions have enabled for causing an exponential rise in the output efficiency of the work performed as compared to the human-involved sytems, which typically is involved with several erroneous aspects causing output ineffieciency. Nonetheless, it is believed that further study and advancements are required to solve the limits of these smart technologies in advanced farming. The delayed and continuous release of the active ingredients made possible by nanoencapsulation can help enhance the application of herbicides by enhancing penetration. Hence, nanotechnology can offer an eco-friendly, effective, and green solution for managing insect pests in agriculture. In the forthcoming era of agricultural mechanization, the latest nanotechnologies tools and procedures can enhance the way agriculture is seen and have a bright future. Nanoparticles have a tremendous deal of promise as "magic bullets" that target particular plant tissues to discharge their charge to the targeted region of the plant to produce the intended effects. These particles might be loaded with herbicides, fungicides, nutrients, fertilizers, or nucleic acids. Modern agricultural methods like precise nitrogen and fertilizer administration and early disease identification have a bright and optimistic future thanks to biotechnological advancements and the quicker and more accurate diagnosis tools made possible by the use of nanomaterials.

ACKNOWLEDGMENT

Authors thank the Department of Biotechnology, Sathyabama Institute of Science and Technology, Chennai, and the International Research Centre, Centre for Nanoscience and Nanotechnology, Sathyabama Institute of Science and Technology, Chennai for carrying out the work.

Conflict of Interest: The authors declare that they have no competing interests of any kind among themselves.

REFERENCES

Akyildiz, I. F., & Kasimoglu, I. H. (2004). Wireless sensor and actor networks: Research challenges. *Ad Hoc Networks*, *2*(4), 351–367. doi:10.1016/j.adhoc.2004.04.003

An, C., Sun, C., Li, N., Huang, B., Jiang, J., Shen, Y., Wang, C., Zhao, X., Cui, B., Wang, C., Li, X., Zhan, S., Gao, F., Zeng, Z., Cui, H., & Wang, Y. (2022). Nanomaterials and nanotechnology for the delivery of agrochemicals: Strategies towards sustainable agriculture. *Journal of Nanobiotechnology*, *20*(1), 1–19. doi:10.118612951-021-01214-7 PMID:34983545

Ang, J. B. (2019). Agricultural legacy and individualistic culture. *Journal of Economic Growth*, *24*(4), 397–425. doi:10.100710887-019-09171-5

Ayaz, M., Ammad-Uddin, M., Sharif, Z., Mansour, A., & Aggoune, E.-H. M. (2019). Internet-of-Things (IoT)-based smart agriculture: Toward making the fields talk. *IEEE Access : Practical Innovations, Open Solutions*, *7*, 129551–129583. doi:10.1109/ACCESS.2019.2932609

Balafoutis, A. T., Van Evert, F. K., & Fountas, S. (2020). Smart farming technology trends: Economic and environmental effects, labor impact, and adoption readiness. *Agronomy (Basel)*, *10*(5), 743. doi:10.3390/agronomy10050743

Banaseka, F. K., Katsriku, F., Abdulai, J. D., Adu-Manu, K. S., & Engmann, F. N. A. (2021). Signal propagation models in soil medium for the study of wireless underground sensor networks: A review of current trends. *Wireless Communications and Mobile Computing*, *2021*, 1–12. doi:10.1155/2021/8836426

Bannerjee, G., Sarkar, U., Das, S., & Ghosh, I. (2018). Artificial intelligence in agriculture: A literature survey. *International Journal of Scientific Research in Computer Science Applications and Management Studies*, *7*(3), 1–6.

Bennett, E., Carpenter, S., Gordon, L., Ramankutty, N., Balvanera, P., Campbell, B., Cramer, W., Foley, J., Folke, C., & Karlberg, L. (2014). Toward a more resilient agriculture. *Solutions*, *5*(5), 65–75.

Bharati, S., & Mondal, M. R. H. (2021). 12 Applications and challenges of AI-driven IoHT for combating pandemics: A review. *Computational Intelligence for Managing Pandemics*, *5*, 213–230. doi:10.1515/9783110712254-012

Bhatnagar, V., Poonia, R. C., & Sunda, S. (2019). State of the art and gap analysis of precision agriculture: A case study of Indian farmers. [IJAEIS]. *International Journal of Agricultural and Environmental Information Systems*, *10*(3), 72–92. doi:10.4018/IJAEIS.2019070105

Bibri, S. E. (2018). The IoT for smart sustainable cities of the future: An analytical framework for sensor-based big data applications for environmental sustainability. *Sustainable Cities and Society*, *38*, 230–253. doi:10.1016/j.scs.2017.12.034

Bouzembrak, Y., Klüche, M., Gavai, A., & Marvin, H. J. P. (2019). Internet of Things in food safety: Literature review and a bibliometric analysis. *Trends in Food Science & Technology*, *94*, 54–64. doi:10.1016/j.tifs.2019.11.002

Brillante, L., Mathieu, O., Bois, B., Van Leeuwen, C., & Lévêque, J. (2015). The use of soil electrical resistivity to monitor plant and soil water relationships in vineyards. *Soil (Göttingen)*, *1*(1), 273–286. doi:10.5194oil-1-273-2015

Ciocca, F., Lunati, I., Van de Giesen, N., & Parlange, M. B. (2012). Heated optical fiber for distributed soil-moisture measurements: A lysimeter experiment. *Vadose Zone Journal*, *11*(4), vzj2011.0199. doi:10.2136/vzj2011.0199

Corradini, E., De Moura, M. R., & Mattoso, L. H. C. (2010). A preliminary study of the incorparation of NPK fertilizer into chitosan nanoparticles. *Express Polymer Letters*, *4*(8), 509–515. doi:10.3144/expresspolymlett.2010.64

Crosson, P. R. (2016). Sustainable agriculture. In *Global development and the environment* (pp. 61–68). Routledge.

DeRosa, M. C., Monreal, C., Schnitzer, M., Walsh, R., & Sultan, Y. (2010). Nanotechnology in fertilizers. *Nature Nanotechnology*, *5*(2), 91. doi:10.1038/nnano.2010.2 PMID:20130583

Facchinetti, D., Santoro, S., Galli, L. E., Fontana, G., Fedeli, L., Parisi, S., Bonacchi, L. B., Šušnjar, S., Salvai, F., Coppola, G., Matteucci, M., & Pessina, D. (2021). Reduction of pesticide use in fresh-cut salad production through artificial intelligence. *Applied Sciences (Basel, Switzerland)*, *11*(5), 1992. doi:10.3390/app11051992

Fahad, M., Javid, T., Beenish, H., Siddiqui, A. A., & Ahmed, G. (2021). Extending ONTAgri with Service-Oriented Architecture towards Precision Farming Application. *Sustainability (Basel)*, *13*(17), 9801. doi:10.3390u13179801

Farooq, M. S., Riaz, S., Abid, A., Umer, T., & Zikria, Y. (2020). Role of IoT technology in agriculture: A systematic literature review. *Electronics (Basel)*, *9*(2), 319. doi:10.3390/electronics9020319

Gagliardi, G., Lupia, M., Cario, G., Cicchello Gaccio, F., D'Angelo, V., Cosma, A. I. M., & Casavola, A. (2021). An internet of things solution for smart agriculture. *Agronomy (Basel)*, *11*(11), 2140. doi:10.3390/agronomy11112140

Gupta, M., Abdelsalam, M., Khorsandroo, S., & Mittal, S. (2020). Security and privacy in smart farming: Challenges and opportunities. *IEEE Access : Practical Innovations, Open Solutions*, *8*, 34564–34584. doi:10.1109/ACCESS.2020.2975142

Gutiérrez, J., Villa-Medina, J. F., Nieto-Garibay, A., & Porta-Gándara, M. Á. (2013). Automated irrigation system using a wireless sensor network and GPRS module. *IEEE Transactions on Instrumentation and Measurement*, *63*(1), 166–176. doi:10.1109/TIM.2013.2276487

Harris, D. R., & Fuller, D. Q. (2014). Agriculture: definition and overview. Encyclopedia of Global Archaeology, 104–113. Springer

Hatti, V., Raghavendra, M., Singh, Y. V., Sharma, S. K., Halli, H. M., & Goud, B. R. (2020). Application of nano technology in crop production to enhance resource use efficiency with special reference to soil and water conservation: A review. *Journal of Soil and Water Conservation*, *19*(4), 375–381. doi:10.5958/2455-7145.2020.00050.8

Jha, K., Doshi, A., Patel, P., & Shah, M. (2019). A comprehensive review on automation in agriculture using artificial intelligence. *Artificial Intelligence in Agriculture*, *2*, 1–12. doi:10.1016/j.aiia.2019.05.004

Kumar, A., Lee, B. G., Lee, H., & Kumari, A. (2012). Secure storage and access of data in cloud computing. *2012 International Conference on ICT Convergence (ICTC)*, 336–339. 10.1109/ICTC.2012.6386854

Kumar, A. (2012). Citrus greening disease detection using aerial hyperspectral and multispectral imaging techniques. *Journal of Applied Remote Sensing*, *6*(1), 63542. doi:10.1117/1.JRS.6.063542

Kumar, R., Singh, M. P., Kumar, P., & Singh, J. P. (2015). Crop Selection Method to maximize crop yield rate using machine learning technique. *2015 International Conference on Smart Technologies and Management for Computing, Communication, Controls, Energy and Materials (ICSTM)*, (pp. 138–145). IEEE. 10.1109/ICSTM.2015.7225403

Lakhiar, I. A., Jianmin, G., Syed, T. N., Chandio, F. A., Buttar, N. A., & Qureshi, W. A. (2018). Monitoring and control systems in agriculture using intelligent sensor techniques: A review of the aeroponic system. *Journal of Sensors*, *2018*, 2018. doi:10.1155/2018/8672769

Lee, C.-L., Strong, R., & Dooley, K. E. (2021). Analyzing precision agriculture adoption across the globe: A systematic review of scholarship from 1999–2020. *Sustainability (Basel)*, *13*(18), 10295. doi:10.3390u131810295

Lencsés, E., Takács, I., & Takács-György, K. (2014). Farmers' perception of precision farming technology among Hungarian farmers. *Sustainability (Basel)*, *6*(12), 8452–8465. doi:10.3390u6128452

Li, H., Lee, W. S., Wang, K., Ehsani, R., & Yang, C. (2014). 'Extended spectral angle mapping (ESAM)' for citrus greening disease detection using airborne hyperspectral imaging. *Precision Agriculture*, *15*(2), 162–183. doi:10.100711119-013-9325-6

Lin, K., & Hao, T. (2020). Link quality analysis of wireless sensor networks for underground infrastructure monitoring: A non-backfilled scenario. *IEEE Sensors Journal*, *21*(5), 7006–7014. doi:10.1109/JSEN.2020.3041644

Lu, J., Zhou, M., Gao, Y., & Jiang, H. (2018). Using hyperspectral imaging to discriminate yellow leaf curl disease in tomato leaves. *Precision Agriculture, 19*(3), 379–394. doi:10.100711119-017-9524-7

MacDonald, S. L., Staid, M., Staid, M., & Cooper, M. L. (2016). Remote hyperspectral imaging of grapevine leafroll-associated virus 3 in cabernet sauvignon vineyards. *Computers and Electronics in Agriculture, 130*, 109–117. doi:10.1016/j.compag.2016.10.003

Mahmood, I., Imadi, S. R., Shazadi, K., Gul, A., & Hakeem, K. R. (2016). Effects of pesticides on environment. In *Plant, soil and microbes* (pp. 253–269). Springer. doi:10.1007/978-3-319-27455-3_13

Mandal, S. N., & Sinha, P. (2021). Pest and Disease Monitoring and Surveillance for Ensuring Food Security. In *Innovations in Agriculture for a Self-Reliant India* (pp. 293–304). CRC Press. doi:10.1201/9781003245384-19

Mehra, M., Saxena, S., Sankaranarayanan, S., Tom, R. J., & Veeramanikandan, M. (2018). IoT based hydroponics system using Deep Neural Networks. *Computers and Electronics in Agriculture, 155*, 473–486. doi:10.1016/j.compag.2018.10.015

Mendes, J., Pinho, T. M., Neves dos Santos, F., Sousa, J. J., Peres, E., Boaventura-Cunha, J., Cunha, M., & Morais, R. (2020). Smartphone applications targeting precision agriculture practices—A systematic review. *Agronomy (Basel), 10*(6), 855. doi:10.3390/agronomy10060855

Mohanraj, I., Ashokumar, K., & Naren, J. (2016). Field monitoring and automation using IOT in agriculture domain. *Procedia Computer Science, 93*, 931–939. doi:10.1016/j.procs.2016.07.275

Mulla, D. J. (n.d.). *Precision Agriculture: Key Advances and Remaining Knowledge Gaps*. ISPA.

Mutschler, P. (2018). *Threats to Precision Agriculture*. Research Gate.

Partel, V., Kakarla, S. C., & Ampatzidis, Y. (2019). Development and evaluation of a low-cost and smart technology for precision weed management utilizing artificial intelligence. *Computers and Electronics in Agriculture, 157*, 339–350. doi:10.1016/j.compag.2018.12.048

Patrício, D. I., & Rieder, R. (2018). Computer vision and artificial intelligence in precision agriculture for grain crops: A systematic review. *Computers and Electronics in Agriculture, 153*, 69–81. doi:10.1016/j.compag.2018.08.001

Paul, P., & Saavedra, M. (2020). Agro informatics vis-à-vis Internet of Things (IoT) integration & potentialities—An analysis. *Agro Economist-An International Journal, 7*(1), 13–20. doi:10.30954/2394-8159.01.2020.2

Podder, P., Mondal, M., Bharati, S., & Paul, P. K. (2021). *Review on the security threats of internet of things*. ArXiv Preprint ArXiv:2101.05614.

Poonia, R. C., Gao, X.-Z., Raja, L., Sharma, S., & Vyas, S. (2018). *Smart Farming Technologies for Sustainable Agricultural Development*. IGI Global.

Popa, A., Hnatiuc, M., Paun, M., Geman, O., Hemanth, D. J., Dorcea, D., Son, L. H., & Ghita, S. (2019). An intelligent IoT-based food quality monitoring approach using low-cost sensors. *Symmetry, 11*(3), 374. doi:10.3390ym11030374

Rai, V., Acharya, S., & Dey, N. (2012). *Implications of nanobiosensors in agriculture.*

Raut, P., & Ghare, P. H. (2020). Analysis of wireless channel parameters for the different types of soil in the wusns. *2020 IEEE International Students' Conference on Electrical, Electronics and Computer Science (SCEECS)*, (pp. 1–4). IEEE.

Sadeghi, S., Soltanmohammadlou, N., & Nasirzadeh, F. (2022). Applications of wireless sensor networks to improve occupational safety and health in underground mines. *Journal of Safety Research*, *83*, 8–25. doi:10.1016/j.jsr.2022.07.016 PMID:36481040

Selmani, A., Oubehar, H., Outanoute, M., Ed-Dahhak, A., Guerbaoui, M., Lachhab, A., & Bouchikhi, B. (2019). Agricultural cyber-physical system enabled for remote management of solar-powered precision irrigation. *Biosystems Engineering*, *177*, 18–30. doi:10.1016/j.biosystemseng.2018.06.007

Senthil Kumar, P., & Janet Joshiba, G. (2019). Water footprint of agricultural products. In *Environmental Water Footprints* (pp. 1–19). Springer.

Singh, P., Singh, R. P., & Singh, Y. (2022). An optimal energy-throughput efficient cross-layer solution using naked mole rat algorithm for wireless underground sensor networks. *Materials Today: Proceedings*, *48*, 1076–1083. doi:10.1016/j.matpr.2021.07.307

Talaviya, T., Shah, D., Patel, N., Yagnik, H., & Shah, M. (2020). Implementation of artificial intelligence in agriculture for optimisation of irrigation and application of pesticides and herbicides. *Artificial Intelligence in Agriculture*, *4*, 58–73. doi:10.1016/j.aiia.2020.04.002

Tapan, A., Biswas, A. K., & Kundu, S. (2010). Nano-fertiliser-a new dimension in agriculture. *Indian Journal of Fertilisers*, *6*(8), 22–24.

Tarafdar, J. C., Agrawal, A., Raliya, R., Kumar, P., Burman, U., & Kaul, R. K. (2012). ZnO nanoparticles induced synthesis of polysaccharides and phosphatases by Aspergillus fungi. *Advanced Science, Engineering and Medicine*, *4*(4), 324–328. doi:10.1166/asem.2012.1160

Tarafdar, J. C., Raliya, R., & Rathore, I. (2012). Microbial synthesis of phosphorous nanoparticle from tri-calcium phosphate using Aspergillus tubingensis TFR-5. *Journal of Bionanoscience*, *6*(2), 84–89. doi:10.1166/jbns.2012.1077

Tarafdar, J. C., Xiong, Y., Wang, W.-N., Quinl, D., & Biswas, P. (2012). Standardization of size, shape and concentration of nanoparticle for plant application. *Applied Biological Research*, *14*(2), 138–144.

Torbick, N., Chowdhury, D., Salas, W., & Qi, J. (2017). Monitoring rice agriculture across myanmar using time series Sentinel-1 assisted by Landsat-8 and PALSAR-2. *Remote Sensing (Basel)*, *9*(2), 119. doi:10.3390/rs9020119

Udomkun, P., Nagle, M., Argyropoulos, D., Mahayothee, B., & Müller, J. (2016). Multi-sensor approach to improve optical monitoring of papaya shrinkage during drying. *Journal of Food Engineering*, *189*, 82–89. doi:10.1016/j.jfoodeng.2016.05.014

Van Erp-van der Kooij, E., & Rutter, S. M. (2020). Using precision farming to improve animal welfare. *Perspectives in Agriculture, Veterinary Science, Nutrition and Natural Resources*, *15*, 1–10. doi:10.1079/PAVSNNR202015051

Vijayakumar, P. S., Abhilash, O. U., Khan, B. M., & Prasad, B. L. V. (2010). Nanogold-loaded sharp-edged carbon bullets as plant-gene carriers. *Advanced Functional Materials*, *20*(15), 2416–2423. doi:10.1002/adfm.200901883

Vincent, D. R., Deepa, N., Elavarasan, D., Srinivasan, K., Chauhdary, S. H., & Iwendi, C. (2019). Sensors driven ai-based agriculture recommendation model for assessing land suitability. *Sensors (Basel)*, *19*(17), 3667. doi:10.339019173667 PMID:31450772

Wang, Z., Walsh, K. B., & Verma, B. (2017). On-tree mango fruit size estimation using RGB-D images. *Sensors (Basel)*, *17*(12), 2738. doi:10.339017122738 PMID:29182534

Wilson, M. A., Tran, N. H., Milev, A. S., Kannangara, G. S. K., Volk, H., & Lu, G. Q. M. (2008). Nanomaterials in soils. *Geoderma*, *146*(1–2), 291–302. doi:10.1016/j.geoderma.2008.06.004

Yang, C. (2020). Remote sensing and precision agriculture technologies for crop disease detection and management with a practical application example. *Engineering (Beijing)*, *6*(5), 528–532. doi:10.1016/j.eng.2019.10.015

Zambon, I., Cecchini, M., Egidi, G., Saporito, M. G., & Colantoni, A. (2019). Revolution 4.0: Industry vs. agriculture in a future development for SMEs. *Processes (Basel, Switzerland)*, *7*(1), 36. doi:10.3390/pr7010036

Zha, J. (2020). Artificial intelligence in agriculture. *Journal of Physics: Conference Series*, *1693*(1), 12058. doi:10.1088/1742-6596/1693/1/012058

Zhang, B., Han, C., & Yu, X. B. (2015). A non-destructive method to measure the thermal properties of frozen soils during phase transition. *Journal of Rock Mechanics and Geotechnical Engineering*, *7*(2), 155–162. doi:10.1016/j.jrmge.2015.03.005

Zhang, Z., Yu, X., Wu, P., & Han, W. (2015). Survey on water-saving agricultural internet of things based on wireless sensor network. *International Journal of Control and Automation*, *8*(4), 229–240. doi:10.14257/ijca.2015.8.4.23

Chapter 9
Internet of Things-Based Micro Climate Monitoring System for Irrigation Water Optimization

S. N. Kumar
Amal Jyothi College of Engineering, Kottayam, India

Andrew Thomas Jacob
Amal Jyothi College of Engineering, Kottayam, India

Amal Varghese
Amal Jyothi College of Engineering, Kottayam, India

Emma Francis
Amal Jyothi College of Engineering, Kottayam, India

Mibin P. Sabu
Amal Jyothi College of Engineering, Kottayam, India

Neenu Rose Antony
Amal Jyothi College of Engineering, Kottayam, India

ABSTRACT

Precision farming is a relatively new concept that has recently gained popularity. It is based on the idea of using technology to gather and analyze data on various factors that affect crop growth and yield. By analyzing data in real time and responding quickly to changes in the environment, precision farming can help farmers optimize the conditions of their fields and maximize the potential of their crops, leading to increased yields and more efficient use of resources. This research work put forward a study focusing on the role of IoT in smart agriculture. Precision farming offers many benefits, including increased efficiency and accuracy, improved data and analytics, reduced environmental impact, improved compliance with regulations, improved communication with stakeholders, and increased profitability for farmers.

DOI: 10.4018/978-1-6684-7879-0.ch009

INTRODUCTION

Precision farming is a modern agricultural technique that uses technology to optimize crop yields and reduce waste (Chung et al., 2016). It involves the use of precision equipment and data analysis to monitor and manage factors such as soil conditions, weather patterns, and crop growth in real-time. It enables farmers to make more informed decisions about planting, fertilization, irrigation, and pest control, which results in higher crop yields and more efficient resource use. Some examples of precision farming technologies include GPS-controlled tractors, drones, and sensors that monitor soil moisture and temperature. Precision farming technologies include a wide range of tools and systems. Some examples include:

- GPS-controlled tractors and other farm equipment, which can help farmers plant and harvest crops more accurately and efficiently.
- Drones, which can be used to gather data on crop growth and health, as well as to spray pesticides and fertilizers.
- Sensors, which can be used to monitor soil moisture, temperature, and other factors that affect crop growth.

The precision agriculture paradigm is depicted in Figure 1.

Figure 1. Precision agriculture paradigm

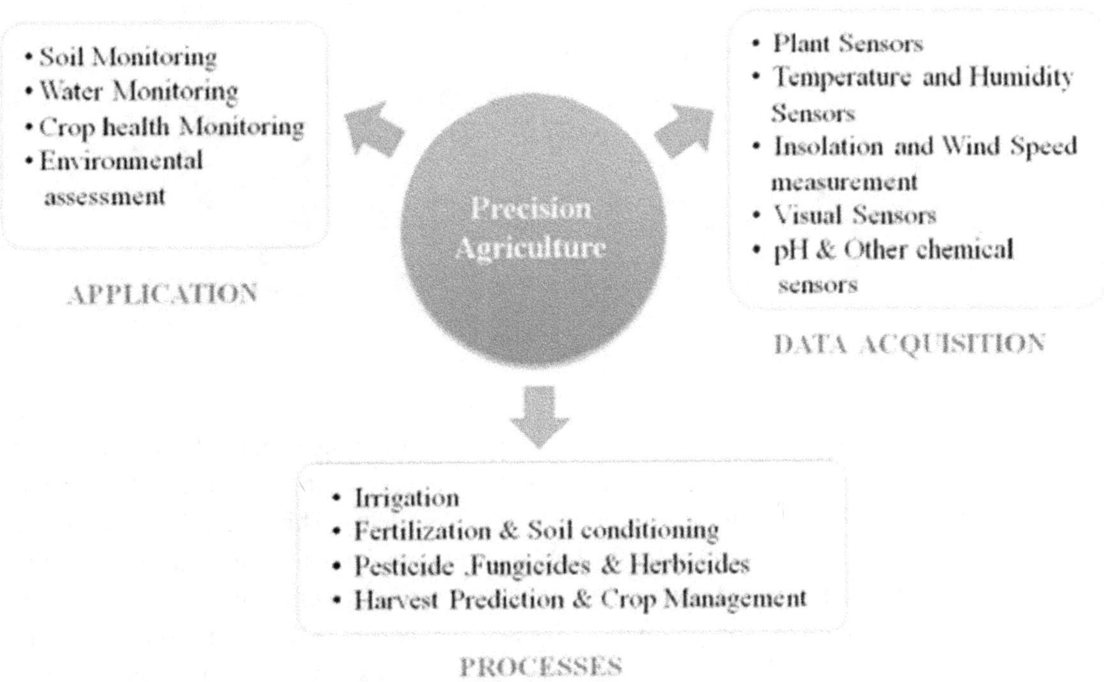

FPGA-based automatic monitoring of parameters in the agriculture field was proposed in Dinesh et al. (2011). Precision irrigation systems can help farmers use water more efficiently by applying it only when and where it is needed. Yield monitoring systems can help farmers track the performance of different varieties of crops and identify areas of the field where yields are low (Mendelsohn, 2007). These systems can track the daily average yields of different crop varieties and compare them with the average yields of the farm to identify the highest-performing varieties. For example, by using precision irrigation systems, farmers can save water and energy, which can reduce their utility bills. Drones and sensors allow farmers to monitor the health of their crops in real-time and detect issues such as pest infestations, water stress, nutrient deficiencies, or disease. By addressing these problems early on, farmers can prevent them from becoming major issues and potentially save the entire crop (Fageria, 1998). This can help them save money on expensive inputs, such as pesticides and fertilizers. The cloud-based management in smart agriculture is depicted in Figure 2.

Figure 2. Cloud-based management in smart agriculture

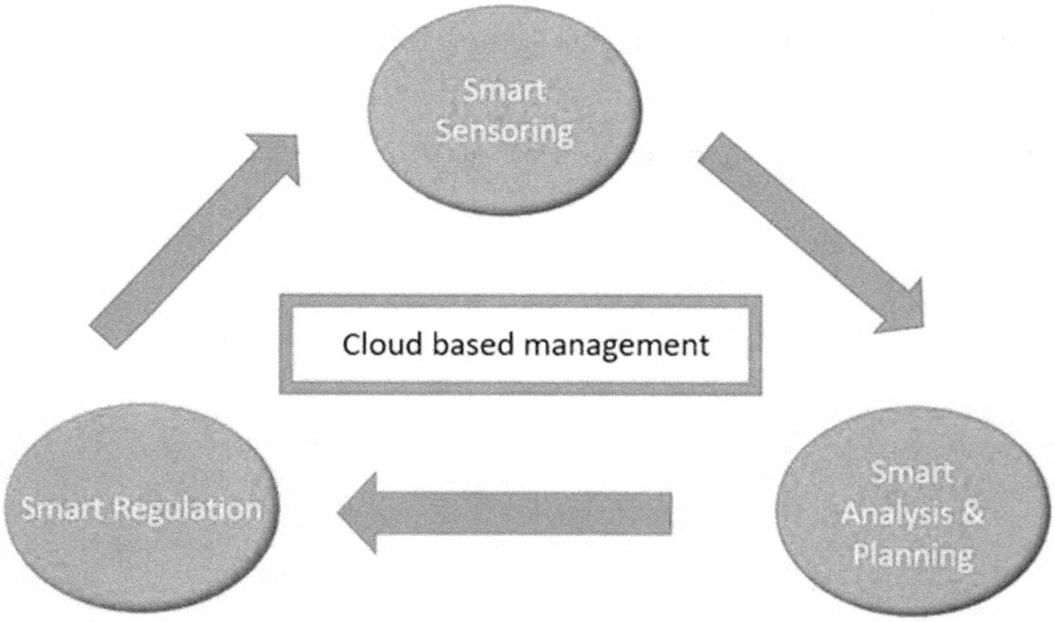

Precision farming also allows farmers to be more responsive to changes in market conditions (Bongiovanni et al., 2004). By using data and analytics, farmers can make adjustments to their crop production in real-time to meet changing demand for different crops or products. This can help them to increase their revenue and improve their bottom line. A further benefit of precision farming is the improvement of crop yields and the reduction of crop failure risks, which can assist farmers in increasing food security. By using technology to optimize crop production, farmers can ensure that they are producing enough food to meet the needs of a growing global population. They can also make sure that they are producing crops that are resilient to changing weather patterns and other environmental factors. The different sectored contribution in India is depicted in Figure 3.

Figure 3. Sectored contribution to economy in India

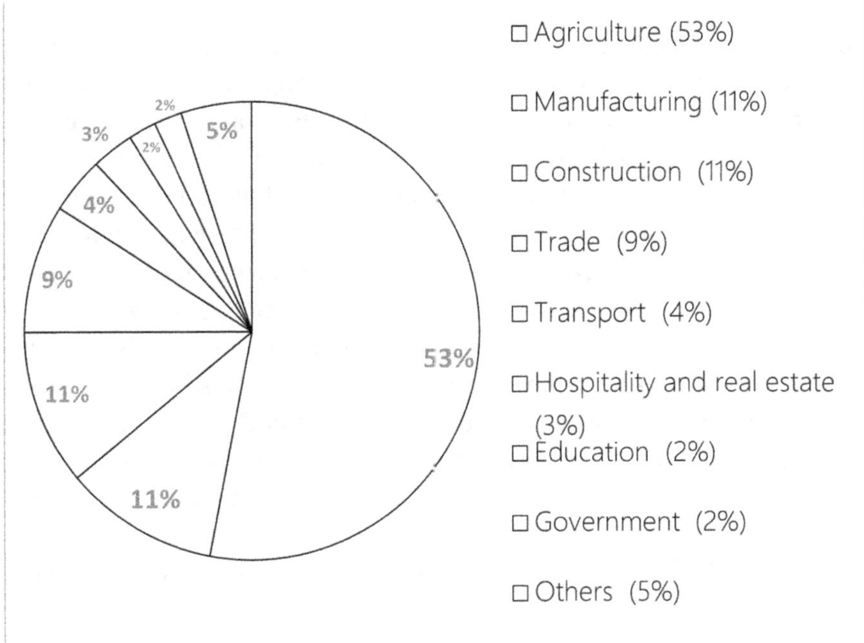

These precision farming techniques can also help farmers to improve the sustainability of their operations. Optimizing crop production with technology can reduce the environmental footprint and make farming more sustainable (Farooq et al., 2019). For example, Precision irrigation and fertilization systems can reduce water and fertilizer usage while also protecting soil health. Precision irrigation systems use sensors to measure the moisture content of the soil and only water the plants when the soil needs it. This ensures that the plants get the right amount of water without any waste, which helps conserve water resources. Precision fertilization systems allow farmers to apply just the right amount of fertilizer needed for the crop. This minimizes the amount of fertilizer that is wasted or runs off into the environment. This can help prevent soil degradation, as well as reduce the number of pollutants that enter our waterways. Precision farming also allows farmers to make more accurate and precise decisions about their operations. The data and information collected by precision farming tools and systems can be used to create detailed maps and models of fields, which can help farmers to identify patterns and trends in crop growth and yields (Ikram et al., 2022). This information can be used to make more precise decisions about planting, fertilization, irrigation, and pest control. In addition, precision farming can reduce the amount of labour farmers need to manage their farms. By using GPS-controlled tractors and other equipment, farmers can automate many tasks, such as planting and harvesting, which can save time and labour costs (Zhang et al., 2012). Drones and other sensors can also help farmers to identify problems in their fields more quickly, which can help them to take action more efficiently (Zhang et al., 2012). It is also possible to improve the quality of crops by using precision farming. Monitoring crop growth and health with technology can help farmers identify problems early and address them before they become serious. This can help to improve crop yields and also improve the quality of crops, which can increase the value of the crops to the farmers.

ROLE OF IoT IN AGRICULTURE

IoT, or the Internet of Things, is the network of physical devices, vehicles, and other items embedded with sensors, software, and connectivity which enables these objects to collect and exchange data. In agriculture, IoT can be used to improve crop yields, reduce water usage, and optimize the use of fertilizers and pesticides (Mat et al., 2018). IoT-enabled sensors can be placed in the field to monitor soil moisture, temperature, and nutrient levels, and to detect pests and diseases (Farooq et al., 2022). This data can be used to optimize irrigation and fertilization, and to take preventative measures to avoid crop loss. Additionally, IoT can be used to track and optimize the movement of livestock and to monitor the health and wellbeing of animals. IoT in agriculture can also be used to improve the efficiency of farming operations. For example, IoT-enabled precision agriculture systems can be used to guide tractors and other farm equipment, allowing for more accurate planting, harvesting, and other tasks. This can reduce the amount of fuel used, and can also reduce the amount of damage to the soil caused by heavy machinery. Another application of IoT in agriculture is in the use of drones and other unmanned aerial vehicles (UAVs). These devices can be equipped with sensors, cameras, and other equipment to collect data on crop growth, soil conditions, and weather patterns (Attaran et al., 2016). This data can be used to make more accurate predictions about crop yields and to identify areas of the field that may need more attention.

As a whole, IoT has the potential to improve efficiency and productivity in agriculture, while also reducing costs and environmental impact. (Verdouw et al., 2016). It can help farmers to make better decisions, increase yields, and produce high-quality crops. IoT in agriculture can also be used to improve the traceability of food products. IoT-enabled sensors and RFID (Radio-Frequency Identification) tags can be used to track food products from the farm to the consumer. This allows for greater transparency in the food supply chain, which can improve food safety and can also enable farmers to improve their marketing efforts. Furthermore, IoT in agriculture can help farmers to better manage their resources and their carbon footprint. For example, by monitoring soil moisture and weather patterns, farmers can reduce water usage and avoid over-irrigation, which can save money and conserve water resources.

IoT can provide farmers with a wealth of data and insights that can be used to improve their operations and make better decisions. For example, by analyzing data on crop growth and weather patterns, farmers can identify which varieties of crops perform best in different conditions and can make more informed decisions about when to plant and harvest (Manickam, 2020). As a result of IoT, the agricultural industry has the potential to improve productivity, reduce costs, and increase sustainability (Kim et al., 2020). It can enable farmers to make more informed decisions and to produce higher-quality food products more efficiently. Another benefit of IoT in agriculture is in the field of Livestock management, IoT technology can be used to monitor the health and wellbeing of animals. This can be done by using sensors to track the movement, temperature, and eating habits of the animals. This can enable farmers to identify and respond to health issues more quickly, which can reduce the risk of disease and increase the overall productivity of the herd. IoT can also be used to automate the feeding and watering of livestock, which can improve the efficiency of the operation and reduce labor costs. By creating a digital farming landscape that is tailored to the needs of individual farmers, precision farming can help farmers take full advantage of their resources, enabling them to optimize their production and reduce their costs. For example, Through IoT-enabled systems, fertilizers and pesticides can be used efficiently, resulting in reduced environmental impact and reduced resource consumption. It is clear that IoT has enormous potential in the agricultural industry, as it can increase efficiency and productivity, reduce costs, and have a global impact on the environment. Farmers can utilize technology to collect real-time data and insights that can help them make more informed decisions, as well as produce high-quality, safe food products on a larger scale. By utilizing the latest technology, farmers can create a more efficient and sustainable agricultural system that is beneficial for both their businesses and the environment.

IoT (Internet of Things) based soil moisture and NPK (Nitrogen, Phosphorus, Potassium) detection refers to the use of IoT devices to monitor and measure the moisture and nutrient levels in soil. This can be accomplished through the use of sensors that measure various soil parameters such as moisture content, pH level, and nutrient levels (Zotarelli et al., 2010). The data collected by these sensors can then be transmitted to a remote location for analysis and monitoring (Namani et al., 2020). This technology can be used in agriculture to optimize irrigation and fertilization practices, leading to improved crop yields and reduced water and fertilizer usage. There are several ways that IoT-based soil moisture and NPK detection can be implemented in agriculture (Gondchawar et al., 2016). By optimizing irrigation schedules and reducing water usage with soil moisture sensors, farmers can make sure their soil is always at its best to produce healthy crops. With pH sensors, farmers can quickly and precisely measure the acidity or alkalinity of their soil, and thereby adjust it accordingly to ensure the best possible yield of their crops. Measurement of soil nitrogen, phosphorus, and potassium levels using nutrient sensors. By accurately measuring the nutrient levels in the soil, farmers can be sure that they are not over-applying or under-applying fertilizers. This will not only reduce costs, but it will also prevent potential environmental problems that can arise from over-fertilization (Prathibha et al., 2017).

Remote monitoring of the data collected by the sensors in real-time through a wireless network and cloud-based platform, this allows farmers and agronomists to access the data from anywhere and make decisions accordingly. The data collected by these sensors can be analyzed using machine learning algorithms to identify patterns and trends, which can then be used to make predictions about future crop yields and soil conditions. Overall, IoT-based soil moisture and NPK detection can help farmers and agronomists to optimize their irrigation and fertilization practices, leading to improved crop yields and reduced water and fertilizer usage (Podder et al., 2021). One of the main benefits of IoT-based soil moisture and NPK detection is the ability to monitor and measure soil conditions in real-time. This allows farmers and agronomists to quickly respond to changes in soil conditions, such as dry spells or heavy rains, which can have a major impact on crop yields. Optimization technique such as genetic algorithm is utilized in (Lejdel, 2022) for efficient utilization of energy in agriculture land.

An added benefit of IoT-based soil moisture and NPK detection is the ability to optimize irrigation and fertilization practices. By using sensors to measure soil moisture and nutrient levels, farmers can ensure that they are applying just the right amount of water and fertilizer to their crops (Jain A et al, 2020). This helps to reduce costs, as well as protect the environment by reducing runoff and preventing unnecessary fertilizer application. Additionally, IoT-based soil moisture and NPK detection can also be used to improve crop yields by providing a more precise way of measuring soil conditions. This can help farmers identify specific areas of the field that may require additional attention and resources, such as irrigation or fertilization. Moreover, IoT-based soil moisture and NPK detection can also be used to identify patterns and trends in soil conditions over time (Suanpang et al., 2019). By analyzing the data collected by the sensors, farmers and agronomists can gain a better understanding of how different soil conditions impact crop yields. This knowledge can be used to make predictions about future crop yields and soil conditions, which can help farmers to make more informed decisions about their irrigation and fertilization practices. A soil moisture and NPK detection system based on IoT can help farmers and agronomists improve crop yields, reduce water and fertilizer use, and help them make more informed fertilization and irrigation decisions. Intelligent techniques are coupled with the IoT for proficient utilization of energy for improving the yield of crops (Ashifuddin Mondal et al., 2018).

IoT-based soil moisture and NPK detection can also be used in precision agriculture. Precision agriculture relies on sensors, drones, and other technologies to monitor soil conditions and track weather patterns, allowing farmers to make data-driven decisions about when and how to apply resources. This allows farmers to use resources more efficiently and reduce their environmental footprint, while still allowing them to maximize yields. By using IoT-based sensors to measure soil conditions in real-time, farmers can create detailed maps of their fields that show variations in soil moisture, pH, and nutrient levels. This information can then be used to create customized irrigation and fertilization plans for different parts of the field, allowing farmers to target specific areas that need more or less water or fertilizer (Dhanaraju et al., 2022; Suhag et al., 2021). Additionally, by using sensors to measure soil conditions and make predictions about future crop yields, farmers can make more informed decisions about when to plant, harvest, and sell their crops. This can help them to increase their profits by selling their crops at the right time and at the right price. IoT-based soil moisture and NPK detection can also be used in combination with other technologies, such as drones and robots, to create more efficient and effective farming systems (Rajakumar et al., 2018). For example, drones (Hasan et al., 2021) equipped with sensors can be used to take high-resolution images of fields, which can then be used to create detailed maps of soil conditions. Robots can then be used to perform tasks such as planting, harvesting, and monitoring crops. It can also be used in combination with other technologies to create more efficient and effective farming systems. The role of IoT in smart agriculture is depicted in Figure 4.

Figure 4. Role of IoT in smart agriculture

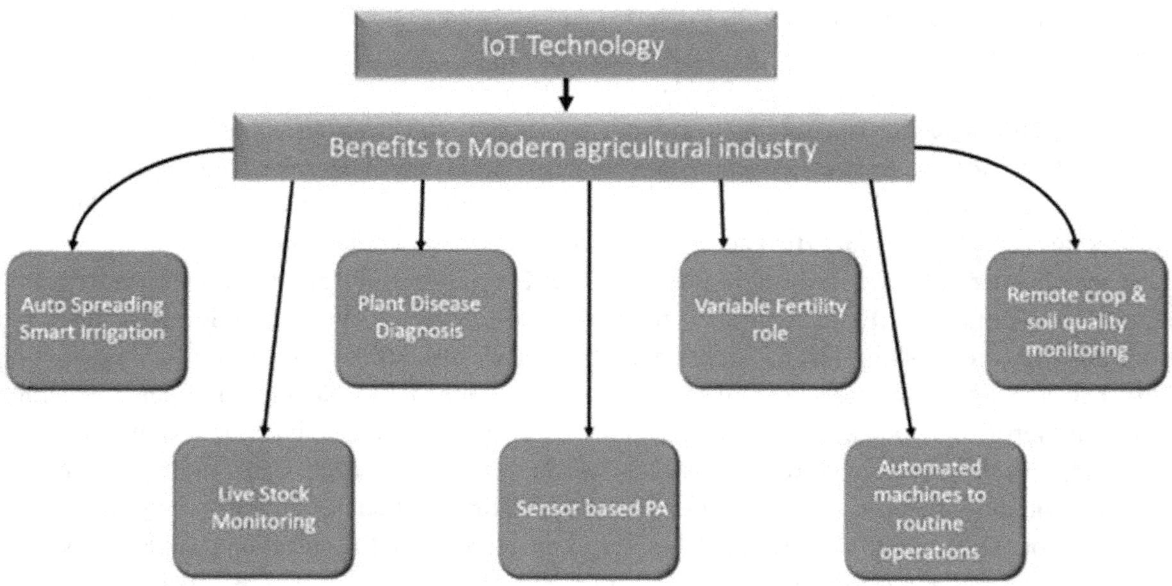

SMART GRID IN AGRICULTURE

A smart grid in agriculture refers to the use of advanced technology and communication systems to manage and optimize the distribution and use of energy on a farm. This includes using sensors and other devices to monitor energy use, and using that data to improve efficiency, reduce costs, and increase the overall productivity of the farm (Hirsch et al., 2019), capacitive soil sensor node gains prominence in generating good yield (Hirsch et al., 2019). Smart grids in agriculture can also include the integration of renewable energy sources, such as solar and wind power, to reduce the farm's dependence on fossil fuels (Khan et al., 2020). Additionally, it can also be integrated with precision agriculture technologies to automate farming operations, such as irrigation and crop management. Smart grids in agriculture can also include the use of advanced control systems to manage the distribution of energy throughout the farm. For example, a smart grid can be used to manage the distribution of electricity to different areas of the farm, ensuring that the power is directed where it is needed most. Smart grids can also be used to optimize the use of energy storage systems, such as batteries, to ensure that power is available when it is needed.

Another important aspect of smart grids in agriculture is the use of communication systems to connect different devices and systems on the farm. This allows for real-time monitoring and control of energy use, as well as the ability to remotely control and adjust systems. The implementation of smart grids in agriculture can also help to increase the resilience of the farm to power outages and other disruptions. For example, by integrating renewable energy sources, a farm can become less dependent on the traditional power grid, and by using advanced control systems, the farm can quickly respond to changes in energy demand. In general, smart grids in agriculture can help to increase the efficiency, productivity, and sustainability of farming operations, while also reducing costs and increasing the resilience of the farm. One of the key benefits of smart grids in agriculture is the ability to collect and analyze large amounts of data on energy use and farming operations. This data can be used to identify patterns and trends, and to make informed decisions about how to optimize energy use and improve overall farm performance. For example, data on energy use can be used to identify areas where energy is being wasted, and to implement changes to improve efficiency. The machine learning-based water optimization for irrigation is depicted in Figure 5. The IoT-based sensors are used to acquire the data and the machine learning algorithms are used to tune the parameters for the water delivery to the agriculture field.

Figure 5. Machine learning-based water optimization for irrigation

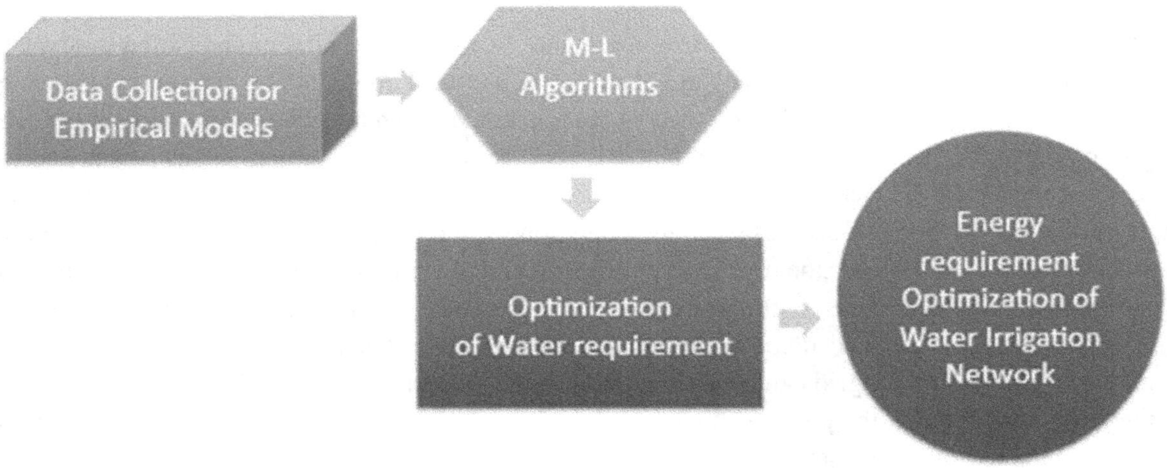

Additionally, smart grids can be used in agriculture as a means to integrate technologies and systems such as precision agriculture, renewable energy sources, and energy storage. This allows for a more holistic approach to managing energy use and farming operations and can lead to significant cost savings and improvements in overall farm performance. Smart grids in agriculture also have the potential to support the development of new business models and revenue streams, such as the sale of excess energy to the grid, or the use of data and analytics services to improve farm performance, energy efficient edge fog cloud architecture for smart agriculture was proposed in Alharbi et al. (2021). Moreover, smart grids in agriculture can also play a role in supporting the adoption of sustainable farming practices. For example, by integrating renewable energy sources and reducing the farm's dependence on fossil fuels, smart grids can help to reduce greenhouse gas emissions and support the transition to a more sustainable food system. Therefore, smart grids in agriculture can provide a wide range of benefits, including increased efficiency, productivity, and sustainability, as well as new business opportunities and support for sustainable farming practices.

CONCLUSION AND FUTURE WORKS

In summary, A precision farming system can also enhance farmers' competitiveness. By using technology to optimize crop production and reduce costs, farmers can produce more crops at a lower cost than traditional farming methods. This can help farmers to be more competitive in the global market and increase their revenues. It also offers many benefits, including increased efficiency and accuracy, reduced labour costs, improved crop quality, increased competitiveness, and improved profitability for farmers. Another important aspect of precision farming is the use of data and analytics. By collecting and analyzing data from various sources such as weather, soil, and crop growth, farmers can make more informed decisions about their operations. This data can be used to create detailed maps and models of fields, which can help farmers to identify patterns and trends in crop growth and yields. Additionally, with the use of machine learning algorithms, farmers can predict crop growth and yields, soil conditions, and weather patterns. This allows farmers to take proactive measures to improve crop yield and quality and to make more accurate predictions of crop yields and market prices.

REFERENCES

Alharbi, H. A., & Aldossary, M. (2021, July 30). Energy-efficient edge-fog-cloud architecture for IoT-based smart agriculture environment. *IEEE Access : Practical Innovations, Open Solutions*, *9*, 110480–110492. doi:10.1109/ACCESS.2021.3101397

Ashifuddin Mondal, M., & Rehena, Z. (2018). Iot based intelligent agriculture field monitoring system. In *2018 8th International Conference on Cloud Computing, Data Science & Engineering (Confluence)* (pp. 625-629). IEEE.

Attaran, A., & Rashidzadeh, R. (2016, July 11). Chipless radio frequency identification tag for IoT applications. *IEEE Internet of Things Journal*, *3*(6), 1310–1318. doi:10.1109/JIOT.2016.2589928

Bongiovanni, R., & Lowenberg-DeBoer, J. (2004, August). Precision agriculture and sustainability. *Precision Agriculture*, *5*(4), 359–387. doi:10.1023/B:PRAG.0000040806.39604.aa

Chung, S. O., Choi, M. C., Lee, K. H., Kim, Y. J., Hong, S. J., & Li, M. (2016). Sensing technologies for grain crop yield monitoring systems: A review. *Journal of Biosystems Engineering*, *41*(4), 408–417. doi:10.5307/JBE.2016.41.4.408

Dhanaraju, M., Chenniappan, P., Ramalingam, K., Pazhanivelan, S., & Kaliaperumal, R. (2022, October 21). Smart Farming: Internet of Things (IoT)-Based Sustainable Agriculture. *Agriculture*, *12*(10), 1745. doi:10.3390/agriculture12101745

Dinesh, M., & Saravanan, P. (2011, June). FPGA based real time monitoring system for agricultural field. *International Journal of Electronics and Computer Science Engineering.*, *1*(3), 1514–1519.

Fageria, N. K. (1998, January). Optimizing nutrient use efficiency in crop production. *Revista Brasileira de Engenharia Agrícola e Ambiental*, *2*, 6–16. doi:10.1590/1807-1929/agriambi.v02n01p6-16

Farooq, M. S., Riaz, S., Abid, A., Abid, K., & Naeem, M. A. (2019, October 25). A Survey on the Role of IoT in Agriculture for the Implementation of Smart Farming. *IEEE Access : Practical Innovations, Open Solutions*, *7*, 156237–156271. doi:10.1109/ACCESS.2019.2949703

Farooq, M. S., Sohail, O. O., Abid, A., & Rasheed, S. (2022, January 13). A survey on the role of iot in agriculture for the implementation of smart livestock environment. *IEEE Access : Practical Innovations, Open Solutions*, *10*, 9483–9505. doi:10.1109/ACCESS.2022.3142848

Gondchawar, N., & Kawitkar, R. S. (2016, June). IoT based smart agriculture. *International Journal of Advanced Research in Computer and Communication Engineering*, *5*(6), 838–842.

Hasan, M., Uddin, K. N., Sayeed, A., & Tasneem, T. (2021). Smart agriculture robotic system based on internet of things to boost crop production. In *2021 2nd International Conference on Robotics, Electrical and Signal Processing Techniques (ICREST)* (pp. 157-162). IEEE. 10.1109/ICREST51555.2021.9331091

Hirsch, C., Bartocci, E., & Grosu, R. (2019). Capacitive soil moisture sensor node for IoT in agriculture and home. In *2019 IEEE 23rd International Symposium on Consumer Technologies (ISCT)* (pp. 97-102). IEEE. 10.1109/ISCE.2019.8901012

Ikram, A., Aslam, W., Aziz, R. H., Noor, F., Mallah, G. A., Ikram, S., Ahmad, M. S., Abdullah, A. M., & Ullah, I. (2022, May 17). Crop Yield Maximization Using an IoT-Based Smart Decision. *Journal of Sensors*, *2022*, 2022. doi:10.1155/2022/2022923

Jain, A., & Kumar, A. (2020). Smart agriculture monitoring system using IoT. *International Journal for Research in Applied Science and Engineering Technology*, *8*(7), 366–372. doi:10.22214/ijraset.2020.7060

Khan, A., Aziz, S., Bashir, M., & Khan, M. U. (2020). IoT and wireless sensor network based autonomous farming robot. In *2020 International Conference on Emerging Trends in Smart Technologies (ICETST)* (pp. 1-5). IEEE. 10.1109/ICETST49965.2020.9080736

Kim, W. S., Lee, W. S., & Kim, Y. J. (2020, December). A review of the applications of the internet of things (IoT) for agricultural automation. *Journal of Biosystems Engineering*, *45*(4), 385–400. doi:10.100742853-020-00078-3

Lejdel, B. (2022). *Optimization of energy in smart farms using a genetic algorithm. In Driving Factors for Venture Creation and Success in Agricultural Entrepreneurship*. IGI Global.

Mat, I., Kassim, M.R., Harun, A.N., & Yusoff, I.M. (2018). Smart agriculture using internet of things. In *2018 IEEE conference on open systems (ICOS)* (pp. 54-59). IEEE.

Mendelsohn, R. (2007, March). What causes crop failure? *Climatic Change*, *81*(1), 61–70. doi:10.100710584-005-9009-y

Namani, S., & Gonen, B. (2020). Smart agriculture based on IoT and cloud computing. In *2020 3rd International Conference on Information and Computer Technologies (ICICT)* (pp. 553-556). IEEE. 10.1109/ICICT50521.2020.00094

Podder, A. K., Al Bukhari, A., Islam, S., Mia, S., Mohammed, M. A., Kumar, N. M., Cengiz, K., & Abdulkareem, K. H. (2021, April 1). IoT based smart agrotech system for verification of Urban farming parameters. *Microprocessors and Microsystems*, *82*, 104025. doi:10.1016/j.micpro.2021.104025

Prathibha, S.R., Hongal, A., & Jyothi, M.P. (2017). IoT based monitoring system in smart agriculture. In *2017 international conference on recent advances in electronics and communication technology (ICRAECT)* (pp. 81-84). IEEE.

Rajakumar, G., Sankari, M. S., Shunmugapriya, D., & Maheswari, S. U. (2018, April). IoT based smart agricultural monitoring system. *Asian J. Appl. Sci. Technol.*, *2*, 474–480.

Suanpang, P., & Jamjuntr, P. (2019). A smart farm prototype with an Internet of Things (IoT) case study: Thailand. *Technology, 5*(12), 15.

Suhag, S., Singh, N., Jadaun, S., Johri, P., Shukla, A., & Parashar, N. (2021). IoT based soil nutrition and plant disease detection system for smart agriculture. In *2021 10th IEEE International Conference on Communication Systems and Network Technologies (CSNT)* (pp. 478-483). IEEE. 10.1109/CSNT51715.2021.9509719

Verdouw, C., Wolfert, S., & Tekinerdogan, B. (2016). Internet of Things in agriculture. *CABI Reviews, 14*, 1-2.

Zhang, C., & Kovacs, J. M. (2012, December). The application of small unmanned aerial systems for precision agriculture: A review. *Precision Agriculture*, *13*(6), 693–712. doi:10.100711119-012-9274-5

Zotarelli, L., Dukes, M.D., & Morgan, K.T. (2010). Interpretation of Soil Moisture Content to Determine Soil Field Capacity and Avoid Over-Irrigating Sandy Soils Using Soil Moisture Sensors: AE460/AE460. *EDIS*, (2).

Chapter 10
Precision Agriculture and Farming Using Cyber-Physical Systems:
A Systematic Study

C. V. Suresh Babu
Hindustan Institute of Technolgy and Science, India

K. Yadavamuthiah
Hindustan Institute of Technology and Science, India

ABSTRACT

The ideals of precision farming, agronomy, and automation in the food supply chain have also been established because of robotization and agricultural advancements. The rigidity, networking, standardization, robustness, skill requirements, absence of real-time data, and actuation are some of the areas where present systems still have serious flaws. The development of robotics and independent systems was prompted using artificial intelligence, and intelligent ICT robotic systems in cyber-physical systems will be necessary. These systems can communicate, compute, track, and control physical elements and processes in a variety of operations. This system uses ICT robots equipped with the necessary sensors to monitor the entire agricultural area, transferring data to edge nodes. While the procedure was still in progress, it began updating the field and storing the information in the cloud, and it transferred updates to the farmers. This chapter explores the various developments of robotics and autonomous systems in the agricultural field.

INTRODUCTION

Modern agriculture faces enormous obstacles to creating a sustainable future worldwide. Global issues such as climate change, urbanization, environmental degradation, increasing demand for animal protein, and population growth threaten the ability of the agricultural sector to meet global food demands. By

DOI: 10.4018/978-1-6684-7879-0.ch010

2050, with a population of nearly 9 billion, there will be a greater demand for different types of food, including meat, fruits, vegetables, and higher-quality and healthier food. However, increasing food production to sustain the growing population may lead to competition for land and water. To ensure sustainability, agricultural innovation is necessary, with the environment, biodiversity, and public health as key pillars. The relationship between agriculture and the environment presents challenges and opportunities for technological advancements. Although intensive agriculture provides benefits to humanity, it harms ecosystems by introducing harmful substances, such as phosphorus and nitrogen, to the environment. The natural resource base also faces structural issues, including water shortages, salinization, watercourse contamination, and ecosystem deterioration. Soil degradation is a widespread problem that threatens soil productivity and serves as the foundation of agriculture. Inadequate fertilizer use leads to soil degradation and loss of fertility, whereas excessive fertilizer use can harm the ecosystem. A commitment to sustainable agriculture is necessary to address these challenges, and a logical conceptual model is needed to emphasize the relationships between natural and human systems in the agricultural system (Wiyon et al., 2023).

BACKGROUND

India's farmers face several challenges such as limited access to high-quality seeds, soil fertility depletion, inadequate irrigation, and low mechanization. Soil exhaustion and reduced productivity are caused by traditional farming practices. Mechanization in agriculture is limited, with manual labor being predominant and decreasing in efficiency. To ensure sustainable agricultural growth, the government needs to invest in research and development, encourage private investment, establish seed banks and subsidies, promote mechanization technologies and water conservation practices, and improve water use efficiency. Addressing these challenges and adopting modern agricultural practices could improve productivity, promote sustainable development, and ensure food security for the Indian population.

TRADITIONAL FARMING VS. MODERN FARMING

Traditional Farming

Indians use the phrase "traditional farming" to describe farming that is done with the help of ingrained knowledge and wisdom accumulated over many years of experience. Visual and aural instructions are used to convey the majority of the information. It frequently performs a good job of adjusting to the environment. The gene pool used was created internally and provided a good understanding of the regional climate. It is claimed that more independence exists. By contrast, traditional agricultural methods have received less attention in terms of record keeping. Therefore, it was utterly unknown to the contemporary educated population, which prevented them from appreciating it. Regardless of the activity, it adopts a holistic approach and is woven into the daily lives of those who engage in it (He, 2023). Agriculture, as practiced in the traditional agricultural system, is defined by its utter lack of modernism and commercialization. Labor and seeds are all produced using traditional techniques. It was a very old-fashioned method of producing food that had been around for a very long time, without the aid of modern technology or innovation. Subsistence farming accounts for a vast majority of this type of farming. Plows are one of the many enduring agricultural implements still in use today (He, 2023).

Advantages

- When substituting natural fertilizers like vermicompost and cow dung for synthetic ones, we can lower the price of fertilizers.
- There is a reduction in the amount of water needed since there are no chemical fertilizers used.
- The cultivable region is perfectly suited for the multi-crop type of agriculture because we only use natural fertilizers.
- The old method requires fewer resources, which lowers the cost of production.
- The practice of organic farming ensures that the food is totally safe to consume.

Drawbacks

- Contrary to high-tech farming, which just takes a few hours, traditional farming requires farmers to spend about 15 hours harvesting their crops.
- Because harvesting takes some time, it must be sold for a higher price to make up for the time and money spent allowing the crops to mature.
- Conventional farming uses soil; therefore, breakdown takes down the vast majority of the time that is available.
- Furthermore, as a result of this, the crops are at a higher risk of contracting soil disease.
- Pesticides are used on crops to prevent pests from attacking them.
- Plants are not in good health as a result (He, 2023).

Modern Farming

A type of farming known as "modern" farming disregards this entrenched wisdom and positions itself as being superior to traditional methods of production. Moreover, modern culture seems to be more capital-intensive and substantial in style, making little use of India's enormous gene pool. It is a monotony that does not take into account regional resources, context, climate, or other considerations, and it stands out in the neighborhood like a sore thumb. The process of farming using modern techniques and technology is referred to as "modern farming." Farming is made feasible by modern science as well as a number of other aspects of discoveries and improvements. Both new capital investment sources and contemporary irrigation systems are being used in modern agriculture. With the use of modern technology for cultivation and harvesting, new methods, as well as high-yielding hybrid HYV seeds, are being used. Today's farmers have a commercialized career path and a variety of work options available in agriculture. These include drip irrigation, spray irrigation, canal systems, and modern technologies. The contemporary agricultural system provides enough resources and opportunity for supplementary enterprises as well as for any operations that provide services, in addition to supporting the primary activity (Raj et al., 2021).

Advantages

- Ensuring that everyone has access to food without running out of food grains.
- The area is being utilized to its fullest extent (crop intensity of more than 300 percent)
- The outcome has been a rise in food grain production.
- Even our mangoes are available in the United States, and we import California apples as well as various goods, according to the global market.
- Due to the expansion of companies and the utilization of machinery, this industry is growing and producing fantastic jobs.

Disadvantages

- The salinization of the soil and loss of micronutrients are caused by excessive fertilizer application.
- A fallacious perception of safety is produced as a result of it.
- Most people cannot afford modern equipment due to its high cost.
- The item's distinctiveness is losing ground (Raj et al., 2021).

PROBLEM STATEMENT

Although more people live in India than ever before, less of the country's land is arable. Of all the industries, the food industry is the most dominant. Agriculture provides the land for the food industry, but according to the current situation, despite the need for food, agriculture is not currently being grown as effectively as other industries (Yoganandan et al., 2022).

Table 1. India: Historical population data

Year	Population	Growth Rate
2023	1,428,627,663	0.81%
2022	1,417,173,173	0.68%
2021	1,407,563,842	0.80%
2020	1,396,387,127	0.96%
2019	1,383,112,050	1.03%
2018	1,369,003,306	1.09%
2017	1,354,195,680	1.16%

The FAO defines arable land (measured in hectares) as land that is either temporarily under temporary crops (double-cropped areas are counted once), temporarily under temporary meadows for mowing or pasture, momentarily under market or kitchen gardens, or temporarily fallow. Land that was abandoned as a result of relocating cultivation is not included (Goyal et al.,2016).

India had 155,369,076 acres of arable land in 2020, 2021, and 2022, which is a 0% growth from 2019.

India had 155,369,076 acres of arable land in 2019, which is a 0.07% decrease from 2018.

India had 155,477,911 acres of arable land in 2018, which is a 0.27% decrease from 2017.

India had 155,895,136 acres of arable land in 2017, which is a 0.27% decrease from 2016 (Goyal et al., 2016).

What Are the Issues That Farmers Face and What Are the Primary Reasons That Farmers Leave Agriculture?

Seeds

In India, the lack of access to high-quality seeds is a significant challenge faced by small and marginal farmers who cannot afford the high costs of such seeds. The distribution of guaranteed quality seeds is as important as their production. Several factors contribute to this problem, including high production costs, a lack of awareness among farmers, and inadequate distribution networks. Addressing this issue requires a multi-pronged approach, including government investment in R&D, policies to encourage private investment, and the establishment of seed banks and subsidies to make high-quality seeds more affordable. Ensuring access to quality seeds is critical for the sustainable growth of Indian agriculture, and concerted efforts from various stakeholders are required to address issues of high cost, lack of awareness, and inadequate distribution networks (Dev et al.,2023).

Manures and Fertilizers

The depletion of soil in India is a significant challenge for the sustainability of agriculture, as traditional farming practices have led to soil exhaustion and decreased productivity. Increasing the application of manure and fertilizers can help restore soil fertility and increase crop yields. However, the indiscriminate use of fertilizers can lead to soil and water pollution, making it essential to judiciously use fertilizers and adopt soil-testing methods to determine the appropriate type and quantity of fertilizers needed for crops. Sustainable agricultural practices, such as crop rotation and intercropping, can also help maintain soil fertility and ensure the sustainability of Indian agriculture. As agriculture is the primary livelihood for almost 60% of the Indian population, it is crucial to take urgent action to address the depletion of soil and promote sustainable farming practices (Nalluri et al., 2023).

Lack of Irrigation

Inadequate irrigation coverage is a significant challenge for agriculture in India, with only one-third of the crop area under irrigation, despite being the world's second-largest irrigated country after China. Bringing more than half of the planted area under guaranteed irrigation is crucial to mitigate the risks of crop failure due to erratic rainfall and to increase crop yields. The government can invest in irrigation infrastructure, promote water conservation practices, and improve water-use efficiency. Farmers can adopt alternative irrigation methods, such as drip irrigation, sprinklers, and water-harvesting techniques, to reduce water waste. A collaborative effort between the government, private sector, and civil society organizations is necessary for sustainable agriculture growth in India (Nair et al., 2022).

Lack of Mechanization

In India, despite some regions adopting extensive mechanization in agriculture, the majority of agricultural operations are still conducted manually using traditional tools and implements. Hand labor is an essential source of employment for many rural households, but it can be physically demanding and time-consuming, limiting agricultural efficiency. Encouraging the adoption of appropriate mechanization technologies and providing training and support to farmers can help improve productivity, reduce drudgery, and enhance the overall efficiency of agriculture in India. The cost of mechanization and limited access to modern equipment are key challenges that need to be addressed to promote the adoption of mechanization in agriculture in India (Haider & Joshy, 2022).

Soil Erosion

Soil erosion caused by wind and water is a significant problem in India, leading to the loss of fertile topsoil and reduced crop yield. To restore soil fertility, farmers can adopt soil conservation measures, such as contour plowing, terracing, and the use of cover crops and mulching to prevent further erosion. Applying organic and inorganic fertilizers can also help improve soil structure and enhance nutrient availability, promoting sustainable agriculture and food security in the country (Sathiyamurthi et al., 2023).

Agricultural Marketing

The lack of dependable marketing infrastructure, inadequate market information, and weak bargaining power are some of the significant challenges faced by farmers in rural India in marketing agricultural products. To address these issues, the government has initiated several programs to improve the marketing infrastructure in rural areas, such as the establishment of Agricultural Produce Market Committees and the Electronic National Agriculture Market, and promoting farmer-producer organizations. These measures are crucial to empower farmers and ensure they receive fair prices for their produce, which is essential for improving the status of agricultural marketing in rural India (Garg et al., 2023).

SURVEY PAPER

Conservation Techniques for Reducing Sediment Load in Ungauged Agricultural Watersheds: Spatial Optimization

Conservation practices (CPs) are used in agricultural watersheds to reduce soil erosion and improve water quality, resulting in long-term management of natural resources. This is crucial because larger populations and climate change are putting additional strain on agricultural systems. Nonetheless, given their complicated, nonlinear, and location-dependent responses, it is still difficult to optimize the application of these approaches. In this study, CPS were localized and characterized at the field scale using GIS, and watershed modeling was performed using an annualized agricultural non-point source model. The activities under investigation are related to the development of sediment basins, crop rotations, riparian buffers, and the conservation reserve program. A total of 33 conservation scenarios were created and their effects on reducing sediment erosion were quantified. The Mississippi River Basin Program, which aims to reduce one of the largest aquatic dead zones on the planet, used this strategy in a watershed that was not gauged. The reduction in sediment load at the watershed scale was significantly affected by the focused strategy, according to the simulation results. Among the many methods evaluated, riparian buffers were the most effective at reducing silt. This study also serves as a guide for future research projects that aim to construct decision-support systems and optimize the placement of CPs in agricultural watersheds (Bingner et al., 2023).

Agriculture 4.0: A Systematic Literature Analysis on the Digitization of the Agricultural Industry

Short- and medium-chain chlorinated paraffins in sediments from China's rivers, their spatial distribution, and the risk they pose Chlorinated paraffins (CPs) were generated in large quantities for a variety of industrial uses, but due to poor handling practises and subsequent environmental releases, they contaminated the entire world. In the current study, the presence and geographical distribution of short- and medium-chain chlorinated paraffins (SCCPs and MCCPs) were examined in 171 sediment samples taken from urban rivers across China that had a strong black stench. The concentrations of total SCCP and MCCP were not detected at 1.0 106 (median: 1.3 104) ng/g DW and 8.3 to 9.4 104 (median: 1.1 103) ng/g DW, respectively. Although urban rivers with a strong smell were polluted by point sources of SCCPs and MCCPs, no obvious spatial distribution of these pollutants was found. Substantial positive associations were found between SCCPs and total organic carbon, as well as between SCCPs and other persistent organic compounds, such as antibiotics, plasticizers, polychlorinated biphenyls, and polybrominated diethyl ethers. The average proportions of MCCPs to SCCPs in the majority of samples fell between the ranges of 11 and 16, which suggests that at least two different categories of CP technical mixtures are produced and used in China. Similar to their commercial versions, SCCPs and MCCPs have the same mineral composition. The Federal Environmental Quality Guidelines and the Risk Quotient were both used in ecological risk assessments, and they both concluded that the presence of SCCP and MCCP in surface sediments poses an ecological concern (Abbasi et al., 2022).

Environmental Implication

Due to their widespread production, propensity for persistence, long-distance transmission, bioaccumulation potential, and toxicity, SCCPs and MCCPs can be regarded as "hazardous materials". In order to better understand the features of SCCP/MCCP pollution, this research carried out a thorough investigation of SCCP/MCCP in urban sediment samples across China that had a strong black stench. This investigation highlighted their environmental risk (Abbasi et al., 2022).

Unmanned Aerial Vehicle (UAV) Geometrical Assessment of Hazelnut Trees in an Intensive Orchard for Precision Agricultural Applications

The precise maintenance of a hazelnut orchard depends greatly on knowledge of tree size. Research has demonstrated that site-specific crop management enables the best possible management and input-use efficiency. Most of the time, manual approaches that require a lot of labor and are not very precise are used to measure the parameters of trees. The purpose of this study was to propose, assess, and test a straightforward and novel method for canopy characterization in an intensive hazelnut orchard using photos captured by an unmanned aerial vehicle (UAV). Radius (Rc), canopy height (hc), tree height (htree), and trunk height (htrunk) were the factors considered. Using UAV photos, two distinct approaches were employed to estimate the canopy volume. The effectiveness of the approach was assessed by comparing manual and UAV data and utilizing the Pearson correlation coefficient and root mean square error (RMSE). Rc, hc, and htree showed significant increases in correlation values, whereas htrunk showed a significantly decreased correlation. It was a wise idea to use the suggested approach for volume computation (Vinci et al., 2023).

Smart Agriculture: Crop Protection Utilizing Cyber-Physical Systems and Machine Learning

In the fields of artificial intelligence (AI), cyber-physical systems (CPS), and global digital transformation (GDT), the world is currently undergoing the fourth generation of revolution. Moreover, cyber-physical systems will be able to function more skillfully, enthusiastically, and cooperatively by applying artificial intelligence. Crop control in agriculture is a procedure to prevent the crop from becoming infected by various pests by protecting it as much as possible. Any living species that has the potential to harm an agricultural crop is considered a pest, including mantodea, poultry, weeds, pathogens, and other creatures. An average of 20–30% of the world's potential agricultural yield is lost each year as a result of pests and other diseases. In order to minimize these losses, crop protection is necessary. This chapter examines the CPS concept, which uses machine learning and cyber-physical systems to increase crop productivity and safeguard the crop from wild animals. Crop suggestion, automated irrigation and fertilization, crop protection, plant monitoring, and harvesting are all part of the CPS's smart agricultural process (Choubisa et al., 2022).

The process of moving from manual production to machine production is referred to as the "industrial revolution." "Industry 4.0" refers to the latest industrial revolution. Industry 4.0 involves the automation and sharing of information in manufacturing industries using cutting-edge trends like the Cyber Physical System (CPS), smart agriculture, the Internet of Things, machine learning, cloud computing, artificial intelligence, block chain technology, and many more.

The first industrial revolution was characterized by the employment of mechanical equipment and steam engine power. Industrial rebellion in the second category was then followed by the use of electricity for mass production and assembly lines. Information technology and electronic chips were the driving forces behind the third industrial revolution. In the present, CPS is related to Industry 4.0.

Industry 4.0 proposes new guidelines for the manufacturing sector, such as:

By integrating sensor-generated signals with virtual reconstruction and plant models, we may virtualize assembly processes;

- Enhanced availability through greater assembly availability
- Smart and dispersed planning for making decisions.
- Gathering information in real time.

Exploring the Application of Industry 4.0 Technologies in the Agricultural Food Supply Chain: A Systematic Literature Review

In order to address the rising worldwide demand for food products and worries about food safety and security, Industry 4.0 (I4.0) is required throughout the "agricultural food supply chain (AFSC)." Additionally, this change is made feasible by recent developments in I4.0 technologies, such as "the Internet of Things (IoT), Blockchain, Big Data (BD), Information and Communication Technology (ICT), Cloud Computing (CC), and Cyber-Physical Systems (CPS). Each of these technologies plays a particular role in enabling the AFSC ecosystem to be as smart as possible to handle the difficulties facing the world today. Hence, to comprehend their applications and current trends, this study analyzed these key AFSC technologies. This study relied on 146 articles from the "Scopus and Web of Science'' databases. Peer-reviewed journals from 2010 to 2020 were included in the articles. In AFSC, the applications of the assessed technology are analyzed in accordance with five research dimensions: traceability and food safety, information system management, food waste, control and monitoring, decision-making and agribusiness, and other irrelevant applications. According to this study, integrating the technologies under consideration can be more beneficial for AFSC in terms of delivering affordable solutions and enhancing sustainability. In addition, blockchain has the potential to revolutionize the way food security and safety are ensured. To further spur scholars to advance this field, the present problems and the planned research agenda in the relevant areas were also identified. This is the first study to provide an overview of the six technologies' most recent advancements in AFSC research (Yadav et al., 2022)

CROP: A Novel Framework for Automatic Crop Damage Estimation in Smart Agriculture

Agriculture is affected by natural disasters. Destruction of crops causes farmers to suffer significant losses. Natural calamities or incidents caused by climate or weather occur frequently and result in losses of billions of dollars. To compensate for these losses, crop insurance provides financial security for the agricultural sector. A crop insurance payout is a lengthy process, and processing claims requires time. In this study, we present a proof-of-concept for a revolutionary crop damage estimation approach, eCrop, which is a component of our proposed agricultural cyber-physical system. Grid-based techniques were used in eCrop. In addition, we presented a novel technique for detecting crop damage. This serves as the foundation for the eCrop. It is a model built on a Convolutional Siamese Neural Network (CSNN). A meta-learning methodology is used to train the model. There was a 92.86% improvement in the accuracy. To estimate crop damage automatically, we can adapt our eCrop method to agricultural insurance claim processing. Any cropland size and crop can be grown on it (Mitra et al., 2022).

Cyber Security for Detecting Distributed Denial of Service Attacks in Agriculture 4.0: Deep Learning Model

Hackers exploit Internet of Things (IoT) networks, which interconnect industrial equipment to the Internet more extensively. Recent deployments of potent deep learning (DL) models have allowed the creation of network intrusion detection systems (NIDSs) that can secure Agriculture 4.0 networks. To detect DDoS attacks, it is possible to develop an effective and flexible intrusion detection system using long short-term memory (LSTM) and a convolutional neural network combined with long short-term memory

(CNN-LSTM) architecture. A proposal for identifying various DDoS attack types was developed using the CIC-DDoS 2019 dataset. The CICFlowMeter-V3 network was utilized to develop the dataset. The development of deep learning techniques was evaluated using the common network traffic dataset, which includes benign packets such as regular NetBIOS, Portmap, Syn, UDPLag, and UDP. The performance of the model was evaluated using various metrics, including precision, recall, F1-score, and accuracy. The proposed technology achieves 100% accuracy, which is a high level of precision. With regard to every evaluation metric, CNN-LSTM has a rating of 100%. To assess the performance of our model, we built it using a deep learning technique and compared it with current systems. Additionally, we believe that this proposed paradigm offers Agriculture 4.0, the highest possible defense against cyberthreats (Aldhyani et al., 2023).

An Intelligent Agro-Cloud Framework for Efficient Crop Cultivation

Agriculture is thought to support the Indian economy. However, despite these factors—water shortages, flooding, storms, lack of labor, and bad agromarketing—farmers are still unable to make a reasonable profit. In addition, ignorance, high costs, and small land size prevent the effective application of contemporary technologies in farming. The proposed cloud-based system consists of three essential layers: (i) the input layer, (ii) the storage and processing layer, and (iii) the application layer. The input layer gathers information from a variety of Internet of Things (IoT) devices, including sensors, meters, cameras, robots (for spraying and weeding), drones, and satellite data. The required information from the IoT devices is stored and processed periodically by the agro-cloud. Crop development, soil monitoring, irrigation control, and pesticide and fertilizer management are all aspects of smart farming aided by layer 2. The right irrigation method can be chosen depending on the type of soil, crop, water quality, and rainfall by using soft computing technologies such as artificial intelligence, neural networks, and machine learning algorithms. Applications such as graphical user interface systems (GUI), SMS notifications, and IoT-based mobile applications are some of the ways in which the application layer communicates with farmers. Based on the suggested framework for effective agricultural cultivation, a farmer adviser mobile app was created that gives farmers timely advice on how to plan their land, choose their seeds, plant their crops, and safeguard and ensure their plants. Hence, the proposed research helps farmers increase agricultural productivity by utilizing less labor and resources (Anitha et al., 2023).

The Development of AgriVerse: Past, Present, and Future

To achieve environmentally friendly agriculture, the agricultural metaverse (AgriVerse) seeks to optimize the production chain by reducing costs, enhancing efficiency, and shattering information silos. Despite the fact that it is characterized by the virtual-real interaction of agriculture-related activities based on massive datasets, knowledge, and models, the relationship between AgriVerse and extensively researched plant modeling is hazy. The decentralized science (DeSci) of the plant model, the decentralized agricultural organization (DAO), and a brief overview of the research on plant modeling were used to examine the ongoing shift in the epoch of artificial intelligence (AI) and to foresee the future of AgriVerse. The application scenarios for the three AgriVerse modules are shown. AgriVerse's prospects and difficulties are highlighted. AgriVerse practitioners from all backgrounds are expected to join the community as a result of this study, which is also intended to identify important research issues facing the field.

Towards the Fifth Innovation Era in Agricultural Innovation

In contrast to the COVID-19 pandemic and its effects, agriculture is experiencing numerous difficulties. Threatened biodiversity, the enclosure of arable land, climate variability, a lack of personnel, and diseases of farm animals are just a few of the 20 major issues we have discovered. There are three basic options to overcome these obstacles: development, replication, and innovation. We contend that one can be influenced by problems in the following ways. This is demonstrated by a case study from Australia, where COVID-19 issues led to the development of robotics in the fruit business. Throughout history, there have been periods of innovation in agriculture, and one such period is about to begin. Modern crop and cultivation concepts, field robotics, fossil-free energy and goods, smart connected systems, and animal welfare technology systems are some traits that define this new era. Since there are many problems in the world right now, innovation is the only thing that seems to have any chance of solving them (Frankelius et al., 2023).

Smart Monitoring of Potato Crop: A Cyber-Physical System Architecture Model in the Field of Precision Agriculture

Advanced mechatronic systems have undergone a significant transition over the past 20 years in cyber-physical systems (CPS). In the realm of precision farming, CPS is crucial, and it is anticipated that production will increase to feed the world's population and stop starvation. It is necessary to develop techniques, tools, hardware, and software according to transversal methods in the area of precision agriculture to accelerate the manifestation of CPS. Prototypes and testbeds should also be used to validate the underlying concepts. In this respect, this study provides an integrated system architecture for precision agriculture management based on CPS design technology (Rad et al., 2015).

Increasing the Effectiveness and Adoption of Agricultural Phosphorus Management Strategies to Minimize Water Quality

Optimal agricultural production depends on phosphorus (P), but when it is lost through erosive processes such as erosion, runoff, or leaching (which releases soluble reactive phosphorus (SRP)), water quality suffers. P is impacted by the use of conservation measures (CP) at the edge of water resources, during transit, or at the source (by avoiding or managing P trapping). The implementation of CP frequently involves trade-offs. As an illustration, studies by numerous researchers have demonstrated that crop rotation reduces total P by over 50%, while improving SRP by as much as 40%. Because SRP is more permeable than particulate P, conservation tillage may result in an increase in water quality degradation. Conservation practices, such as P treatment with conventional tillage and a fluvial buffer, must be implemented as a system of practices to increase redundancy and address all loss channels. To implement practices in crucial source areas and offer the finest remediation for the least amount of money, planning and adaptation must also take place at the watershed scale. Farmers must participate in riparian planning, which should incorporate financial support and outreach into the community. For be employed more efficiently to reduce and delay off-site P losses. To better regulate phosphorus leaching, manage the legacy storage of phosphorus in soil, and reduce higher phosphorus losses anticipated as a result of climate change, new and inventive CPs are required. Climate change will cause an increase in P losses, which will lead to a decline in water quality if quick modifications to CP implementation are not made.

At a watershed level, these adjustments must be done in a deliberate, transparent, and intelligent manner (Sharpley et al., 2019).

Principal Concepts

- All P routes must be under the control of conservation actions that reduce phosphorus.
- Systems must be used to implement phosphorus-reducing conservation strategies.
- To enhance control of P, new and creative conservation techniques are required.
- When putting conservation practices into effect, farmer decision-making must be taken into account.
- The planning of watersheds and the implementation of conservation practices must be deliberate.

Automation and Robotics: The Trend Towards Cyber Physical Systems in Agriculture Business

The implementation of Cyber Physical Systems (CPS), which will promote new ideas for future farming systems, is one of the enormous opportunities provided by digital transformation in Europe. The transformation of mobile devices into cyber-physical systems with almost no communication barriers is driven by extremely rapid advancements in communication and information technology. Because of the enormous potential for efficiency increases at all levels, automation is the most significant development in the creation of agricultural mobile equipment (machines, processes, and farm operations). CPS technologies provide innovative systemic and regional solutions to support real-time and strategic decisions and significantly increase system transparency and predictability within the sustainability triangle (Herlitzius, 2017).

A Robot-Based Intelligent Management Design for Agricultural Cyber-Physical Systems

In this study, an intelligent management system based on robots is proposed that can control the growth of crops in their initial surroundings. The proposed robot uses an FPGA device as its computing architecture, which is different from that of most other agricultural robots. Binarized neural networks (BNNs) were used in this robot's design to detect target crops in real time and with high accuracy, and cryptographic hardware functionalities were used to ensure the secrecy of sensor file transfers in an autonomous environment. The robot can regulate actuators to maintain an appropriate development environment for crops based on a patterned and virtualizable design strategy. It can also dynamically change the functionality of its cryptographic hardware to satisfy a variety of requirements. According to experiments, the processing time of cryptographic operations may be reduced in the proposed robot by 99.46% to 99.62% compared to the traditional microprocessor-based architecture, whereas the BNN inference can be sped up by a factor of 13,359.6, on average. The robot design can cut 37.4% of slice LUTs and 29.81% of slice registers in a Xilinx Zynq XC7Z020 device, resulting in a 13.6% reduction in power consumption in order to implement all the AES, DES, and 3DES operations (Huang et al., 2021).

DIGITAL FARMING

Sensors

Sensors come in a variety of forms, and each one serves a specific purpose in the industry. As sensors produce data straight from the field, they are essential to the development of digital agricultural solutions. These sensors are set up all over the area and are able to produce the raw data. As the acquired data is transmitted to the cloud, it is processed and analyzed, and several of the most recent breakthroughs are then presented to the farmer (King, 2017).

Rain Gauge

A rain gauge, also known as udometer, pluvia metior, pluviometer, ombrometer, or hyetometer, is a tool used by meteorologists and hydrologists to collect and measure the amount of liquid precipitation over a specific area and time period. It enables the determination of rainfall depth in millimeters and provides precise measurements of precipitation levels. This scientific instrument is essential in determining the water cycle, climate patterns, and water resource management. With its ability to accurately quantify the volume of rainfall, the rain gauge is an important tool for weather forecasting, agriculture, and various other applications that rely on precipitation data (King, 2017).

Lux Meter

A lux meter is an instrument that measures the intensity of light in units of lux falling on a surface. In agriculture, lux meters can be used to measure the amount of light that reaches crops in different areas of a field, allowing for the optimization of crop placement and orientation to ensure optimal growth and development. Lux meters are also useful in measuring the amount of artificial light in greenhouses and other controlled environments, ensuring that crops receive the appropriate amount of light for their growth stage. Additionally, lux meters can help monitor light pollution in agricultural areas, which can negatively impact plant growth and development. By utilizing lux meters in agriculture, crop yields can be improved, resource use can be optimized, and sustainable agricultural practices can be achieved (King, 2017).

Leaf Wetness

Leaf wetness is a crucial factor in agriculture as it refers to the amount of moisture on a plant's leaves. Monitoring leaf wetness is essential for disease control as many plant diseases thrive in moist conditions. Farmers can identify periods of high humidity by monitoring leaf wetness and take preventive measures to prevent disease spread. Additionally, leaf wetness helps determine when plants have received enough moisture from rain or irrigation, optimizing water usage and preventing over- or under-watering. Monitoring leaf wetness is also vital for pesticide application as some pesticides require a specific amount of time to dry on the plant before they can be effective. Farmers use sensors or observations to monitor leaf wetness, and advanced systems that use artificial intelligence can analyze data from sensors to provide real-time recommendations for disease prevention, irrigation, and pesticide application (King, 2017).

Temperature, Humidity, and Air Pressure

The use of sensors to collect data on temperature, humidity, and air pressure is a crucial aspect of modern agriculture. This information can assist farmers in making informed decisions about crop management, irrigation, and harvesting. For example, if the temperature is too high, farmers may need to adjust their irrigation schedules to avoid water stress in plants. High humidity levels can also necessitate preventative measures to avoid the development of fungal diseases. Moreover, changes in air pressure can significantly impact crop growth and development. Although the data collected by these sensors may be raw and unprocessed, analysis of the data can reveal patterns and trends, leading to improved agricultural practices and better crop yields. By adopting technology, farmers can increase efficiency, make informed decisions, and optimize agricultural production (King, 2017).

Irrigation

Irrigation is a critical component of modern agriculture, and the use of soil moisture sensors has made it more efficient and effective. These sensors are placed throughout the field to measure soil moisture levels regularly. When the moisture level falls below a certain threshold, the irrigation pump installed on the farm is alerted. The pump sends a message to the farmer's mobile device via radio signals, requesting authorization to begin irrigation. If the farmer approves the request, the pump will begin to irrigate the field automatically, and it will continue to do so until it receives a signal from the soil moisture sensor to stop. This technology allows for precise and targeted irrigation, reducing water waste and preventing overwatering, which can lead to waterlogging and crop damage. By leveraging technology to automate irrigation, farmers can save time and resources while also improving their crop yields (King, 2017).

Drones

Drones make it possible to collect data and apply nutrients in a resource-efficient manner, facilitating the forecasting of agricultural output and evidence-based planning. Emerging technologies: In order to mainstream emerging technologies like yield estimation or insurance, drones can be a useful facilitator (Sundar et al., 2023).

In specific industries, drones can address a variety of problems. It has the potential to reveal a variety of intriguing insights that support the farmer's decision-making. Let's see how useful drones are in the world of agriculture:

- Crop monitoring
- Soil and field analysis
- Weed identification
- Pest and disease identification
- Crop spraying
- Crop health assessment
- Livestock management

Weather Data

Weather data plays a critical role in modern agriculture, providing farmers with the information they need to make informed decisions about their crops. Digital technology and mobile devices, farmers can access accurate weather information that includes factors like temperature, precipitation, and wind speed. This data is essential for preventing losses in agriculture, allowing farmers to anticipate extreme weather events and take proactive measures to protect their crops. These measures might include adjusting irrigation schedules, installing protective covers or shelters, or harvesting crops early. By leveraging technology to access weather data, farmers can optimize their agricultural production and minimize the risks associated with a changing climate (Sridhara et al., 2023).

CYBER PHYSICAL SYSTEMS FOR AGRICULTURE DOMAIN

The agricultural sector has seen a significant shift towards mechanization in recent years, with the use of robotics and autonomous systems aided by the applications of Artificial Intelligence (AI) and embedded systems. While drones have been used in the past, the introduction of a new generation of intelligent ICT robotic systems is expected to revolutionize the field. These systems are capable of communication, computation, monitoring, and control of physical components and processes in various applications. They can also be networked, interconnected, interactive, and autonomous. Designed as intelligent sensor networks of autonomous systems with unique sensing and actuation capabilities, these systems are vital for acquiring data in interactions with the real world, particularly those involving users.

The integration of sensors and actuators into cyber systems has led to the correlation between the physical and digital worlds, making it impossible to separate them into distinct categories. Cyber-physical systems (CPS) were born from this concept, where cyber technologies could use real-time control from conventional embedded systems, making them responsive to the physical world. A computational physical system (CPS) integrates computation with physical processes, using sensors and actuators to connect computational systems to the real environment. However, incorporating physical elements in CPS requires dependability and safety standards that differ significantly from those in general-purpose computing, necessitating the adaptation of computing abstractions. The challenge of CPS lies in the inherent concurrency of physical components, which is qualitatively different from object-oriented software components. Hence, it is crucial to evolve beyond the current state of computing technology. Despite the integration of computation, control, sensors, and networking into every aspect, the behaviour of the system's cyber and physical components working together is critical in CPSs. Components behaviour must be compatible and trustworthy, considering the embedded systems, wireless sensor networks, the Internet of Things, communication protocols, and their interactions, which are the most critical components of CPS. CPS is a new development in systems exhibiting autonomous behaviour. The ability to make systems cooperate with one another is a crucial challenge in the development of robots for the future. In contrast to traditional robot automation involving industrial robots performing repetitive tasks within fenced areas, precision agriculture is a vastly different scenario that will be the target of disruptive developments in this context. This is crucial because of the world population's increase, which is impacting our access to fertile land due to climate change. Agriculture must adapt to these challenges to avoid a potential shortage of food in the future. Therefore, developing autonomous robots that can cooperate to address these real-world challenges is of utmost importance. The agricultural sector has seen significant changes in recent years, with the introduction of robotics and autonomous systems driven by the applications of AI and embedded systems. CPS, which integrates computation with physical processes, is crucial to the development of intelligent ICT robotic systems that are capable of communication, computation, monitoring, and control of physical components and processes in various applications. The future of robots lies in their ability to cooperate with each other, making precision agriculture an essential area of development. With a global population increase and the effects of climate change on fertile land, developing autonomous robots that can work together to address these challenges is of utmost importance (Udutalapally et al., 2021).

Process

The agriculture sector is a significant contributor to the economy. However, with the increasing population and demand for food, conventional farming techniques face challenges in meeting the rising demand while maintaining crop quality and productivity. The Agriculture Cyber-Physical System (A-CPS) is an innovative solution to enhance crop quality and productivity while utilizing minimum cropland. The system leverages Artificial Intelligence (AI) for the automatic detection of plant diseases (Suresh Babu, 2023). Plant diseases, particularly those caused by microbial pathogens, pose a significant threat to crop yield and quality. With the constantly mutating pathogens, it is challenging for farmers to keep track of the diseases and respond promptly. The A-CPS addresses this challenge by leveraging AI to detect plant diseases automatically. The AI system relies on image analysis of crop images captured by a health maintenance system to detect diseases. The health maintenance system consists of a solar sensor node responsible for capturing images and a developed soil moisture sensor. The system is designed to continuously sense and automate the process of monitoring plant health. The system's high longevity compared to its peers makes it ideal for long-term monitoring of plant health. The captured images are analyzed using a Convolutional Neural Network (CNN) model that is trained to recognize disease symptoms.

The CNN model is trained on a large dataset of crop images labeled with disease symptoms. The model learns to recognize patterns in the images and associate them with specific diseases. During training, the model adjusts its internal weights to minimize the difference between its predictions and the actual disease labels. The training process ensures that the model can accurately recognize diseases in unseen images.

Once trained, the CNN model is deployed in the A-CPS system to analyze images captured by the health maintenance system. The system is designed to be used in real-time, allowing farmers to monitor their fields remotely using a smartphone application. The captured images are analyzed by the CNN model, which detects any disease symptoms present in the images. The system then provides the farmer with a diagnosis and recommendations for treatment.

The A-CPS system has undergone rigorous testing to evaluate its performance. The system's performance was evaluated by monitoring crop health and productivity over a period of three months. The system demonstrated robust performance by remaining rust-free and sustaining varied weather conditions.

The system's accuracy in predicting plant diseases is remarkable, achieving an accuracy of 97.34%. This high accuracy is due to the CNN model's ability to learn and recognize patterns in images that are associated with specific diseases. The high accuracy of the system allows farmers to respond promptly to disease outbreaks, reducing crop loss and increasing productivity. The Agriculture Cyber-Physical System is an innovative solution to the challenges faced by conventional farming techniques. The system leverages AI to provide automatic detection of plant diseases, enhancing crop quality and productivity while utilizing minimum cropland. The system's accuracy and reliability make it a valuable tool for farmers in responding promptly to disease outbreaks and increasing crop productivity. The use of CNN models for disease prediction provides an accurate and reliable means of predicting plant diseases, leading to a more efficient and effective agriculture system.

Figure 1. Architecture of the AI-based agriculture cyber-physical system (A-CPS)
Source: Udutalapally et al. (2021)

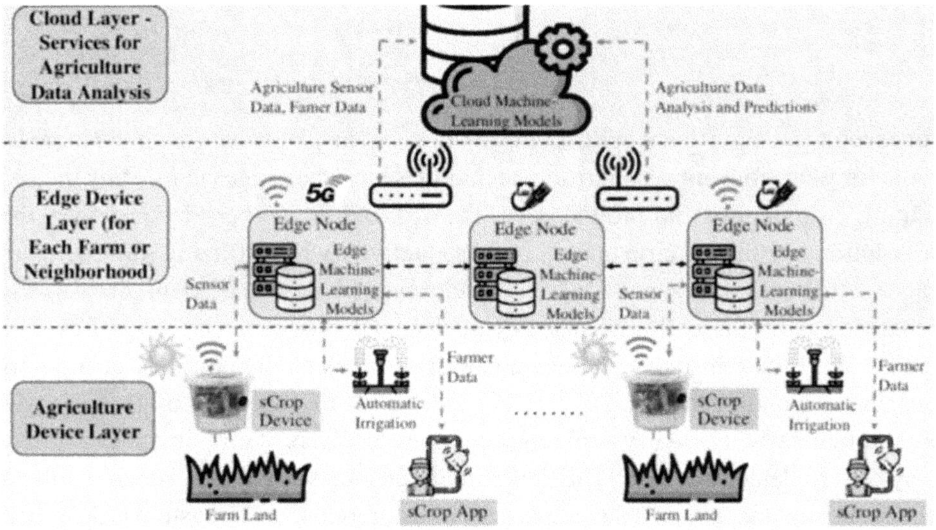

CONCLUSION

Modern agriculture must overcome enormous obstacles to create a sustainable future across the globe. On a worldwide scale, these issues include, for instance, population growth, urbanization, energy scarcity, environmental degradation, rising animal protein consumption, and, of course, climate change. Global efforts will need to be managed to ensure that the capabilities of the numerous farming sector, including crops, poultry, aquaculture, and forestry are not undermined in order to supply the world's food needs. Agricultural processes can be made more efficient by using all of the technologies discussed in this article. Although technology and automation can help lessen the load on humans required to achieve this, many agricultural practices must also be changed (for example, the reduction of fertilizers or chemicals in conjunction with the actual implementation of alternative tactics). In hopes of helping to develop and acknowledge successful enhancement and research actions with a genuine societal impact, the trend towards multiple disciplines dialogue with the involved stakeholders is growing rapidly, and the need to refine human and system/machine interfaces and interrelations is being taken into account.

REFERENCES

Abbasi, R., Martinez, P., & Ahmad, R. (2022). The digitization of agricultural industry – a systematic literature review on agriculture 4.0 smart agricultural technology Volume-2. *Technology in Society*, *69*, 100042.

Aldhyani, T. H. H., & Alkahtani, H. (2023). Cybersecurity for detecting distributed denial of service attacks in Agriculture 4.0: Deep learning model. *Mathematics*, *11*(1), 233. doi:10.3390/math11010233

Anitha, Muthusamy, & Thiruvenkadam. (2023). An Intelligent Agro-Cloud Framework For Efficient Crop Cultivation. *Journal of Pharmaceutical Negative Results*, 754–763.

Choubisa, M., & Doshi, R. (2022). Crop Protection Using Cyber-Physical Systems and Machine Learning for Smart Agriculture. In P. M. Rizwan (Ed.), *Handbook of Research on Cyber-Physical Systems for Smart Agriculture* (pp. 211–237). IGI Global.

Dev, P., Khandelwal, S., Yadav, S. C., Arya, V., Mali, H. R., Poonam, & Yadav, K. K. (2023). Conservation Agriculture for Sustainable Agriculture. *International Journal of Plant and Soil Science*, *35*(5), 1–11. doi:10.9734/ijpss/2023/v35i52828

ElKadiri, R., Momm, H. G., Bingner, R. L., & Moore, K. (2023). Spatial optimization of conservation practices for sediment load reduction in ungauged agricultural watersheds. *Soil Systems*, *7*(1), 4. doi:10.3390oilsystems7010004

Frankelius, P., & Rose, M. V. (2023). Towards the fifth innovation era in agricultural innovation. In Agricultural Bioeconomy (pp. 205-229). doi:10.1016/B978-0-323-90569-5.00003-2

Garg, M., Singhal, S., Sood, K., Rupeika-Apoga, R., & Grima, S. (2023). Price Discovery Mechanism and Volatility Spillover between National Agriculture Market and National Commodity and Derivatives Exchange: The Study of the Indian Agricultural Commodity Market. *Journal of Risk and Financial Management*, *16*(2), 62. doi:10.3390/jrfm16020062

Goyal, Prabha, Rai, & Singh. (2016). Indian Agriculture and Farmers-Problems and Reforms. In Indian Agriculture and Farmers (pp. 79-87). Poddar Publication.

Haider, R., & Joshy, J. K. (2022). An effect of farm mechanization on Indian agriculture with respect to rice yield. *South Asian Food and Agriculture Research and Policy*, 1-16.

He, Y. (2023). Knowledge and Technology System of Traditional Agriculture. In The Agricultural Civilization of Ancient China. doi:10.1007/978-981-19-8469-3_4

Herlitzius, T. (2017). *Automation and Robotics - The Trend Towards Cyber Physical Systems in Agriculture Business.* SAE Technical Paper 2017-01-1932.

Huang, C.-H., Chen, P.-J., Lin, Y.-J., Chen, B.-W., & Zheng, J. X. (2021). A Robot-Based Intelligent Management Design For Agricultural Cyber-Physical Systems. *Computers and Electronics in Agriculture*, *181*, 105967. doi:10.1016/j.compag.2020.105967

Kang, M., Wang, X., Wang, H., Hua, J., de Reffye, P., & Wang, F.-Y. (2022). The Development of AgriVerse: Past, Present, and Future. *IEEE Transactions on Systems, Man, and Cybernetics. Systems*. Advance online publication.

King, A. (2017). Technology: The Future of Agriculture. *Nature*, *544*(7651), S21–S23. doi:10.1038/544S21a PMID:28445450

Latha, S., Gomathi, T., Pavithra, S., Sudha, P. N., Nalluri, A., & Bhardwaj, P. (2023). Chitosan-Based Porous Carbon Materials for Agriculture and Agro-waste Applications. In A. N. Grace, P. Sonar, P. Bhardwaj, & A. Chakravorty (Eds.), *Handbook of Porous Carbon Materials. Materials Horizons: From Nature to Nanomaterials* (pp. 1–20). doi:10.1007/978-981-19-7188-4_28

Mirkouei, A. (2020). A cyber-physical analyzer system for precision agriculture. *Environmental Science Current Research*, *3*(1), 1–8. doi:10.24966/ESCR-5020/100016

Mitra, A., Singhal, A., Mohanty, S. P., Kougianos, E., & Ray, C. (2022). eCrop: A Novel Framework for Automatic Crop Damage Estimation in Smart Agriculture. *SN Computer Science*, *3*(4), 319.

Nair, J., & Thomas, B. K. (2022). Why is adoption of micro-irrigation slow in India? A review. *Irrigation and Drainage*, *71*(3), 413–427.

Rad, C.-R., Hancu, O., Takacs, I.-A., & Olteanu, G. (2015). Smart monitoring of potato crop: A cyber-physical system architecture model in the field of precision agriculture. *Agriculture and Agricultural Science Procedia*, *6*, 73–79. doi:10.1016/j.aaspro.2015.08.041

Raj, E. F. I., Appadurai, M., & Athiappan, K. (2021). Precision Farming in Modern Agriculture. In A. Choudhury, A. Biswas, T. P. Singh, & S. K. Ghosh (Eds.), *Smart Agriculture Automation Using Advanced Technologies*. Transactions on Computer Systems and Networks. doi:10.1007/978-981-16-6124-2_4

Sathiyamurthi, S., Ramya, M., Saravanan, S., & Subramani, T. (2023). Estimation of soil erosion for a semi-urban watershed in Tamil Nadu, India using RUSLE and geospatial techniques. *Urban Climate*, *37*, 101424. doi:10.1016/j.uclim.2023.101424

Sharpley, A. N., Duncan, E. W., & Hoag, D. L. K. (2019). Increasing the effectiveness and adoption of agricultural phosphorus management strategies to minimize water quality impairment. *Journal of Environmental Quality*, *48*(5), 1204–1217. doi:10.2134/jeq2019.03.0114 PMID:31589706

Sridhara, S., Manoj, K. N., Gopakkali, P., Kumar, V., & Krishna, K. R. (2023). Evaluation of machine learning approaches for prediction of pigeon pea yield based on weather parameters in India. *International Journal of Biometeorology*, *67*(1), 165–180. doi:10.100700484-022-02396-x PMID:36323951

Sundar, C. B. V., & Asokhan, M. (2023). Pivotal Role of Drones in Agriculture – a Review. *Asian Journal of Agricultural Extension. Economia e Sociologia*, *41*(4), 131–140.

Suresh Babu, C. V. (2023). *Introduction to Data Science* (1st ed.). Anniyappa Publications.

Suresh Babu, C. V. (2023). *Artificial Intelligence and Expert System* (1st ed.). Anniyappa Publications.

Udutalapally, V., Mohanty, S. P., Pallagani, V., & Khandelwal, V. (2021). sCrop: A Novel Device for Sustainable Automatic Disease Prediction, Crop Selection, and Irrigation in Internet-of-Agro-Things for Smart Agriculture. *IEEE Sensors Journal, 21*(16), 17525–17538. doi:10.1109/JSEN.2020.3032438

Vinci, A., Brigante, R., Traini, C., & Farinelli, D. (2023). Geometrical characterization of hazelnut trees in an intensive orchard by an unmanned aerial vehicle (UAV) for precision agriculture application. *Remote Sensing (Basel), 15*(2), 541. doi:10.3390/rs15020541

Wiyono, S., Suwardi, & Syakir, A. M. (2023). International Conference on Modern and Sustainable Agriculture (ICOMSA): Building Future Agriculture for a Better World. *IOP Conference Series: Earth and Environmental Science, 1133*, 011001.

Yadav, V. S., Singh, A. R., Raut, R. D., Mangla, S. K., Luthra, S., & Kumar, A. (2022). Exploring the application of Industry 4.0 technologies in the agricultural food supply chain: A systematic literature review. *Computers & Industrial Engineering, 169*, 108304. doi:10.1016/j.cie.2022.108304

Yoganandan, G., Rahman, A. A. A., Vasan, M., Senthilkumar, S., & Sathya, S. (2022). Evaluating agripreneurs' satisfaction: Exploring the effect of demographics and emporographics. *Journal of Innovation and Entrepreneurship, 11*(2), 2. doi:10.118613731-022-00193-9

Chapter 11
Smart Cyber-Physical System-Based Plant Disease Detection for Agriculture

R. Karthickmanoj
Academy of Maritime Education and Training, India

S. Aasha Nandhini
SRM Institute of Science and Technology, India

T. Sasilatha
Academy of Maritime Education and Training, India

D. Lakshmi
Academy of Maritime Education and Training, India

ABSTRACT

Agriculture cyber-physical systems are becoming increasingly significant in improving crop quality and output while employing the least amount of farmland possible. Many agricultural components and methods have been automated to produce faster and superior-quality items. As a result, several methods and techniques to help prevent or reduce plant diseases have been created. Imaging analysis tools and gas sensors are increasingly being included in innovative cyber agribusiness for plant disease monitoring. This chapter develops intelligent cyber-physical system-based plant disease detection for successfully preventing and managing plant diseases and decision-making. To extract the important features linked to the sick region, a novel multi-statistical feature extraction technique has been developed. The proposed methodology is implemented using a Raspberry Pi board running Stretch OS. Metrics like detection and classification accuracy are used to assess the efficacy of the multi-statistical feature extraction methodology on the plant disease detection system.

DOI: 10.4018/978-1-6684-7879-0.ch011

SMART AGRICULTURE

Plant diseases cause significant losses to farmers in agricultural fields. According to a study, crop losses due to diseases, pests, animals, and weeds account for 20 to 40% worldwide agricultural production. The conventional approach of identifying illnesses by visually evaluating specific properties of leaves, such as texture, color, and shape, is only sometimes practical. As a result, most farmers worldwide engage expert agriculturists to identify illnesses in their crops on huge farms. However, it is a time-consuming and costly process.

Smart agriculture produces around 26% of India's GDP and employs over 60% of the population. Farmers estimate that global food output will more than treble by 2050. Based on acquired data on humidity, temperature, sunshine, CO2, and moisture level, our innovative agriculture solution with an AI integration model can precisely forecast plant growth's favorable or unfavorable state. This IoT-enabled system helps significant commercial farming operations reduce costs and improve food security by facilitating early diagnosis and control of plant disease.

Some farmers use a conventional system that needs more current tools for automating plant disease identification and categorization. Farmers on big farms must detect plant diseases, which will result in a major drop in the number of quality of agricultural goods. As a result, automated plant disease detection systems are critical for farmers to boost crop output by continuously monitoring plant disease with little effort in remote agricultural locations.

INTRODUCTION

Crop diseases detection is critical in agriculture because workers must continuously check whether the crop harvested is adequate. Taking an critical since it can create serious plant issues that influence quality of the product, volume, and performance. Plant diseases generate disease epidemics on a regular basis, resulting in widespread death and a large economic effect. Computer vision techniques may be used to do image-based automated inspection. Manual identification takes longer, requires greater precision, and can only be done in small amounts. This technique can identify early plant infections, and infection management measures may be utilized to handle pest issues while decreasing dangers to the environment and public people. Because visual monitoring plants for illnesses is less accurate and time-consuming, machine learning automates diagnosing diseases based on features extracted from photos. Image acquisition and extraction of feature are carried out at the field to be watched. Experts classify the monitoring station. Finally, the farmers are advised on how to solve the situation via text messages or a mobile app. The farm's camera sensor captures images of the plant leaves. The idea is then further processed to identify and segment the affected leaf section.

The agricultural specialist extracts features from the segmented section and sends them to him for study. The farmer could appropriately apply pesticides to the affected areas if he had precise knowledge of where the illness had spread, resulting in economic and environmental benefits. Pre-processing and segmentation are critical components of a vision-based disease diagnosis system. Attribute extraction from either the split image is required for effective classification. The process of grouping or splitting an image into various portions is known as image segmentation Khirade (2015). Different techniques can be used to segment an image, ranging from simple to complex segmentation procedures.

To extract features, color, texture, and area are used. Experts use classifiers to categorize data. The researchers Singh (2016) suggested unique image segmentation techniques for detecting as well as categorizing crop leaves illnesses automatically. The images were segmented using the genetic algorithm and classified using.

The author Dhakate (2015) suggested a method based on artificial neural networks for identifying diseases in pomegranate plants. The diseases that will be demonstrated. With a 90% accuracy rate, the proposed method produced satisfactory results. The author Aasha Nandhini (2018) presents a proposed method to detect and identify illnesses in crop using compressed sensing (CS). For several plants, the plan was tested using simulation and experimental analyses. The data were classified using SVM, and the results showed 98.5 percent accuracy.

LITERATURE SURVEY

Researchers Shoaib (2022) developed a way for identifying tomato infections using a deep learning-based system that leverages visual data from plant leaves in this consideration. In this work, newly built convolutions trained on 18,161 segmented and unsegmented tomato leaf pictures were used to identify and diagnose several tomato illnesses using the Inception Net model. We employed a neural network-based deep learning architecture. In this area of research, two cutting-edge semantic segmentation models are utilized to detect and classify illness areas. The model ranked binary plant leaf pixels as environment and region of interest.

A unique system Pandian (2022) was proposed in this study to recognize plant leaf diseases. The dataset's class sizes were balanced using data augmentation techniques. One thousand training epochs of the suggested deep convolutional neural system were conducted in an environment containing several graphics processing units. The most appropriate hyper-parameter values were chosen randomly for improved training performance using the coarse-to-fine searching strategy.

Throughout this process, artificial intelligence has been employed to comprehensively examine the different methods used in crop disease diagnosis. In computer vision, deep learning has become increasingly important for improving performance results for identifying plant diseases. Due to the application of deep learning advancements to various disciplines, there has been substantial development in the machine learning and computer vision domains. Several research studies have compared the performance and applicability of the two methodologies to establish the advantage of the deep model over the algorithm for machine learning. Computational intelligence technology has the ability to decrease farming losses dramatically. In this performance of ML as well as DL in identifying citrus illnesses is compared in this article Jackulin (2022). The accuracy of disease categorization achieved through trials is excellent since the DL technique surpasses the ML method.

To segregate the target region of the rice grain image from the background, the suggested approach employs a modified K-Means segmentation algorithm. Following segmentation, we extract features based on three characteristics: color, shape, and smoothness. The researcher Sujatha (2021) suggested novel density color extracting features approach was used to extract color features. At the same time, texture characteristics were retrieved using bit pattern characteristics. Grayscale co-occurrence matrices are defined by the shape features listed below: Determined by measuring the afflicted part's area and diameter. After that, the photographs are categorized using a novel support vector machine-based deep learning architecture that recognizes distinct feature values.

By using improved K-means segmentation technique, the targeted region of the rice grain image is separated from the external environment. Following the extraction of attributes based on aspects, a novel intensity-based color image retrieval approach for obtaining textural properties is given in this article Archana (2022). The dimension and width of the afflicted sections, on the other hand, establish structural properties. The images are then categorised to identify distinct attribute values that used the ground-breaking support vector computer bayesian neural network technique.

The author Pantazi (2019) proposed technique demonstrates an automated method for agricultural diagnostic testing that use for feature extraction and Classifier to categorize varied leaf sample images from various crop kinds. When employing the proposed technique, every crop health phase uses its own One Description. The grape leaf-trained systems demonstrated flawless generalization when tested in other crops. Whenever ambiguous test data can belong with one or more criteria, a novel technique based on resolving conflict among Each Systems – determines the correct state.

This article Ferentinos (2018) used deep learning techniques that build neural network convolutional models to recognise and detect plant ailments employing basic leaves images of normal & sick plants. The models were programmed would use an available set of images that included 25 different plants in 58 different pairings, including plants grown. Numerous model architectures are developed, having the most effective system recognising plants with a high degree of success.

Throughout this study, the utility of which were previously CNN-based features against traditional, hand-crafted descriptors for identifying varied color textures in a variety of imaging situations is compared. This work Bello (2019) contains 46 groups of 23 color texture datasets, 68 picture descriptors, and ten experimental settings. Overall, the findings reveal that deep networks are significantly superior, especially when dealing with non-stationary textures and when the acquisition settings are changed often. On the other hand, manually constructed descriptors outperformed CNN-based features in terms of resistance to picture rotation and the ability to differentiate stationary textures under steady imaging settings.

Integrates encoded and decoded information to improve model performance even when complex backgrounds and barriers are present. The adversarial training approach uses long-range spatial labels to overcome the problem of building objects being obscured by existing occlusions, such as trees, vehicles, and shadows, resulting in high-quality building segmentation results under challenging settings. The quantitative data acquired by the proposed method Abdollahi (2020) may improve outcomes by finding and correcting the disagreement between the segmentation model output and the reference map.

Using three datasets comprising images of distinct types of insects. Thenmozhi, K (2019) established an outstanding deep model for classifying species of insects. The following methodology was assessed and compared to previously train deep learning systems for insect's categorization. To improve the pre-trained systems, transfer learning was applied.

To enhance categorization among broadleaf species, the author Le V.N.T.,(2020) devised an unique technique that combines filtered data recovered by linked Local Binary Pattern operators with characteristics recovered using seedlings contour filters. The use of morphological operators that open and close was used to minimize noise in Crop images. But use a testbed setup, images were obtained at four phases of development.

Tomatoes are among the most widely consumed vegetables, because tomatoes may be used in a variety of ways in any household, independent of culture. However, as a result of several infections, the quality and quantity of tomato harvests deteriorate. As a result, Agarwal (2020) outlines a deep learning-based strategy for detecting disease. For sickness diagnosis and categorization, a Convolution Neural

Network-based approach is applied. The suggested model outperforms the other pre-trained models, as demonstrated by the testing results.

This study Javidan (2023) suggests using a faster suggested model to detect healthy tomato leaves. To begin, in order to understand more about the illness, we substitute VGG16 with a residual depth network for image feature extraction. Finally, employing three distinct types of feature extraction networks, we do a k-means experiment. The revised technique for agricultural leaf disease detection outperformed. According to the testing results, the earliest Faster Classifier outperformed the original in terms of identification precision and detection time by 2.71%.

The author Azgomi (2023) proposed a novel deep residual convolutional neural network of layers in this study for identifying various plant leaf diseases. ResNet197 was built with six-layer blocks. ResNet197 was trained and tested using a pooled leaf disease in plants image dataset. ResNet197's layers and hyper parameter values were optimized using the evolutionary search technique. It attained an average accuracy of classification of 99.58 percent on the testing dataset. These experimental findings outperformed both conventional ResNet topologies and newer transfer learning approaches.

Early diagnosis of banana illnesses is critical in order to discover the diseases before they do too much damage to the plants. This paper Haridasan, (2023) proposed a novel based technique for early identification of banana illnesses. Five deep-learning architectures were employed to create models for banana disease detection, with all models achieving high accuracies. The created technique was capable of detecting illnesses in 99% of the collected leaves from the actual world. This technology will assist smallholder farmers in detecting banana illnesses early and increasing their yield.

This research Divyanth, (2023) focuses on the fundamental problems and obstacles associated with classifying leaf diseases. There is also a comparison of several methods based on an agricultural product, technique, efficiency, benefits, and drawbacks. The review research looks at the frequently employed ML algorithms, finding that the Support Vector Machine is frequently employed for disease classification. A evaluation of various approaches and their related accuracy is also undertaken, and it is discovered that features paired with Linear SVM offer the greatest accuracy.

Deep learning is frequently used to identify plant leaf fungal diseases. However, it cannot analyze fuzzy data on its own. To manage this data, a deep neuro-fuzzy network is used. However, owing to disadvantages such as requiring an appropriate training set and necessitating retraining when a new challenge occurs in this approach Meng, X(2023), the deep neuro-fuzzy network is only suitable for some-shot tasks. At the same time, the meta-baseline with the deep neuro-fuzzy network as the backbone is an excellent choice since it combines a deep neuro-fuzzy network with meta-learning to generalize from such a small group of cases to a large set of new data. Experiment data indicate the model's benefits. This model is excellent for detecting single-shot plant leaf detection.

This Luo, (2023) work provides a novel approach to detecting illegal opium poppy growing using drone aerial imagery. The exact segmentation of the poppy plant from the acquired picture is critical to this endeavor. Target regions must be recovered based on unique morphological aspects of the poppy plant in order to build a segmentation mask that is close to actual data, and intricate environmental influence must be eliminated. By augmenting too-small samples with a hybrid approach technique, the poppy plant's bitmap qualities are improved. Second, the proposed method with the self-attention mechanism is improved to segment the bigger dataset. To cope with the special physical characteristics of poppy species, the multi-head self-attention module is modified in this way by applying relative location encoding. The findings showed that the suggested approach could accurately segment the poppy plant.

The authors of this paper, Vallabhajosyula (2022), created a novel approach for autonomous plant disease diagnosis. Transfer learning is used to improve previously taught models. The research provides a comprehensive taxonomy of the efficacy of many pre-trained neural nets, as well as the efficacy of a weighted ensemble among those systems vital for identifying plant leaves sickness. Moreover, the suggested method's effectiveness is examined using just a publicly available crops community data that includes classifications generated on fourteen crops. The effectiveness testing of the proposed model shows it is efficient at systems can aid kinds of plant ailments and beats pre-trained models.

In this researcher Ashwinkumar, S(2022) to determine the damaged areas of the leaf image, the suggested thresholding-based image segmentation is used. As an image retrieval strategy, mobile net techniques are applied, with hyper parameters tuned using the emperor penguin optimization algorithm to increase crop disease diagnosis rate. To offer suitable majority class to a plant images, a novel-based classifier is applied. A large amount of simulations were done to illustrate the increased efficiency of the proposed model. The outcomes of the experiments showed that the OMNCNN model outperformed earlier region techniques in terms of maximum efficiency.

Harakannanavar, (2022) samples of tomato leaves with faults are studied in this study. Early warning indications will allow farmers to detect illnesses. They were testing tomatoes leaf disease samples. Tomato salad leaves are stretched to be subjected to Histogram. Equalization is used to enhance the quality of tomato leaf samples. K-means sorting is used to split data space. The margins of leaf samples are extracted via shape tracing. To identify the informative properties of both the leaves information, many descriptors are employed. ML algorithms are used to classify the collected characteristics. SVM is used to assess the correctness of a prediction.

Fungal diseases are among the factors influencing agricultural productivity. Early detection of plant diseases reduces crop loss and assists farmers in reducing risk while improving profitability. As a result, automated leaf illness identification is important in limiting the disease transmission as well as reducing harm. The goal of this study is to look into a method for studying and identifying plant ailments that makes use of image analysis technologies and ensembles of machine learning algorithms. Rani, (2021) the experimental findings reveal that ensemble classifiers outperform some other baseline classifiers in terms of prediction accuracy.

Chen, Z., (2022) recommended strategies were applied to improve the network model's target recognition impact. Throughout the experimental stage, to evaluate the model's performance, sample images were randomly picked from the resulting trees illness data to construct test and training sets. The test findings revealed that perhaps the upgraded YOLOv5 network's mean average accuracy was higher. The upgraded YOLOv5 network outperformed both the version of network models in regards to overall detection accuracy. The enhanced model accurately recognizes plant illnesses in natural settings and serves as a technical guide for crop disease treatment and prevention.

This article Orchi (2022) provides a current summary of research done over the last year in the area of illness diagnosis of various crops. A comparison evaluation of numerous crop disease detection systems was also conducted. Moreover, this article examines the many problems to be faced as well as potential solutions. Following that, numerous solutions to these problems are proposed. Furthermore, this study presents a future viewpoint that is a highly helpful and beneficial resource for agricultural disease diagnosis investigators.

Three distinct rice illnesses are considered in this Sharma, (2022) study. A careful comparison of the data shows that transfer learning approaches outperform traditional methodologies for machine learning. InceptionResNetV2 is shown to produce the most significant results, followed by XceptionNet. The research may be integrated into supporting growers with early identification of rice illness, allowing for timely action to be taken. In the future, efforts should be devoted to working with larger datasets in order to generalize the experiment's findings.

On the offered Turkoglu (2022) subject, the six cutting-edge networks in the suggested models are assessed separately as well as an ensemble. Eventually, the results of various pairings utilizing the proposed concepts are shown. We gathered images of 15 different types of diseases and pests widespread in Turkey to assess the validity of the suggested method. The results demonstrate that the suggested approaches achieve or outperform region performance in this setting.

Babu's paper (2022) discusses a DL technique for identifying paddy grains disorders, as well as the common leaf diseases that this study discovered. Deep learning algorithms are used to identify and diagnose rice plant disease. This approach of early identification of grain illnesses could be employed as a prophylactic as well as a diagnostic technique. When compared to current region procedures, the suggested methodology has a higher accuracy of 99.45%.

Bhati (2022) analyzed the most commonly used methodologies for identifying plant diseases throughout this study. It includes a variety of approaches for identifying crop. Subsequently, researchers offered a hybrid approach to identifying crop sickness, and we expect superior prediction with this suggested model.

Hassan (2022) proposed special framework on the inception layer and residue link in this work. To limit the number of variables, a depth - wise separable convolution was utilised. Three separate plant pathogens datasets were used to create and assess the conceptual model. In terms of precision, the proposed approach beats region deep learning algorithms.

Several soft computing approaches for automating the identification of crop diseases have been developed. Vishnoi (2021) review summarizes numerous aspects of such research, as well as their advantages and disadvantages. Modern feature extraction algorithms are investigated in order to find those which seem to operate effectively over many harvest leaves. The findings might aid academics in better understanding the use of artificial intelligence in agricultural disease tracking.

In this study, Abbas (2021) provide a method for detecting plant disease by utilizing the Probabilistic suggested system to produce synthetic pictures of tomatoes leaf tissue. The model is then trained on synthetic and natural pictures using transfer learning to categorize leaf spot data into ten different disorders. Using the already available data, the research approach was fully tested & trained. The proposed approach achieved a high degree of precision in tomato leaf image categorization. The proposed solution clearly outperforms the present system.

In this work, the hybrid model proposed by Bedi, P (2021) was used to diagnose Bacteria Spots illness in fruit trees utilizing leaf images, but it might also be often used identify other diseases in plants. A suggested framework necessitates smaller training variables than initially disclosed methods. As just a consequence, the time required to train the systems for independent crop detection of pathogens and diagnostic testing in crops that used the learning algorithm is greatly decreased.

A computational intelligence strategy is suggested to recognize and categorize bean plant diseases utilizing a public dataset containing leaf images and the open-source framework Tensor Flow. In this research, Elfatimi, (2022) developed a method for classifying bean crop diseases as well as identifying and describing the most efficient network design. Furthermore, we analyzed the outcomes of each architecture after applying it independently to determine the optimal architectural configuration for categorizing beans leaf illnesses and their outcomes. Furthermore, to meet the categorization objectives, they were trained using the Method was used in order framework in controlled environments such as

Feature selection methods to test whether we could achieve faster training timelines, higher accuracy, and easier retraining.

The author Suhag, S (2021) offered a strategy for the Internet of Things. Ground sensors, such as proximal ground sensors to test soil fertility that help in the analysis of the situation of ground agricultural water, ploughing, fluid, or land for post-harvest, are used to collect tree data in the form of images at various intervals. These devices also identify crop diseases and ground micronutrients. Also utilized are temperature sensors. Detectors for groundwater resources are used to continuously check the water's quality. The gardener will receive all information via IoT. Binarization thresholding is used for picture categorization. The robot does picture classification and recognition during harvesting.

The latest research Maray (2022) describes an artificial intelligence (AI) powered classification and detection algorithm for coconut disease for sustainable farming. The proposed method described here seeks to classify oil palm tree diseases in a precision agriculture setting to increase crop yield. The Proposed method begins by using a noise removal method based on median filtering. The Nave Bayes fuzzy swarm segmentation is then used to find the impacted leaf regions. Another application of the capsule required to advance is as a feature representation. The goal of this work is to identify diseases in coconut trees using the proposed model. The developed framework was used in the experimental investigation, and the findings were outstanding.

Farmland factors like humidity and temperature soil moisture content, and plant diseases are tracked by the planned Somasundareswari (2022) plan using categorization techniques. The system uses a Raspberry Pi to automate tasks and detect diseases. Through a Web page, the farmers can see the sensor information. The farmer is automatically emailed information about any illnesses discovered in the plants.

We uncovered the most prevalent and damaging illnesses impacting wheat crop yields Liu, Z (2022), as well as 45 researches upon that identification and categorization of bacterial infections that have plagued six cereal grains in previous years in this study. We also located and compiled a significant number of publicly available datasets, which we highlighted as one of the key hurdles in investigating the use of machine learning in agricultural production detection. According to the results of this investigation, deep convnets based on spectral images are the most effective method for detecting early-stage illnesses, and ensemble learning is one of the successful and widely used training techniques.

Manually crop monitoring is a labor-intensive, sluggish, and error-prone operation in conventional agriculture. Plant disease-related economic and production losses can be reduced with continuous plant monitoring. By automating the real problem of leaf disease identification using deep deep learning as well as the Internet - Of - things, the current solution Zamani (2022) significantly advances precision farming. Image acquisition mechanization with Raspberry Pi and smart line tracking robot, leaf identifier to YoloV3 CNN model, disease spread with image analysis, disease detection with InceptionV3, and fully automated pesticide sprayer are all covered in this paper.

Jiang, P. (2019) presents a deep learning technique towards real-time detection for diseases of apple leaves utilizing improved neural networks. The dataset for apple plant leaf, which would be made up from experimental shots and complicated imagery acquired in the environment, was created in this study utilizing data mining methods and image processing and analysis tools. On this basis, the GoogleNet Inception structure and Rainbow concatenation are utilized to develop a novel apple leaf disease detection model based on deep models. Finally, utilizing a collection of pictures of sick apple leaves from the hold-out test set, the proposed system is trained to identify these five frequent apple leaf illnesses.

Ozguven (2019) suggested an automated diagnosis of leaf spot infection in sugar beets in this paper. Based on the test findings, the technique presented for diagnosing sickness severity using imaging-based expert systems was trained and tested using 155 images, and the overall accurate classification rate was judged to be 95.48%. Moreover, the proposed approach revealed how adjusting CNN settings in accordance with the image and regions to be recognized might increase the Fast R-CNN architecture's success. Considering important parameters, this proposed strategy outperforms all current methods mentioned in previous research. As a consequence, the technology is projected to reduce the time spent detecting sugar beet leaf spot disease in big production areas, as well as human error and time spent establishing the cause.

Atila, (2021) proposed the Efficient Net deep learning architecture for plant disease detection in this research, and its performance was compared to that of other cutting-edge deep learning models. The Plant Village dataset was used to develop the models. The basic and upgraded information were used to train all of the algorithms. The Efficient Net architecture and other deep learning models were trained via transfer learning. In the transfer learning, all layers of the models were made to be trainable. When compared to other deep learning models in the original and supplemented datasets, the Efficient Net architecture achieved the highest values in the test dataset.

This research Sardogan, (2018) describes proposed algorithm-based technique for detecting and classifying tomato leaf disease. The collection comprises five hundred images of tomato leaves with four disease signs. We created a CNN model for automated extracting features and categorization. Hue data is workers in the world in leaf disease of plants research. The effects in our model are used on three channels depending on Spectral information. For network training, the LVQ was fed the output feature vector of the convolution component. These research results demonstrate that the proposed method accurately identifies four types of tomatoes plant leaves.

Pooja, V (2017) research provides an illness classification and detection approach that utilizes machine learning principles and image analysis technologies. First, the diseased region is identified & captured, and afterwards picture processing is conducted. Following that, the sections are acquired, the region of interest is identified, and an extracted feature is performed. Ultimately, all gathered data were submitted via SVM Classifiers to get the results. The Support Vector Machines outperform the job of illness categorization; their findings suggest that the method proposed in this study produces significantly better results than formerly utilized sickness detection methods.

The purpose of this research Padol (2016) is to help inside the identification and classification of grape disease identification using the SVM classification method. The sick region is first identified using segmented by K-means grouping, and afterwards color - shape characteristics being retrieved. Lastly, another categorization approach is employed in order to identify the kind of plant leaf. The suggested approach detects and classifies the tested illness with an efficiency of 88.89%.

In this article, researchers Bouali, E.T (2021) shed further light on the ICT components of to Smart agriculture and describes the details of a real-world testbed implementation. The overall architecture, wireless communication infrastructure, and processing underlying the control unit are all presented. The latter is a Fuzzy logic-based intelligent system. Furthermore, we provide venues for integrating renewable energy into Smart agriculture and argue for the use of batteries for efficient energy consumption.

This Chapter proposed a unique multi-feature extraction approach that improves classification accuracy. The objective is to use the image-splitting approach to extract several characteristics from the picture image. The suggested technique's performance is measured in terms of detection and classification accuracy. A comparison with the present K-means classification technique is also provided. The following is an outline of the Chapter. The second section goes more into the suggested framework.

Section 3 goes further into evaluation process. Section 4 concludes with a conclusion and future study recommendations.

MULTI-STATISTICAL FEATURE EXTRACTION FRAMEWORK FOR CLASSIFICATION

Figure 1. Proposed framework

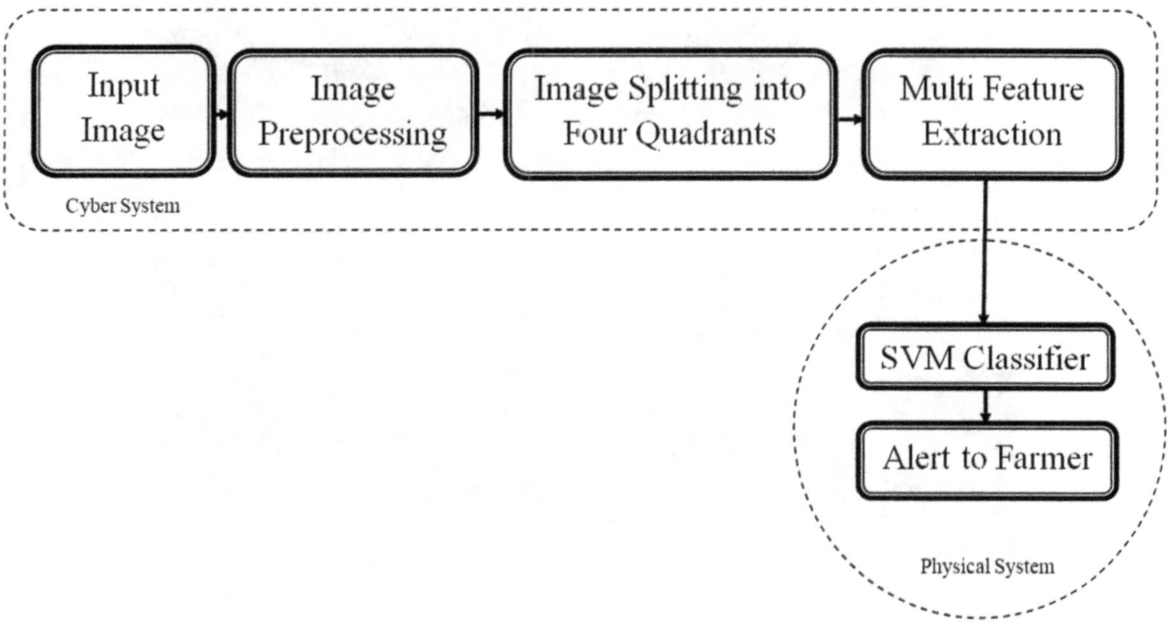

A novel multi-statistical feature extraction (MSFE) technique is proposed for the leaf disease detection system. The system is used to monitor plant leaves in the field from afar. This technique makes it easier to identify the illness at an early stage. The system will collect the image, pre-process it, then split it into four quadrants to collect multiple features. This technique is advantageous since it reduces visual monitoring by farmers in person in fields and can detect the early onset of diseases as they emerge on crop leaves. Figure 1 depicts an overview of the suggested framework. The framework's process flow is as follows: The video sensor captures images of the leaves, which are then pre-processed to improve image quality. The images are acquired from the field at regular intervals. To increase the quality, the image is enhanced before being translated into the HSI model.

Figure 2. Input pomegranate leaves: (a) Alternaria alternata, (b) Anthracnose, (c) Cercospora leaf spot, (d) bacterial blight

Figure 3. Image splitting done on H component of HSV transformed images: (a) Alternaria alternata, (b) Anthracnose, (c) Cercospora leaf spot, and (d) bacterial blight

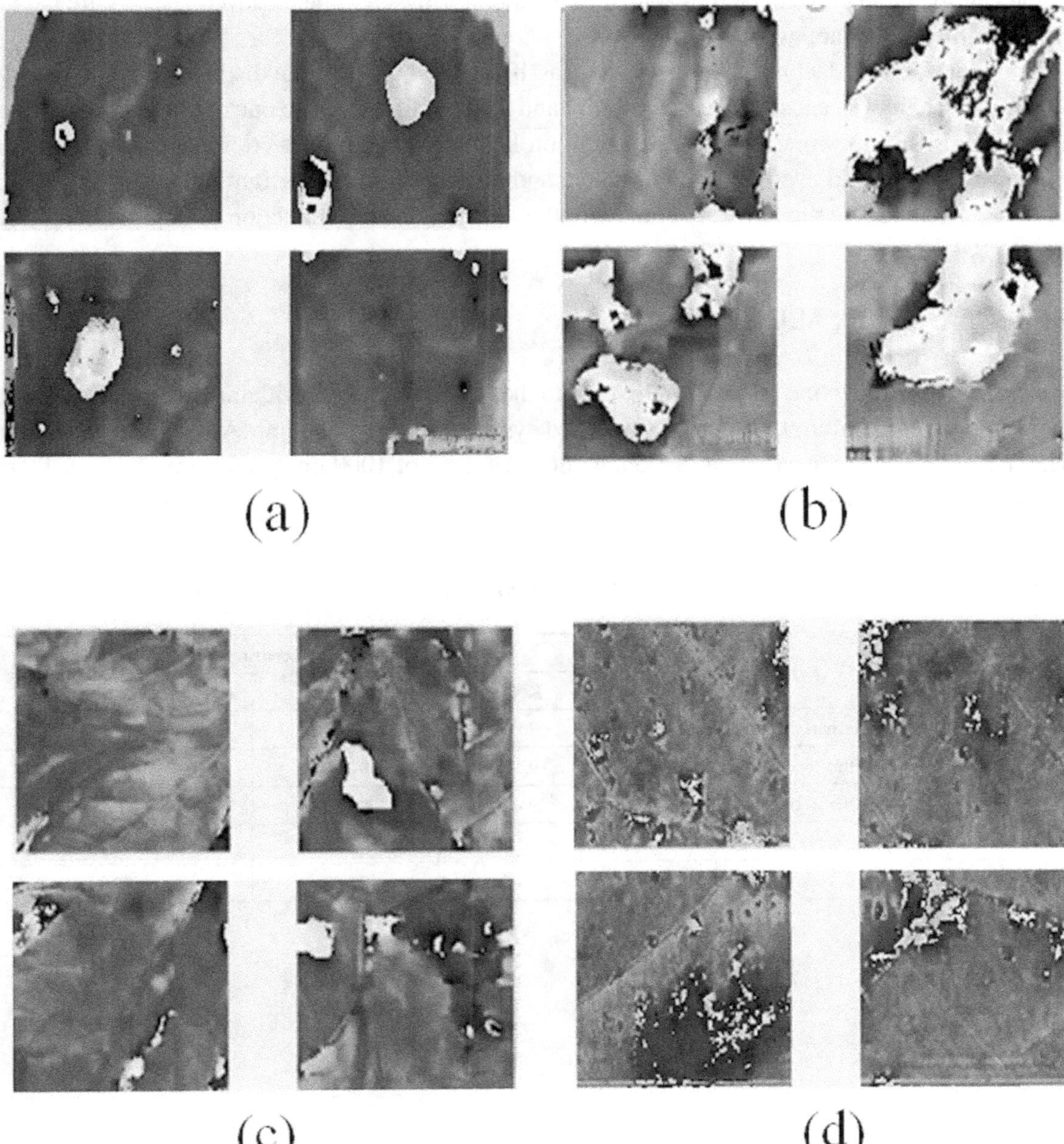

The H component of the HSI transformed image is then spitted into four quadrants. Different features are extracted from each quadrant. For feature extraction, GLCM features and ORB descriptors are employed. The first and third quadrants are used to extract GLCM characteristics. The second and fourth quadrants are used to extract ORB characteristics. The Grey-level co-occurrence matrix is used to extract texture-based GLCM properties such as energy, contrast, homogeneity, entropy, and correlation. ORB is a hybrid of the FAST key point detector and the BRIEF descriptor, with a few modifications. An ORB detector is ideal replacements for SURF and SIFTS detectors. Once features are extracted, the mean is computed and fixed as the threshold. The threshold will be compared with the extracted features to select the significant features to be transmitted to the monitoring site. The retrieved features are fed into the classification algorithm as training data.

The Support Vector Machine is a supervised classification algorithm that distinguishes between two groups by using a line. Because of its accessibility and adaptability, it is also one of the most extensively used algorithms. In this work the main contribution lies in the feature selection part hence improved accuracy can be achieved even by using a simple and efficient SVM algorithm. Following classification, the disease is sent to the farmers, along with the solution, for further action to improve crop yield.

PERFORMANCE EVALUATION

After classification, the found disease is delivered to the farmers, together with the treatment, for further action to improve agricultural yield. The suggested system was tested using MATLAB 2019 software. Pomegranate leaves were used for validation, while a database of 1000 images for training and 400 test images was used for assessment.

Table 1. Comparison with various machine learning algorithm

Crop Considered	Disease Name	Detection Accuracy (%)		
		KNN	RFC	Proposed System
Pomegranate	Alternaria Alternata	96.2	99.12	99.50
	Anthracnose	92.0	96.78	97.38
	Bacterial blight	93.0	96.61	97.20
	Cercospora leaf spot	94.3	97.87	98.9
Overall Accuracy		**93.875**	**97.57**	**98.3**

Figure 4. Comparison with various machine learning algorithms

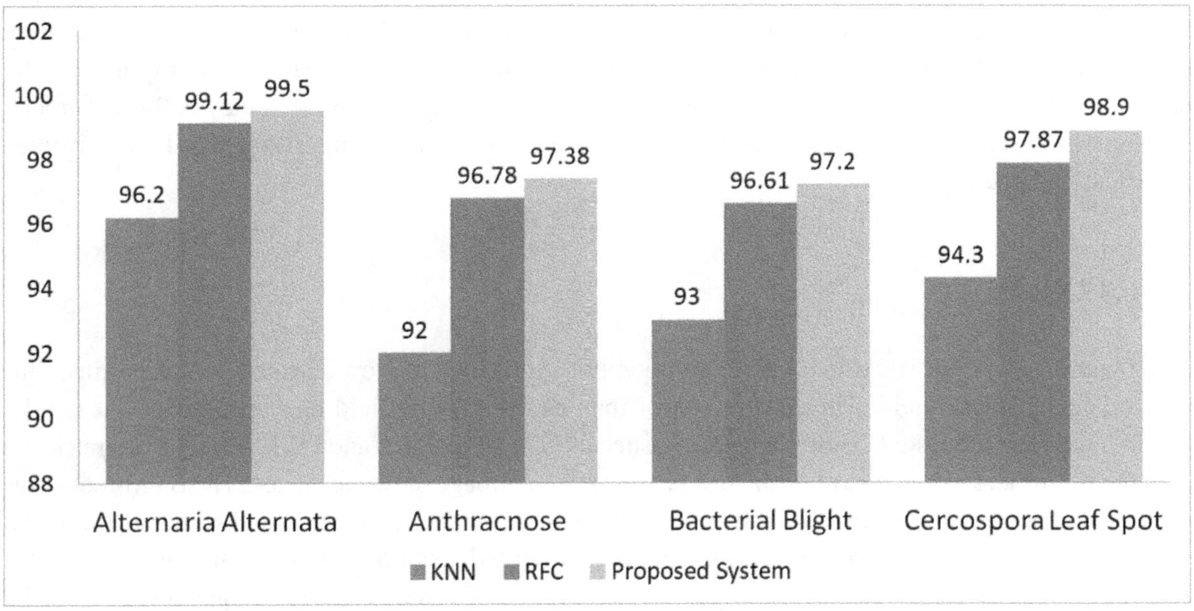

Kaggle was used to create the database. Among the diseases that will be displayed are Alternaria alternata, Anthracnose, Bacterial Blight and Cercospora leaf spot. The suggested system's performance is evaluated in terms of classification and detection accuracy. Figure 2 depicts the input images for the training dataset for various diseases. Figure 3 shows the image splitting of the pre-processed images.

The system's detection accuracy (Acc) is computed using (1).

$$Acc=P/T*100\% \quad (1)$$

Where P is the number of correctly identified images, and T represents the total amount of images included for testing. The K-means clustering algorithm is used for comparison with the proposed technique in order to assess the framework's performance.

Table 2. Shows how accurate the system is for four pomegranate leaf diseases

Leaf Disease	*Alternaria alternata*	Anthracnose	Bacterial Blight	Cercospora Leaf Spot	Accuracy
Alternaria alternata	397	1	0	2	99.25
Anthracnose	1	398	0	1	99.5
Bacterial Blight	0	0	400	0	100
Cercospora Leaf Spot	0	1	0	399	99.75
Average					9.63

Table 1 shows that the suggested approach achieves an average accuracy of 98.3 percent, which is higher than the existing KNN and RFC machine learning algorithm described in the literature. The accuracy of each disease's categorization is tabulated in Table 2, which considers the four different pomegranate leaf diseases. Classification accuracy determines the rate of correctly classifying specific diseases from test images. According to Table 2, the overall categorization accuracy is 99.75 percent. In comparison to previous work, the proposed MFE technique and SVM algorithm have higher average classification accuracy

CONCLUSION

Crop diagnostic system is a critical agricultural application since it allows farmers to detect infections early on and increase yield. The camera sensor may be used in the field to record images at regular intervals or when requested. Multiple features, such as GLCM features and ORB features, are retrieved from the pre-processed picture after the pre-processing and image-splitting stages. The SVM method is trained using photos of pomegranate leaves with four unique diseases. The suggested system is simulated in MATLAB, and the results demonstrate that it has an overall detection accuracy of around 98.3 percent and a classification accuracy of approximately 99.63 percent. The suggested technique outperforms the k-means clustering algorithm.

REFERENCES

Aasha Nandhini, S., Hemalatha, R., Radha, S., & Indumathi, K. (2018). Web enabled plant disease detection system for agricultural applications using WMSN. *Wireless Personal Communications*, *102*(2), 725–740. doi:10.100711277-017-5092-4

Abbas, A., Jain, S., Gour, M., & Vankudothu, S. (2021). Tomato plant disease detection using transfer learning with C-GAN synthetic images. *Computers and Electronics in Agriculture*, *187*, 106279. doi:10.1016/j.compag.2021.106279

Abdollahi, A., Pradhan, B., Gite, S., & Alamri, A. (2020). Building footprint extraction from high-resolution aerial images using generative adversarial network (GAN) architecture. *IEEE Access : Practical Innovations, Open Solutions*, *8*, 209517–209527. doi:10.1109/ACCESS.2020.3038225

Agarwal, M., Singh, A., Arjaria, S., Sinha, A., & Gupta, S. (2020). ToLeD: Tomato leaf disease detection using convolution neural network. *Procedia Computer Science*, *167*, 293–301. doi:10.1016/j.procs.2020.03.225

Archana, K. S., Srinivasan, S., Bharathi, S. P., Balamurugan, R., Prabakar, T. N., & Britto, A. (2022). A novel method to improve computational and classification performance of rice plant disease identification. *The Journal of Supercomputing*, *78*(6), 8925–8945. doi:10.100711227-021-04245-x

Ashwinkumar, S., Rajagopal, S., Manimaran, V., & Jegajothi, B. (2022). Automated plant leaf disease detection and classification using optimal MobileNet-based convolutional neural networks. *Materials Today: Proceedings*, *51*, 480–487. doi:10.1016/j.matpr.2021.05.584

Atila, U., Uçar, M., Akyol, K., & Uçar, E. (2021). Plant leaf disease classification using EfficientNet deep learning model. *Ecological Informatics*, *61*, 101182. doi:10.1016/j.ecoinf.2020.101182

Azgomi, H., Haredasht, F. R., & Motlagh, M. R. S. (2023). Diagnosis of some apple fruit diseases by using image processing and artificial neural network. *Food Control*, *145*, 109484. doi:10.1016/j.foodcont.2022.109484

Babu, S., Maravarman, M., & Pitchai, R. (2022). Detection of Rice Plant Disease Using Deep Learning Techniques. *Journal of Mobile Multimedia*, 757-770.

Bedi, P., & Gole, P. (2021). Plant disease detection using hybrid model based on convolutional autoencoder and convolutional neural network. *Artificial Intelligence in Agriculture*, *5*, 90–101. doi:10.1016/j.aiia.2021.05.002

Bello-Cerezo, R., Bianconi, F., Di Maria, F., Napoletano, P., & Smeraldi, F. (2019). Comparative evaluation of hand-crafted image descriptors vs. off-the-shelf CNN-based features for color texture classification under ideal and realistic conditions. *Applied Sciences (Basel, Switzerland)*, *9*(4), 738. doi:10.3390/app9040738

Bhati, H., & Rathore, M. 2022. Diagnosis and Detection of Plant Diseases Using Data Mining Techniques. In Proceedings of Data Analytics and Management: ICDAM 2021, Volume 1 (pp. 451-458). Springer Singapore. doi:10.1007/978-981-16-6289-8_38

Bouali, E. T., Abid, M. R., Boufounas, E. M., Hamed, T. A., & Benhaddou, D. (2021). Renewable energy integration into cloud & IoT-based smart agriculture. *IEEE Access : Practical Innovations, Open Solutions*, *10*, 1175–1191. doi:10.1109/ACCESS.2021.3138160

Chen, Z., Wu, R., Lin, Y., Li, C., Chen, S., Yuan, Z., Chen, S., & Zou, X. (2022). Plant disease recognition model based on improved YOLOv5. *Agronomy (Basel)*, *12*(2), 365. doi:10.3390/agronomy12020365

Dhakate, M., & Ingole, A. B. "Diagnosis of pomegranate plant diseases using neural network." In *Computer Vision, Pattern Recognition, Image Processing and Graphics (NCVPRIPG), 2015 Fifth National Conference on* (pp. 1-4). IEEE. 10.1109/NCVPRIPG.2015.7490056

Divyanth, L. G., Ahmad, A., & Saraswat, D. (2023). A two-stage deep-learning-based segmentation model for crop disease quantification based on corn field imagery. *Smart Agricultural Technology*, *3*, 100108. doi:10.1016/j.atech.2022.100108

Elfatimi, E., Eryigit, R., & Elfatimi, L. (2022). Beans leaf diseases classification using mobilenet models. *IEEE Access : Practical Innovations, Open Solutions*, *10*, 9471–9482. doi:10.1109/ACCESS.2022.3142817

Ferentinos, K. P. (2018). Deep learning models for plant disease detection and diagnosis. *Computers and Electronics in Agriculture*, *145*, 311–318. doi:10.1016/j.compag.2018.01.009

Harakannanavar, S. S., Rudagi, J. M., Puranikmath, V. I., Siddiqua, A., & Pramodhini, R. (2022). Plant Leaf Disease Detection using Computer Vision and Machine Learning Algorithms. *Global Transitions Proceedings*, *3*(1), 305–310. doi:10.1016/j.gltp.2022.03.016

Haridasan, A., Thomas, J., & Raj, E. D. (2023). Deep learning system for paddy plant disease detection and classification. *Environmental Monitoring and Assessment, 195*(1), 120. doi:10.100710661-022-10656-x PMID:36399232

Hassan, S. M., & Maji, A. K. (2022). Plant disease identification using a novel convolutional neural network. *IEEE Access : Practical Innovations, Open Solutions, 10*, 5390–5401. doi:10.1109/ACCESS.2022.3141371

Jackulin, C., & Murugavalli, S. (2022). A comprehensive review of the detection of plant disease using machine learning and deep learning approaches. Measurement. *Sensors (Basel), •••*, 100441.

Javidan, S. M., Banakar, A., Vakilian, K. A., & Ampatzidis, Y. (2023). Diagnosis of grape leaf diseases using automatic K-means clustering and machine learning. *Smart Agricultural Technology, 3*, 100081. doi:10.1016/j.atech.2022.100081

Jiang, P., Chen, Y., Liu, B., He, D., & Liang, C. (2019). Real-time detection of apple leaf diseases using deep learning approach based on improved convolutional neural networks. *IEEE Access : Practical Innovations, Open Solutions, 7*, 59069–59080. doi:10.1109/ACCESS.2019.2914929

Khirade, S. D., & Patil, A. B. (2015). Plant Disease Detection Using Image Processing. In *Computing Communication Control and Automation (ICCUBEA), 2015 International Conference on* (pp. 768-771). IEEE.

Le, V. N. T., Ahderom, S., Apopei, B., & Alameh, K. (2020). A novel method for detecting morphologically similar crops and weeds based on the combination of contour masks and filtered Local Binary Pattern, operators. *GigaScience, 9*(3), giaa017. doi:10.1093/gigascience/giaa017 PMID:32129847

Liu, Z., Bashir, R. N., Iqbal, S., Shahid, M. M. A., Tausif, M., & Umer, Q. (2022). Internet of Things (IoT) and Machine Learning Model of Plant Disease Prediction–Blister Blight for Tea Plant. *IEEE Access : Practical Innovations, Open Solutions, 10*, 44934–44944. doi:10.1109/ACCESS.2022.3169147

Luo, Z., Yang, W., Gou, R., & Yuan, Y. (2023). TransAttention U-Net for Semantic Segmentation of Poppy. *Electronics (Basel), 12*(3), 487. doi:10.3390/electronics12030487

Maray, M., Albraikan, A. A., Alotaibi, S. S., Alabdan, R., Al Duhayyim, M., & Al-Azzawi, W. K. (2022). Artificial intelligence-enabled coconut tree disease detection and classification model for smart agriculture. *Computers & Electrical Engineering, 104*, 108399. doi:10.1016/j.compeleceng.2022.108399

Meng, X., Tian, X., Wu, Q., Chen, Y., Pan, J., & Hang, Y. (2023). Meta-baseline based on the deep neuro-fuzzy network for few-shot plant leaf fungal disease recognition. *Imaging Science Journal*, 1–13.

Orchi, H., Sadik, M., & Khaldoun, M. (2022). On using artificial intelligence and the internet of things for crop disease detection: A contemporary survey. *Agriculture, 12*(1), 9. doi:10.3390/agriculture12010009

Ozguven, M. M., & Adem, K. (2019). Automatic detection and classification of leaf spot disease in sugar beet using deep learning algorithms. *Physica A, 535*, 122537. doi:10.1016/j.physa.2019.122537

Padol, P. B., & Yadav, A. A. (2016, June). SVM classifier based grape leaf disease detection. In *2016 Conference on advances in signals processing (CASP)* (pp. 175-179). IEEE. 10.1109/CASP.2016.7746160

Pandian, J. A., Kumar, V. D., Geman, O., Hnatiuc, M., Arif, M., & Kanchanadevi, K. (2022). Plant disease detection using deep convolutional neural network. *Applied Sciences (Basel, Switzerland)*, *12*(14), 6982. doi:10.3390/app12146982

Pantazi, X. E., Moshou, D., & Tamouridou, A. A. (2019). Automated leaf disease detection in different crop species through image features analysis and One Class Classifiers. *Computers and Electronics in Agriculture*, *156*, 96–104. doi:10.1016/j.compag.2018.11.005

Pooja, V., Das, R., & Kanchana, V. (2017, April). Identification of plant leaf diseases using image processing techniques. In *2017 IEEE Technological Innovations in ICT for Agriculture and Rural Development (TIAR)* (pp. 130-133). IEEE.

Rani, J. V., Jakka, A., & Kanuru, H. (2022). Disease Detection in Crop Management Using Ensemble Machine Learning. *Sentimental Analysis and Deep Learning Proceedings of ICSADL, 2021*, 907–916.

Sardogan, M., Tuncer, A., & Ozen, Y. (2018, September). Plant leaf disease detection and classification based on CNN with LVQ algorithm. In *2018 3rd international conference on computer science and engineering (UBMK)* (pp. 382-385). IEEE. 10.1109/UBMK.2018.8566635

Sharma, M., Kumar, C. J., & Deka, A. (2022). Early diagnosis of rice plant disease using machine learning techniques. *Archiv für Phytopathologie und Pflanzenschutz*, *55*(3), 259–283. doi:10.1080/03235408.2021.2015866

Shoaib, M., Hussain, T., Shah, B., Ullah, I., Shah, S. M., Ali, F., & Park, S. H. (2022). Deep learning-based segmentation and classification of leaf images for detection of tomato plant disease. *Frontiers in Plant Science*, *13*, 1031748. doi:10.3389/fpls.2022.1031748 PMID:36275583

Singh, V., & Misra, A. K. (2016). Detection of plant leaf diseases using image segmentation and soft computing techniques. *Information Processing in Agriculture*.

Somasundareswari, D., KH, M.S., & Santhiya, A. (2022). Design and Implementation of Smart Farming System. *International Journal of Progressive Research in Science and Engineering*, *3*(5), 184–187.

Suhag, S., Singh, N., Jadaun, S., Johri, P., Shukla, A., & Parashar, N. (2021, June). IoT based soil nutrition and plant disease detection system for smart agriculture. In *2021 10th IEEE International Conference on Communication Systems and Network Technologies (CSNT)* (pp. 478-483). IEEE. 10.1109/CSNT51715.2021.9509719

Sujatha, R., Chatterjee, J. M., Jhanjhi, N. Z., & Brohi, S. N. (2021). Performance of deep learning vs. machine learning in plant leaf disease detection. *Microprocessors and Microsystems*, *80*, 103615. doi:10.1016/j.micpro.2020.103615

Thenmozhi, K., & Reddy, U. S. (2019). Crop pest classification based on deep convolutional neural network and transfer learning. *Computers and Electronics in Agriculture*, *164*, 104906. doi:10.1016/j.compag.2019.104906

Turkoglu, M., Yanikoğlu, B., & Hanbay, D. (2022). PlantDiseaseNet: Convolutional neural network ensemble for plant disease and pest detection. *Signal, Image and Video Processing*, *16*(2), 301–309. doi:10.100711760-021-01909-2

Vallabhajosyula, S., Sistla, V., & Kolli, V. K. K. (2022). Transfer learning-based deep ensemble neural network for plant leaf disease detection. *Journal of Plant Diseases and Protection*, *129*(3), 545–558. doi:10.100741348-021-00465-8

Vishnoi, V. K., Kumar, K., & Kumar, B. (2021). Plant disease detection using computational intelligence and image processing. *Journal of Plant Diseases and Protection*, *128*(1), 19–53. doi:10.100741348-020-00368-0

Zamani, A. S., Anand, L., Rane, K. P., Prabhu, P., Buttar, A. M., Pallathadka, H., Raghuvanshi, A., & Dugbakie, B. N. (2022). Performance of Machine Learning and Image Processing in Plant Leaf Disease Detection. *Journal of Food Quality*, *2022*, 2022. doi:10.1155/2022/1598796

Chapter 12
Effective Path Planning of Cyber-Physical Systems for Precision Agriculture

M. Sundarrajan
SRM Institute of Science and Technology, India

Akshya Jothi
https://orcid.org/0000-0001-9976-2133
SRM Institute of Science and Technology, India

Mani Deepak Choudhry
https://orcid.org/0000-0001-8519-2416
KGiSL Institute of Technology, India

Biji Rose
Dr. N.G.P. Institute of Technology, India

T. Jayapratha
Sri Eshwar College of Engineering, India

V. Nithya
KGiSL Institute of Technology, India

ABSTRACT

The usage of cyber-physical systems is well known to play a vital role in the development of precision agriculture, which in turn raises overall output. When a network of Unmanned Aerial Vehicle (UAV) is deployed for these applications, they are considered CPS. One of the most crucial things that needs to be handled is covering the area that needs to be travelled with the least number of UAV traversals possible. UAV traversal is a crucial part of precision agriculture because it enables farmers to gather crucial information and make informed decisions that may lead to higher crop yields and cheaper costs. Thanks to improved algorithms and techniques, UAVs can traverse agricultural land more effectively and precisely. As a result, farming activities become more productive and environmentally friendly. This chapter explores the various approaches that might be used to determine the UAV's best route when it is deployed over farmland.

DOI: 10.4018/978-1-6684-7879-0.ch012

INTRODUCTION

There is great promise for Smart Farming and Precision Agriculture applications to benefit from emerging technology like Cyber-Physical Systems (CPS), which makes it possible to collect and analyse environmental data in real-time. CPS for short, are systems that have the capacity to connect the real world with the virtual world (Dafflon et al., 2021). This ability is directly related to the information processing that takes place in both worlds. CPS can exist in a number of different contexts, including the agriculture industry. It is anticipated that the use of CPS will play a significant part in the development of precision agriculture, which will, in turn, increase overall productivity (Sharma et al., 2020). It is necessary to develop methods, tools, hardware, and software components based upon transdisciplinary approaches in order to validate the principles via prototypes and test beds in order to expedite and accelerate the realisation of CPS in the field of precision agriculture. Unmanned Aerial Vehicles (UAVs) when collectively working together are termed CPS and they can be widely used in many crop management applications because of their ability to take pictures with a high spatial and temporal resolution (del Cerro et al., 2021). These advancements are projected to have a profound impact on the agricultural industry by speeding up the decision-making process from weeks to days, hence reducing costs and increasing yields. UAVs are a type of aerial robot used and deployed in a wide range of real-world scenarios. The usage of UAVs is becoming increasingly important in many fields, including aerial photography (Martínez-Carricondo et al., 2020), search and rescue (Akshya et al., 2020), disaster management (Akshya et al., 2020), agriculture (Adão et al., 2017), etc. Most modern UAVs are envisioned as intelligent systems, with the goal of facilitating collaboration and the sharing of information between multiple components for the sake of a common goal. Limitations abound in the deployment of a single UAV for a given task. Optimal path planning is a good example of a limitation of this type.

Benefits of UAVs on Agriculture

Any use of unmanned aerial vehicles (UAVs) necessitates a solution to the Coverage Path Planning (CPP) problem. Effectively maximising the coverage area is the goal, and CPP determines the best way to get there. When utilising UAVs, one of the most challenging limits to work on is expanding coverage areas. When plotting a route for a UAV, many other complications can develop as well (Priyadarsini et al., 2021). Taking into account no-fly zones and avoiding collisions with immobile objects are two examples of the difficulties. The algorithm can perfectly or approximately dissect the whole area to develop ideal pathways based on the size and complexity of the area where the UAV is planned to be deployed. Depending on the specifics of the task at hand, several CPP methods address the utilisation of a single UAV or a fleet of them. The number of UAVs deployed in a mission is determined by the mission's unique goals. It is important to include a buffer zone when flying with many UAVs to prevent accidents and facilitate smooth coordination among them (Aggarwal et al., 2020). Any method for determining the best route to follow should aim to maximise coverage while minimising travel time. Within the scope of this article, one UAV is used to monitor a certain region for smart agriculture.

Applications for UAVs in Agriculture

There are several ways that UAVs can be used in agriculture. Crop monitoring and mapping is one of the most often used applications as depicted in Figure 1. UAVs are able to gather high-resolution imagery and data that can be used to evaluate crop health, spot stressed regions, and track vegetation changes over time. UAVs can also be used to precisely and effectively treat crops with fertilisers, herbicides, and other treatments. UAVs can be used to map and analyse soil, giving farmers important information on the quality and makeup of the soil.

Figure 1. Use of UAV over agricultural land

In agriculture, UAVs serve a variety of purposes. Covering the region that needs to be traversed with the least amount of traversal possible for the UAV is one of the most important things that need to be handled. This chapter discusses the various possible ways that could be used to find an optimal path for the UAVs when it is deployed over agricultural land, given the parameters that describe the UAV's coverage capabilities and the supplied geographic area. Path Planning is performed such that the UAV traverses the quickest route to cover the entire agricultural area. The number of UAV turns should be kept to a minimum, and the route should be planned so that the UAV returns to its starting position after the full traverse. A method for determining the best possible path through the area of interest represented as a closed curve. The algorithm divides the region of interest into a series of irregular rectangles, taking into account the UAV's flying altitude and any potential impediments it may meet. After creating a graph from the rectangular partitioning, the problem is turned into the Traveling Salesman Problem. The problem is effectively approximated using the Minimum Spanning Tree (MST) approximation algorithm and Ant Colony Optimization, both of which are presented here (ACO) (Priyadarsini et al., 2022). Another approach for computing the path of the UAVs is discussed where the UAV needs to traverse the area which concaves polygons. Path planning for UAVs is complicated when considering geographic areas as polygons because of the presence of concave polygons. As such, it has been discussed about the Field of View-based Sweep-Line method (FoVSL) to find the best route through a concave polygon. The mentioned approach improves upon the established Sweep-Line method of locating the best route within a polygonal region. The method makes use of the convex polygonal decomposition methodology since it is easier to achieve the answer when the polygons are convex. The number of UAV turns was cut in half by using the Boustrophedon approach. The chapter proposes various algorithms and methods for deploying UAVs over a given agricultural land and investigates the real-world difficulties that may crop up during the design, testing, and operation phases of such systems and describe the different ways in which computational models might contribute.

BACKGROUND

Recently, unmanned aerial vehicles (UAVs) have seen significant usage across a wide range of fields. Despite UAVs' widespread use, there are a number of considerations that must be made before deployment may be made. UAV limitations include flight time, best route, and power usage. Different UAVs are built to accomplish varying goals, and some can even do several at once. UAV deployment becomes more effective when numerous objectives may be solved at once. When it comes to using unmanned aerial vehicles (UAVs) effectively and autonomously, the area-coverage problem is a critical issue that needs constant attention. There have been many suggestions and applications of algorithms to address some of these problems, especially in precision agriculture (PA). Conesa-Muñoz et al. (2016) has utilised a Genetic Approach (GA) to solve path planning to minimise computing time in agricultural contexts, Particle Swarm Optimisation (PSO) algorithm to optimise flight time in 3D areas, taking into account the aircraft's geometry (Li et al., 2016). Pham et al. (2017) employed GA to address the path-finding difficulties of aerial robots in PA scenarios with obstacles. Both Sunny et al. (2017) and Noguchi et al. (1998) attempted to seize control of an ANN that had been taught to navigate a semi-structured agricultural environment with static barriers in a three-dimensional space. This calls for the use of training data inputs, which may be difficult to come by in other contexts. Using the Dubins algorithm and the Rapidly Random Exploring Tree Star (RRT) algorithm, Hao et al. (2018) were able to find the UAVs' safest path through an obstacle-filled agricultural area by prioritising obstacle avoidance over other attributes like efficiency and performance. Path-planning approach for an autonomous multi-rotor UAV based on Dijkstra's algorithm for two-dimensional (2D) settings to visit and cover numerous agricultural regions in the least amount of time (Rozantsev et al., 2017). In addition, Yuan et al. (2019) have used the Ant Colony Optimisation (ACO) algorithm to avoid obstacles and create ideal paths for covering agricultural areas. Validating the routes taken by tractors and harvesters equipped with autonomous steering and navigation systems is a major aid in the agricultural sector. Path discovery in agricultural area coverage was accomplished using a Voronoi Diagram (VD) in two-dimensional space by Nolan et al. (2017). The writers, however, are not looking for the quickest solution. Though this work made significant progress in addressing UAV path planning in PA, more research is needed into the problem.

UAV autonomous guiding, UAV trajectory planning in an environment with obstacles, and specialised applications such as remote sensing with UAVs in precision agriculture are all covered in a variety of synthesis works. Choset published a comprehensive review of CPP techniques for mobile robots, dividing them into two groups based on their methodology. The robots are programmed to follow very straightforward guidelines, however, this heuristic approach to problem-solving does not ensure complete area coverage. However, the methods that employ cellular decomposition to ensure coverage are comprehensive. The author also discusses the flying time, which can be reduced by employing many robots and cutting down on the number of turns. Decomposition approaches, unmanned aerial vehicle CPP methods, multi-unmanned aerial vehicle CPP methods, as well as energy-saving algorithms. Current CPP method surveys that take into account unmanned ground vehicles (UGV) and CPP method surveys that use UAVs expand upon the CPP methodologies used by UGVs. However, when employing UAVs on CPP techniques designed for UGVs, several elements must be taken into account, such as the weight of the sensors, the flight endurance, the direction of the wind, and the severity of the wind (Theile et al., 2020). Kumar et al. (2022) proposed a method that is based on the idea that numerous UAVs will be used to look for victims and rescuers in a disaster zone. The Drones have sensors that can identify and pinpoint their targets. The objective is to reduce search time and maximise efficiency by coordinating the movements of the UAVs as they scan the region. In Ju et al. (2018), the paper presents an overview of the advantages of employing UAVs in agriculture, including enhanced productivity, reduced expenses, and higher yields. The difficulties of employing UAVs in agriculture, such as the necessity for precise mapping and the creation of reliable control systems, are also addressed. The deployment of the multiple UAV system is then described in detail, including the use of a ground control station (GCS), a communication network, and a number of UAVs carrying cameras. It's worth noting that multi-UAV coverage algorithms have been developed, which cut down on the amount of time spent on area coverage. How an area is covered by autonomous robots is algorithm-specific. Algorithms for CPP have been described in the literature using a variety of approaches (including grids, graphs, and neural networks) and can do their calculations either in real-time or in advance. This chapter's structure will look like this for the rest of its length. Various algorithms that are used for partitioning any given agricultural land are discussed in Section 3, Section 4 examines the approaches used by UAVs for path planning over agricultural lands. Section 5 discussed the overall conclusion of making use of UAVs over agricultural land efficiently.

ALGORITHMS FOR PARTITIONING THE AGRICULTURAL LAND

This section explains the methodology used by outlining the partitioning methods and the computational geometry tools used for cell decomposition. By separating configuration spaces in this new way, one is able to extract guidelines for coverage planning.

No Decomposition

Areas with regular shapes and no complexity, such as rectangles, do not require decomposition. To avoid gaps in coverage, a relatively straightforward region can be covered effectively by a pattern with a simple path design, such as a boustrophedon or a square. The boustrophedon method, often known as "the way of the ox," consists of a series of straightforward back-and-forth motions along the polygon's longest side.

The No Decomposition Partition (NDP) algorithm operates on the premise that the farmland can be divided into smaller sub-regions or partitions, and that the UAV may be given a route to follow across each partition. The partitions are laid out to include all of the usable farmland while minimising the amount of overlap between them. The NDP method kicks off with an initial clustering algorithm that divides the farmland into smaller sections as shown in Figure 2. The NDP algorithm then determines a route for the UAV to follow while it traverses each of these partitions. The UAV's flight route is planned to maximise the area covered while minimising the total distance travelled. Among the variables considered by the algorithm are the partition's topography, the position of crops, and any other elements that could affect the UAV's flight. The NDP algorithm has applications in agricultural surveying, where it can be used to plot out routes for unmanned aerial vehicles (UAVs) that cover huge areas with minimal overlap. This can improve the efficiency with which UAVs are used in agricultural surveys, saving both time and effort. In conclusion, the NDP algorithm is a potentially useful method for UAV path planning in agriculture, and more work in this area could lead to greater efficiency and accuracy in UAV-based agricultural land surveying.

Figure 2. No decomposition partitioning

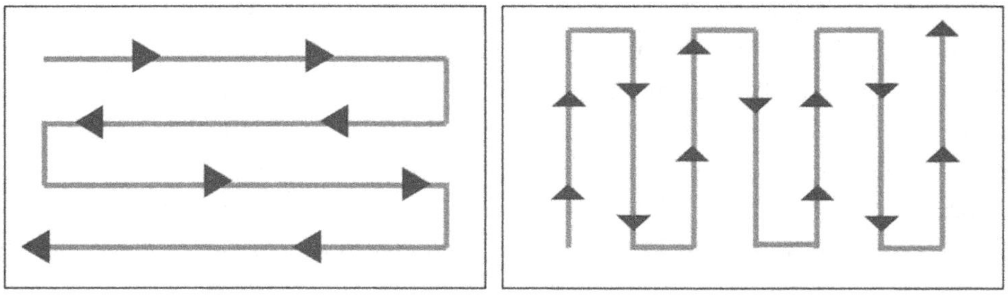

Grid-Mapping Representation

The region of interest is represented by a grid in grid-based path planning, with each grid-cell dimension being constrained by the dimensions of the UAV itself. The area can be grid-mapped using either a 4-neighbours or 8-neighbours method. It's possible for there to be square, rectangular, or hexagonal cells in the grid.

Grid mapping is a method for representing the land in the shape of a grid that is utilised in agricultural field surveying using unmanned aerial vehicles. With this method, the farmland is first divided into a grid of cells of uniform size, and then each cell is given a value based on information gathered by the UAV's sensors and cameras. Producing a precise and comprehensive map of farmland, suitable for uses like crop health monitoring, yield estimate, and soil research is possible thanks to the grid-mapping representation. Big-scale agricultural land surveys benefit greatly from the grid-mapping technique because collecting data manually is both laborious and inefficient for such a large area. Optical, thermal, and multispectral sensors are only some examples of the types of sensors that can be used to apply the grid-mapping technique. Such sensors can monitor the soil moisture, plant health, and pest and disease levels on farmland. Each cell in the grid is given a value based on the information gathered by the sensors. In order to make educated judgements about crop management and other elements of agricultural production, the grid-mapping technique is useful since it provides a very detailed and precise picture of the agricultural area. The grid-mapping method can be used in conjunction with other methods, such as the No Decomposition Partitioning (NDP) algorithm, to determine the most efficient route for the UAV to take while covering the most amount of farmland. Overall, the grid-mapping technique is an exciting new direction for UAV-based agricultural land surveys, with the potential to radically alter the way that agricultural production is monitored and improved. However, the technique needs to be refined and tested in real-world settings before it can be considered fully developed.

Below is a brief summary of the algorithm implemented in UAV-based agricultural land surveying to represent grid-based mapping:

1. A grid of cells of uniform size should be used to divide the farmland.

2. Carry out an aerial survey of the farmland using the UAV's cameras and sensors.
3. Use the sensors and cameras to compile information on the state of the farm's crops, the soil, and other factors.
4. Assign a number to each square in the grid based on readings from the sensors.
5. Incorporate the provided values into a grid-based map of the farmland.
6. The grid-mapping representation can be used for a number of purposes, including keeping tabs on crop health, calculating potential harvests, and assessing the quality of the soil.

Depending on the use case and the peculiarities of the agricultural area, the algorithm can be optimised by employing different types of sensors and cameras, as well as varying grid sizes. The UAV's flight path can be optimised, and all farmland can be reached, by combining this approach with others, such as the No Decomposition Partitioning (NDP) algorithm.

Exact Cellular Decomposition

The exact decomposition of cells could be used on both regular and irregular surfaces. It's common to practise dividing planners' methods into two broad camps: those who work alone, and those who work together. The exact cellular breakdown is used by the single-UAV strategies to calculate their flight paths. On both concave and convex surfaces, they use back-and-forth and spiral techniques. A UAV network, made up of several UAVs, is used to implement cooperative techniques over broad areas. The research employs numerous UAVs to perform back-and-forth flights over a convex polygonal area. Three distinct areas make up the whole. Each pair of consecutive tracks is adjusted to match the UAV's stride. In order to prevent collisions and other mission complications, each UAV plans its route independently, and other routes are developed if necessary as shown in Figure 3. Path planning algorithm for fixed-wing UAV coverage of large complicated polygonal areas with NFZs. Nearby NFZs are of particular interest. The area is broken down into several convex polygons using a technique provided by the authors. The offered path planning method employs a standard boustrophedon technique to map out the routes. To speed up missions when the wind is present, a cost function is put into place. The spiral approach has been used in several works with heterogeneous UAVs.

Figure 3. Exact cellular decomposition

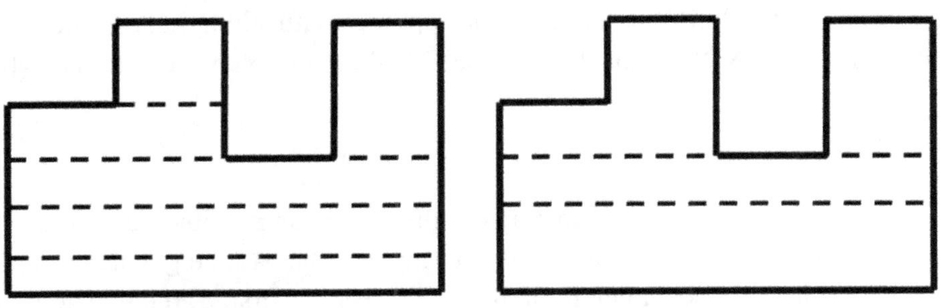

Approximate Cellular Decomposition

The region is roughly subdivided into a collection of uniform cells by means of cellular decomposition. Although typically square, this type of cell can also be depicted as a tetrahedron or a hexagon. Many times, the pathways are designed offline, with the assumption that all relevant data, such as the locations of obstacles and no-fly zones (NFZs), is available. A "-1" is assigned to cells that contain obstacles or the NFZ. The technique offers a grid-based algorithm for determining coverage paths that takes energy efficiency into account. While planning routes through unsteady terrain, they want to minimise energy use. The algorithm proposed in using a grid structure is improved by their work. They use a novel energy cost function in their evaluation of the method.

In unmanned aerial vehicle (UAV) agricultural land surveying, the Approximate Cellular Decomposition (ACD) technique is utilised for path planning. In ACD, the farmland is divided into a grid of smaller, rectangular cells or polygons that the UAV may fly over and collect data from. With the ACD algorithm, there will be a reduction in the UAV's trip time while still inspecting all of the farmland. The ACD algorithm for use by UAVs in agricultural settings entails the following processes, which are briefly described below:

1. Create a grid of smaller rectangles or polygons to use as farmland.
2. Consider the topography, any potential impediments, and any applicable safety rules when establishing the level of accessibility of each cell.
3. Sort the cells into groups according to how easily they could be obtained to them and how close they are to one other.
4. Using an algorithm like the Traveling Salesman Problem, determine the best way to get from one cluster to the next (TSP).
5. Discover the most direct route between each set of nodes to improve the journey's efficiency.
6. Use the UAV to travel along the prescribed route through the farmland.
7. Use the UAV's sensors and cameras to compile information on the land's condition in terms of things like crop yield, moisture levels, and soil composition.
8. Create a model or map of the farmland using data gathered by the UAV.

The performance of the ACD algorithm can be improved by tailoring the size and shape of the cells or polygons to the particulars of the agricultural land under study. The UAV's flight route and data gathering can be further optimised by combining the approach with additional methods, such as the No Decomposition Partitioning (NDP) algorithm and the Grid-Mapping Representation (GMR) algorithm.

GridU Algorithm

The objective of this task is to use a UAV to map out the best route possible through a certain area. In this case, the treated region of interest is a polygon. Using h as the base height and h' as the maximum height, a UAV in India might reach a height of about 400 feet. At first, S, the polygonal region under investigation is divided into squares of uniform size. Within the polygonal region, it can be assumed that a random assortment of static obstructions. The static obstacle locations are more easily obtained once the centroids of the squares are determined. The size of the matching square will not change if there are no obstructions over the area of interest. If the UAV detects a static barrier within the confines of a square, it must raise its altitude to h' to avoid crashing into it. When flying a UAV at a higher altitude, the area it can see and the height it can go to get there both increase. As the size of the area grows, so does the size of the partitions that house the obstructions. Combine the partitions nearest the obstacle-filled square to create a rectangular one. Once the centroids are determined, the closest partition to the obstacle can be determined.

Rectilinear Partitioning Algorithm

Imagine there is a camera mounted on a UAV and it is flying at a height of *h* over a predetermined polygonal region A. Field of View refers to the geographical region that this UAV is able to observe at any given time (FoV). The parameters are taken into account at various stages of the traversal, allowing for the rectilinear division to be performed. The Graham Scan Algorithm determines the smallest possible enclosure area given a polygonal region. In order to ensure that the UAV is able to cover the full polygon, it is now required to partition the area included within the polygon in order to establish an optimal area coverage approach as shown in Figure 4. The flying height *h* of the UAV is assumed to change depending on a number of parameters. An impediment that might appear on the route is one such crucial consideration. In this project, it can be seen that the use of static impediments including buildings, transmission line towers, and trees to test the UAV's ability to fly safely and effectively around them. Particular applications, such as disaster management, require drones to take pictures from low altitudes and then move to a higher one if they encounter a structure or other immovable impediment. Because of variations in UAV altitude.

The field of view (FoV) also shifts, since it is totally height-dependent. So, the area covered by the UAV is proportional to the height at which it is flying. UAVs can survey a larger area the higher up in the air they are. Having the UAVs fly at higher altitudes all the time isn't necessary because the resolution of the photographs they gather could suffer. In order to capture clearer footage, some uses require UAVs to fly at lower altitudes.

Figure 4. Rectilinear partitioning algorithm

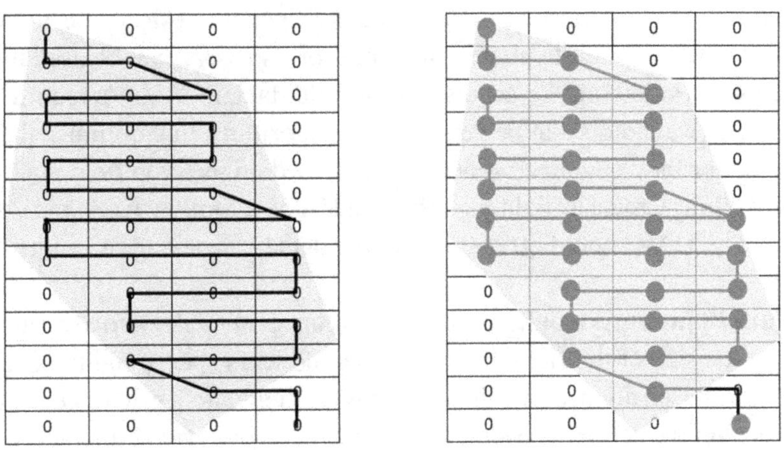

Baseline Area Partitioning Algorithm

For chip layout in the field of VLSI (Very Large Scale Integration) design, the Baseline Area Partitioning Algorithm (BAP) is a heuristic algorithm. To employ the technique, a large rectangular area is divided into many smaller rectangles of roughly the same size. Following is how the algorithm functions:

1. Find the rectangle's preferred orientation by: The "length" of a rectangle is the longer side, while the "width" refers to the shorter side. The proportion of length to breadth establishes the rectangle's orientation.
2. When the partition is first set up, the rectangle is cut in half along its longest length if it is longer than it is broad, or along its shortest side if it is wider than it is long, creating two smaller rectangles of equal area. The first partition is set up in this process.
3. In order to achieve the appropriate partitioning depth, it is necessary to recursively split the rectangles. The number of required rectangles is a common indicator of the required depth.
4. After the recursive subdivision is complete, the algorithm may rebalance the partitions to make sure each rectangle is roughly the same size.
5. In VLSI design, the BAP method is frequently used as a foundation for more advanced partitioning algorithms due to its ease of implementation and high quality first partition. As an added bonus, it can be put to use in other domains like image processing and computer graphics, where rectangular segmentation is a need.

PATH PLANNING ALGORITHMS OF UAVs FOR AGRICULTURAL LAND

Creating an algorithm that can efficiently navigate across crop fields and collect data is necessary for path planning for agricultural UAVs as depicted in Figure 5. Here is an illustration of an agricultural UAV path planning algorithm:

1. **Obtain Field Data:** Accurate field data, such as crop types, soil parameters, elevation maps, and other pertinent information, must be obtained before the path design process can begin. Remote sensing methods, such as satellite photography or drone surveys, can be used to gather this data.
2. **Define Waypoints:** Following the acquisition of field data, the UAV's required waypoints must be determined. The waypoints can be generated by algorithms that optimise the flight path using the available data, or they might be predetermined based on the crop field plan.
3. **Create a Flight Plan:** Create a flight plan that the UAV will follow using the waypoints that have been specified. The UAV should efficiently collect data by covering the entire field according to the flight plan.
4. **Implement Path Planning Algorithm:** The algorithm should take into account a number of variables, including the size of the field, the crop type, the wind speed, and the characteristics of the UAV. Using a grid-based algorithm, which divides the field into a grid of cells and determines the fly path by choosing the cells that need data collection, is one typical method.
5. To ensure obstacle avoidance, the UAV must be able to recognise and steer clear of objects like trees, electricity wires, and buildings while in flight. Ultrasonic or LiDAR sensors that can identify objects in the UAV's route can be used to accomplish this.

6. Real-time monitoring and reporting are essential for ensuring that the path-planning algorithm is operating appropriately. Real-time information on crop health, soil moisture, and other factors can be obtained from the UAV's sensors and used to modify the flight plan as needed.
7. Processing and analysis of the obtained data should be done after the UAV has finished its flight in order to produce insights that can aid farmers in maximising their operations. Information including yield projections, crop health status, and soil factors may be included.

Figure 5. Path planning of UAVs

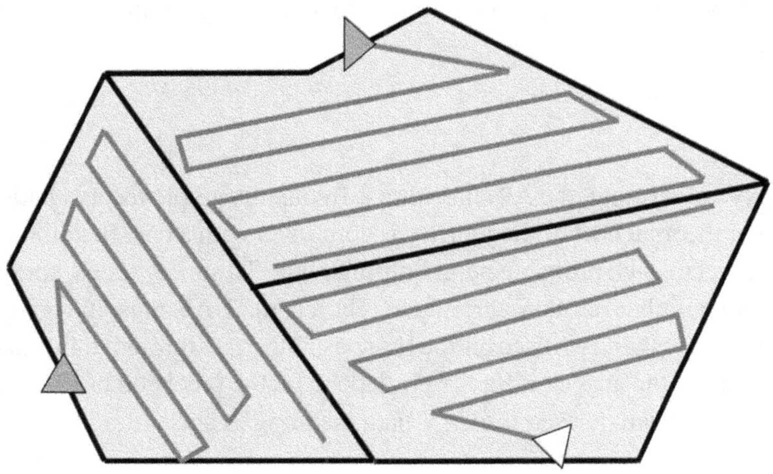

Dijkstra's Algorithm

UAV path planning in farming can benefit from Dijkstra's algorithm. The algorithm uses a breadth-first approach to discover the shortest path between any two nodes in a graph. The search space, which is often a 2D or 3D map of the agricultural area in the case of UAV path planning, is represented by the graph. Dijkstra's algorithm can be used for UAV path planning in farming by following these steps:

1. **Outline the Scope of the Search:** Make a chart to illustrate the agricultural sector. The graph's nodes stand in for the possible locations of the UAV, while the edges denote the corresponding distances.
2. Make use of the calculated distance between each pair of nodes by assigning that value as the edge's weight. The Euclidean distance or some other distance metric to determine the weights.
3. **Initiate the Algorithm:** It is recommended that the use current node as the starting node and that the initial distance is set to 0. Make the distance between every other node infinite.
4. If it is necessary to explore the graph, pick the node with the shortest distance and check it off the list. A new distance from the root node should be added if the current one is smaller than the distance to any of the current node's neighbours.
5. The algorithm ends when the final node is reached. Then find the quickest route by following the path back from the node that needs to be reached to the one you left.

6. Once the shortest path has been determined, the UAV path can be generated by joining together the nodes along the route. Post-processing techniques like path smoothing and obstacle avoidance can further improve the path's efficiency.
7. For agricultural UAVs, Dijkstra's algorithm is a straightforward and effective path-planning technique. However, it fails in situations where the impediments are constantly changing position. Such situations often call for the use of more complex algorithms like RRT or PRM.

In conclusion, an agricultural UAV path planning algorithm should be created to maximise flying effectiveness, minimise obstacles, and gather correct data. Farmers can make knowledgeable decisions about their crops and increase yields by integrating remote sensing methods, in-the-moment monitoring, and data analysis.

A* Algorithm

When it comes to UAVs in farming, the A* algorithm is frequently employed as a path-planning method. A heuristic estimate of the total cost of the journey is combined with a cost function that calculates how much it will cost to travel to each probable neighbouring node. To survey crops, spray with pinpoint accuracy, or carry out any number of other agricultural chores, the UAV must navigate a grid or graph of waypoints, and this method is used to discover the best possible route through this maze. While A* is fast and effective, the quality of the answer it finds may depend on the precision of the heuristic and the cost function. The A* algorithm can be used for UAV path planning in agriculture by following these steps:

1. The issue must be defined. Locate the UAV's beginning and ending points, the grid or graph of waypoints it must follow, and any obstacles or limits it must overcome.
2. Launch the A* algorithm by: Constructing a pair of lists, one open and one closed, to maintain tabs on which nodes have already been assessed.
3. The predicted cost to reach the target and the cost to reach the present node should be calculated using a heuristic function.
4. To create a list of connected nodes accessible from the current node, use the "generate neighbours" function.
5. Compute the distance to each neighbouring node and the anticipated total cost to reach the target through that node.
6. If the nearby nodes have not yet been examined, add them to the open list, and if a better path has been determined, adjust their cost accordingly. Take the current node out of the open set and place it in the closed set.
7. Iterate until the target node is reached or no more nodes remain to be evaluated (steps 3-6).
8. Construct the path by using the parent pointers generated during the search to traverse backwards from the destination node to the initial node.
9. Smooth down the route's sharp curves and extra zigzags to maximise its efficiency.
10. Fly the UAV along the calculated best route to conduct crop surveys or other agricultural operations.

RRT Algorithm

A probabilistically complete motion planning technique that can be utilised for UAV route planning in agriculture is the Rapidly-exploring Random Tree (RRT) algorithm. The idea behind random walk-through (RRT) is to construct a tree of possible paths by gradually expanding a tree of connected nodes in the direction of random samples picked from the search space. The RRT algorithm can be used for UAV path planning in agriculture by following these steps:

1. The problem must be defined. Identify the beginning and ending locations, the search area, and any limits that must be respected.
2. Set up the RRT by making a tree with the initial state as its root node.
3. Create a random location inside the search area.
4. Find the node in the tree that is closest to the random point, and then draw a line between them.
5. Check for obstructions by seeing if the shortest route between the two points violates any constraints or meets any.
6. If the straight-line route is legitimate, the new location is added as a node to the tree structure.
7. Iterate until a solution is found or the time or number of iterations limit is met, whichever comes first.
8. Using the parent pointers generated during the search, construct a path from the goal node to the starting node.
9. Smooth down the route's sharp curves and extra zigzags to maximise its efficiency.
10. Fly the UAV along the calculated best route to conduct crop surveys or other agricultural operations.

While RRT can be helpful for navigating high-dimensional and complex search spaces, it isn't guaranteed to always produce the best solution. Both the sampling method and the NN search algorithm selected might have an impact on the final product's quality.

Evolutionary Algorithms for UAV

Survival of the fittest principles is used by evolutionary algorithms, a type of optimization method, to iteratively refine a pool of potential answers to a problem. These algorithms can be used to plan routes for agricultural UAVs, and they can find avenues that other algorithms might miss. Genetic algorithms (GAs) and particle swarm optimization (PSO) are two popular evolutionary algorithms used for UAV path planning in agriculture (PSO). In order to use evolutionary algorithms for UAV path planning in agriculture, one must take the following measures:

1. The problem must be defined. Identify the beginning and ending locations, the search area, and any limits that must be respected.
2. The first step is to generate a random sample of possible paths, each of which could lead to a solution.
3. Calculate how well each potential path satisfies the problem's criteria, such as minimising the distance, avoiding obstacles, or maximising crop coverage, and this is what is meant by "evaluating fitness."
4. Choose parents: pick the healthiest and most successful people to start the next generation.

5. Reproduce: Through crossover and mutation, generate a new population of potential offspring from the parents.
6. Consider the new population and how well it can adapt to its environment.
7. Shape the next generation by choosing the healthiest members of the population to become parents.
8. Steps 5-7 are to be repeated until a stopping criterion is met, such as when a target fitness level is reached or a predetermined number of generations has been reached.
9. The road from the initial point to the ultimate destination should be built using the most optimal member of the final population.
10. Smooth down the route's sharp curves and extra zigzags to maximise its efficiency.
11. Fly the UAV along the calculated best route to conduct crop surveys or other agricultural operations.

While evolutionary algorithms have the potential to produce high-quality solutions, they are often computationally intensive and require fine-tuning of numerous parameters, including population size, crossover and mutation rates, and selection criteria.

Machine Learning and Deep Learning Algorithms for UAV Path Planning

Machine learning and deep learning algorithms can be used for the path planning of UAVs in agriculture by training models to predict optimal paths based on input data. Some common machine learning and deep learning algorithms for path planning of UAVs include:

1. **Reinforcement Learning:** In order to optimise a reward signal, reinforcement learning algorithms learn to make decisions depending on the outcomes of previous actions and the information gathered from the surrounding environment. UAVs can be taught to navigate their surroundings and avoid danger with the help of these algorithms.
2. **Convolutional Neural Networks (CNNs):** Specifically, CNNs are a sort of deep learning algorithm useful in computer vision and image processing. They can be taught to analyse data from previous flights to determine the most efficient routes for unmanned aerial vehicles.
3. **Recurrent Neural Networks (RNNs):** RNNs are a sort of deep learning technique that may be used to time series and sequence data. Based on historical data and sensor input, they can be used to calculate the best routes for UAVs to take.
4. **Autoencoders:** The autoencoder is a deep learning technique that can learn compact representations of input data without human supervision. UAVs can utilise them to "read" the environment and get recommendations on the best routes to take.
5. **Decision Trees:** It is possible to utilise decision trees for both classification and regression purposes; they are a sort of machine learning method. In the context of unmanned aerial vehicles, they can be used to foresee the best routes for those vehicles to take in light of environmental information.

The steps for using machine learning and deep learning algorithms for UAV path planning in agriculture are as follows:

1. **Collect Data:** Gather information from the UAV's sensors, the surrounding area, and any other sources of data that may be useful.
2. **Pre-Process Data:** Data should be filtered, cleansed, and normalised before being used.

3. **Train Model:** In order to use the data for training a machine learning or deep learning model, it must first be pre-processed.
4. **Test Model:** Put the model through its paces by putting it through a test on a validation set.
5. **Optimize:** Adjust the model's hyperparameters and architecture to see if can be obtained with better results.
6. **Deploy:** UAV's optimal flight paths can be predicted by deploying the trained model and using it to navigate.

High-quality solutions can be generated by machine learning and deep learning algorithms, but they are computationally intensive and often require enormous volumes of training data. The solution's efficacy may also be affected by the details of the models employed and the parameters with which they were trained.

CONCLUSION

UAV path planning can help enable precision farming by supplying farmers with accurate and up-to-date data on the health of their crops, the conditions of their soil, and the patterns of the weather. These data can be put to use to improve resource use, such as water, fertiliser, and pesticide application, which will ultimately result in increased crop yields and decreased expenses. The use of UAVs can make potentially hazardous jobs, such as crop monitoring and spraying, safer for farmers and other workers to do, reducing the likelihood that they would sustain an accident. UAV path planning can assist reduce the usage of resources like water and pesticides, resulting in farming techniques that are more sustainable and better for the environment. Cost savings can be achieved through the use of UAV path planning, which can assist farmers in reducing the costs of labour, fuel, and other expenses related to conventional agricultural methods.

In conclusion, the planning of flight paths for unmanned aerial vehicles (UAVs) used in agriculture is an important area of research that has the potential to raise farmers' production while simultaneously lowering their expenses. The use of unmanned aerial vehicles (UAVs) for precision agriculture necessitates the development of algorithms for efficient and effective path planning that are able to navigate surroundings that are both complex and dynamic while avoiding obstacles and optimising the goals of the mission. A*, RRT, APF, PRM, and D* are only a few of the path-planning algorithms that can be utilised for this purpose. Other options include PRM and APF. The selection of an algorithm will be determined by the particular necessities of the application, and the development of new algorithms is an active area of research that is being conducted currently. The use of unmanned aerial vehicles (UAVs) in agriculture is poised to become an increasingly essential and influential tool for farmers all over the world as a result of the continual progress of UAV technology and path-planning algorithms.

It is possible that future research on the subject of UAV path planning for agricultural land would concentrate on further enhancing the efficacy, accuracy, and resilience of existing algorithms, as well as developing new algorithms that are better suited for certain applications. Additionally, there is a need for study on the integration of several UAVs for the purpose of coordinated path planning and mission execution, since this can further improve productivity and efficiency in precision agriculture. Research is also being done in other areas, such as the development of more advanced sensors and control systems for UAVs as well as the integration of machine learning and artificial intelligence approaches for route planning and decision-making. In the end, ongoing study and development in the field of UAV path planning for agriculture can assist to progress of the field of precision agriculture, which in turn can lead to agricultural methods that are more sustainable and efficient.

REFERENCES

Adão, T., Hruška, J., Pádua, L., Bessa, J., Peres, E., Morais, R., & Sousa, J. J. (2017). Hyperspectral imaging: A review on UAV-based sensors, data processing and applications for agriculture and forestry. *Remote Sensing (Basel)*, *9*(11), 1110. doi:10.3390/rs9111110

Aggarwal, S., & Kumar, N. (2020). Path planning techniques for unmanned aerial vehicles: A review, solutions, and challenges. *Computer Communications*, *149*, 270–299. doi:10.1016/j.comcom.2019.10.014

Akshya, J., & Priyadarsini, P. L. K. (2020). Graph-based path planning for intelligent UAVs in area coverage applications. *Journal of Intelligent & Fuzzy Systems*, *39*(6), 8191–8203. doi:10.3233/JIFS-189140

Akshya, J., & Priyadarsini, P. L. K. (2020). A Comparison of Machine Learning Approaches for Classifying Flood-Hit Areas in Aerial Images. In *International Conference on Innovative Computing and Communications: Proceedings of ICICC 2019* (vol. 1, pp. 407-415). Springer Singapore. 10.1007/978-981-15-1286-5_34

Conesa-Muñoz, J., Bengochea-Guevara, J. M., Andujar, D., & Ribeiro, A. (2016). Route planning for agricultural tasks: A general approach for fleets of autonomous vehicles in site-specific herbicide applications. *Computers and Electronics in Agriculture*, *127*, 204–220. doi:10.1016/j.compag.2016.06.012

Dafflon, B., Moalla, N., & Ouzrout, Y. (2021). The challenges, approaches, and used techniques of CPS for manufacturing in Industry 4.0: A literature review. *International Journal of Advanced Manufacturing Technology*, *113*(7-8), 2395–2412. doi:10.100700170-020-06572-4

del Cerro, J., Cruz Ulloa, C., Barrientos, A., & de León Rivas, J. (2021). Unmanned aerial vehicles in agriculture: A survey. *Agronomy (Basel)*, *11*(2), 203. doi:10.3390/agronomy11020203

Hao, B., & Yan, Z. (2018). Recovery path planning for an agricultural mobile robot by Dubins-RRT* algorithm. *International Journal of Robotics and Automation*, *33*(2). Advance online publication. doi:10.2316/Journal.206.2018.2.206-5443

Ju, C., & Son, H. I. (2018). Multiple UAV systems for agricultural applications: Control, implementation, and evaluation. *Electronics (Basel)*, *7*(9), 162. doi:10.3390/electronics7090162

Kumar, G., Anwar, A., Dikshit, A., Poddar, A., Soni, U., & Song, W. K. (2022). Obstacle avoidance for a swarm of unmanned aerial vehicles operating on particle swarm optimization: A swarm intelligence approach for search and rescue missions. *Journal of the Brazilian Society of Mechanical Sciences and Engineering*, *44*(2), 56. doi:10.100740430-022-03362-9

Li, X., Zhao, Y., Zhang, J., & Dong, Y. (2016, November). A hybrid PSO algorithm based flight path optimization for multiple agricultural UAVs. In *2016 IEEE 28th international conference on tools with artificial intelligence (ICTAI)* (pp. 691-697). IEEE. 10.1109/ICTAI.2016.0110

Martínez-Carricondo, P., Agüera-Vega, F., & Carvajal-Ramírez, F. (2020). Use of UAV-photogrammetry for quasi-vertical wall surveying. *Remote Sensing (Basel)*, *12*(14), 2221. doi:10.3390/rs12142221

Noguchi, N., Reid, J. F., Zhang, Q., & Tian, L. F. (1998). Vision intelligence for precision farming using fuzzy logic optimized genetic algorithm and artificial neural network. American Society of Agricultural Engineers.

Nolan, P., Paley, D. A., & Kroeger, K. (2017, March). Multi-UAS path planning for non-uniform data collection in precision agriculture. In *2017 IEEE Aerospace Conference* (pp. 1-12). IEEE. 10.1109/AERO.2017.7943794

Pham, T. H., Bestaoui, Y., & Mammar, S. (2017, October). Aerial robot coverage path planning approach with concave obstacles in precision agriculture. In *2017 Workshop on Research, Education and Development of Unmanned Aerial Systems (RED-UAS)* (pp. 43-48). IEEE. 10.1109/RED-UAS.2017.8101641

Priyadarsini, P. L. K. (2021, January). Area Partitioning by Intelligent UAVs for effective path planning using Evolutionary algorithms. In *2021 International Conference on Computer Communication and Informatics (ICCCI)* (pp. 1-6). IEEE.

Priyadarsini, P. L. K., Gadupudi, P., & Paladugula, K. (2022, April). Path Planning of Intelligent UAVs using Computational Geometric Techniques. In *2022 International Conference on Electronic Systems and Intelligent Computing (ICESIC)* (pp. 190-193). IEEE.

Rozantsev, A., Sinha, S. N., Dey, D., & Fua, P. (2017). Flight dynamics-based recovery of a UAV trajectory using ground cameras. In *Proceedings of the IEEE Conference on Computer Vision and Pattern Recognition* (pp. 6030-6039). 10.1109/CVPR.2017.266

Sharma, R., Parhi, S., & Shishodia, A. (2020). Industry 4.0 applications in agriculture: cyber-physical agricultural systems (CPASs). In *Advances in Mechanical Engineering: Select Proceedings of ICAME 2020* (pp. 807-813). Singapore: Springer Singapore.

Sunny, M. S. H., Hossain, E., Mimma, T. N., & Hossain, S. (2017, September). An autonomous robot: Using ANN to navigate in a static path. In *2017 4th International Conference on Advances in Electrical Engineering (ICAEE)* (pp. 291-296). IEEE.

Theile, M., Bayerlein, H., Nai, R., Gesbert, D., & Caccamo, M. (2020, October). UAV coverage path planning under varying power constraints using deep reinforcement learning. In *2020 IEEE/RSJ International Conference on Intelligent Robots and Systems (IROS)* (pp. 1444-1449). IEEE. 10.1109/IROS45743.2020.9340934

Yuan, J., Wang, H., Lin, C., Liu, D., & Yu, D. (2019). A novel GRU-RNN network model for dynamic path planning of mobile robot. *IEEE Access : Practical Innovations, Open Solutions*, *7*, 15140–15151. doi:10.1109/ACCESS.2019.2894626

Chapter 13
Analysis of Drone Applications in Precision Agriculture

N. Vanitha
Dr. N.G.P. Arts and Science College, India

S. K. Rakshun Selvaa
Dr. N.G.P. Arts and Science College, India

ABSTRACT

Unmanned Aerial Vehicles (UAV) or drones are a growing technological boon to major fields which possess high complexity in logistics such as defense systems, cargo transportation, mapping geographic locations and terrain, border surveillance, medical emergency shipments, and many other areas where humans find it inaccessible or inefficient to work. Drones can cope with tasks that demand precisive services and tasks that are inefficient under human intervention. Precision agriculture deals with efficiency as a key factor and it uses information technology to keep track of optimum health of the crops and productivity of a particular crop field using precisive approach which can be done with the help of purposely modified drones for precision agriculture. This chapter analyses and evaluates the significance of drones in precision agriculture, challenges possessed in scaling them, and solutions to implement them in large-scale environments. In the future, cognitive analysis and sensory evaluation can be used to further reduce the complexity of accessing drones in various environments.

INTRODUCTION

Precision Agriculture

Since the inception of new key technological improvements after the Agricultural Revolution we're ensured that there is a steady Growth in the field of agriculture by employing various forms of technologies and analysis tools. Precision Agriculture is one of the strategy we employ to maximize crop yield and reduce man labour and dependency, so that efficiency levels of agriculture output can be maximized when compared to regular means of agriculture. Precision Agriculture is a new concept thus needed to

DOI: 10.4018/978-1-6684-7879-0.ch013

be improvised by data processing and management system. Precision Agriculture requires enormous data to be in circulation to analyse and optimize according to the data provided by sensors that detect various forms of information from crop yield to pest infestation and much more important data that are essential to precision agriculture. In Precision Agriculture farmers can deploy systems that can monitor from water to fertilizer management to enhance productivity and ensure that the crop quality is in our expected level. Crop Health Assessment is the foremost duty of any farmer which is a complex mechanism and the complexity involved in this process can by be reduced by deploying High Precisive techniques with help of Precisive Agriculture. Various other variables taken into account in Precision Agriculture includes soil condition, Action of fertilizer and pesticide, irrigation methods and etc. Precision Agriculture data are inconsistent over various time frame that requires Implementation of various Technological Aid which is discussed below in the paper further (Donka et al., 2020).

Significance of Drones in Precision Agriculture

Drones also known as Unmanned Aerial vehicles is one of the intelligent way to reduce the complexity by streamlining the process of obtaining the information processed by the sensors and analyse the information with the help of Topography Analysis. We do employ the principles used in military drone networks in the field of Precision Agriculture such as surveillance, attacking, mapping and etc. Surveillance is done via Drones specifically modified for Precision Agriculture to gather information from crop surveying, pest infestation detection, weed growth analysing, soil inspection, irrigation management and mapping. Attack via Precision Agriculture drone is done purposely to destroy weed and pests that inhibits or stun the growth of plant or reduce the quality of crops. Mapping over inaccessible or large terrain can be done with the help of Precision Agriculture based drones. Crop-fertility can be assessed by the High-Resolution data provided by drone during Surveillance. Natural calamities affected lands which can be challenging to accessible by Humans can deploy drones to access the damage possessed by such calamities. Drones can replace the hefty price tagged satellite technology in image processing while mapping terrain and analysing the vegetation health. Precision Agriculture based data can be obtained to be processed by the farmers and analyse the further parameters to be accounted while utilizing the data during deployment (Petkovics et al., 2017).

Objectives

Our primary objective in deploying drones in Precision Agriculture is to employ key new technologies to reduce the complexity involved in undertaking Precision Agriculture over standard procedures. We need to ensure that such technological improvements can be introduced with the current procedures by not straining the farmers to know deep about such technological hurdles by making more autonomous and self-learning capabilities to such systems that we are employing in Precision Agriculture.

BACKGROUND STUDY

Table 1 provides the various literature reviewed for the analyses of applications of drones in precision agriculture.

Table 1. Background study

S.No	Author Name	Journal	Technique	Observation
1	M V Suhas, S Tejas, Snigdha, Sitaram Yaji, Sanket Salvi	An Agricultural Drone using Internet of Things, Data Analytics and Cloud Computing Features	IOT, Data Analytics, Cloud-Computing	Crop management can transferred between the Drone and the Farmer without any need to inspect them in person
2	Hyun-Sik Son, Deok-Keun Kim, Seung-Hwan Yang, Young-Kiu Choi	Real-Time Power Line Detection for Safe Flight of Agricultural Spraying Drones Using Embedded Systems and Deep Learning	Deep Learning	Using Deep Leaning The Drones can ensure that the flight path is free of obstacles
3	Mansoor Alam, Muhammad Shahab Alam, Muhammad Roman, Muhammad Tufail, Muhammad Umer	Real-Time Machine-Learning Based Crop/Weed Detection and Classification for Variable-Rate Spraying in Precision Agriculture	Machine Learning	Drones can classify the weed detection among crops and can spray the weedicide in varying rate according to the analysed level of Damage
4	Ching-Ju Chen, Ya-Yu Huang, Yuan-Shuo Li, Ying-Cheng Chen, Chuan-Yu Chang,	Identification of Fruit Tree Pests With Deep Learning on Embedded Drone to Achieve Accurate Pesticide Spraying	Deep Learning	Fruit tree pests can be identified by specifically embedded Deep Learning Algorithm and to obtain a accurate Pesticide Spraying rate
5	Giuseppe Potrino, Nunzia Palmieri, Venturino Antonello, Abdon Serianni	Drones Support in Precision Agriculture for Fighting Against Parasites	Distributed Algorithm	Employing Drones in clear a crop cluster to avoid pest infestation
6	Renat N. Abutalipov, Yuriy V. Bolgov, Hamisha M. Senov	Flowering plants pollination robotic system for greenhouses by means of nano copter (drone aircraft)	Nano-Engineering	Nano-Copters can be used as pollinating agents which can be achieved in Greenhouse based farming
7	Kazi Mahmud Hasan, Wida Susanty Suhaili, S. H. Shah Newaz, Md. Shamim Ahsan	Development of an Aircraft Type Portable Autonomous Drone for Agricultural Applications	Portable Drones	Portability is ensured among agricultural drones so they deployed at varied environments and terrain
8	Maanak Gupta, Mahmoud Abdelsalam, Sajad Khorsandroo, Sudip Mittal	Security and Privacy in Smart Farming: Challenges and Opportunities	Authentication and Secure Communication	Privacy is maintained among farmer and the drone
9	Praveen Kumar Reddy Maddikunta, Saqib Hakak, Mamoun Alazab, Sweta Bhattacharya	Unmanned Aerial Vehicles in Smart Agriculture: Applications, Requirements, and Challenges	Agricultural Sensors	Challenges involved in Smart Agriculture are validated
10	Ting-Chu Hsieh, Ming-Chien Hung, Mai-Lun Chiu, Pay-Jiing Wu	Challenges of UAVs Adoption for Agricultural Pesticide Spraying: A Social Cognitive Perspective	Cognitive Analysis	Cognitive perspective in the way of humans can be employed by the drones

Uses of Drones in Precision Agriculture

Precision Agriculture based drones can undertake various task from crop health monitoring to crop growth evaluation, measuring fertility of crops, management of soil inspection, weeds and pest damage detection and control management, even can help in pollinating in plants by modifying the drones as nano-copters. Crop quality is drastically reduced by weed and pests in all major crops that including food crops and cash crops. Such damage analysis and quality control can be effectively managed by drones that are employed over greenhouse based farming. Nano-copters can be a great deal in accessing over plant structures that are smaller than huge field based crops as their tiny size can help in surveillance over small distances with over high resolution in greenhouse based flowering plants. GPS employed drones can cope with large distances fields that are hundreds and thousands of acres in size and can encounter the field area by flying above to capture and monitor the cluster of crops (Vanitha N et al., n.d.). Certain damaged crop clusters can be easily seen via a drone and can be intimated faster than other conventional methods so that the damage including pest infestation can be abided as soon as possible whereas in conventional methods the time of action can be delayed. Drones tethered with a source of water or pesticide can do the task of spraying them autonomously and can be even more precise in saving both water and pesticide while spraying over large fields and they can transmit the data of the crop health and other crop related analysis by IOT and can even employ Artificial Intelligence to even effectively streamline the process of Precision Agriculture (Abutalipov et al., n.d.). Usage of drones in precision agriculture shown in Figure 1.

Figure 1. Applications of drones in precision agriculture

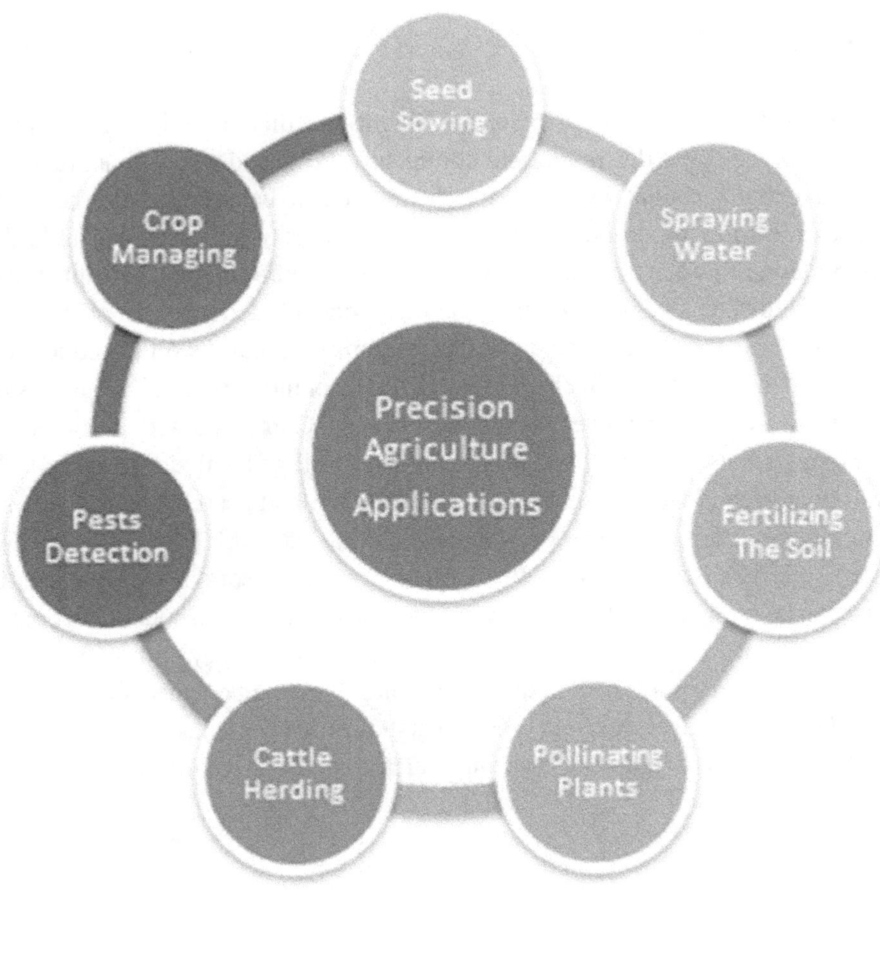

Sowing Drones

Critical issues faced by today's agricultural environment includes worker shortage and low profit cut and high expenditure cost over long time run which is very inefficient and can be made efficient by the adoption of Precision Agriculture. Rice is a crop that is being a staple food for many countries as a drawback of requiring enormous labour hands to sow in the soil. Drones can be made to replace such high work force consuming task by carrying the seeds as a payload and can show them in paddy field. Modular and portable copters specifically designed for carrying bulk amount of seeds over short distances can be employed to meet the demands of sowing seeds without a human and can operate manually with Artificial Intelligence and Deep Learning. Once programmed the drone can be able to learn the pattern of sowing of seeds over fields by the above mentioned technologies and thus can be employed in any fields with variable area size and varied terrain forms. Drones can even monitor soil moisture and can decide the suitable time to sow a seed and can even detect the seed for its quality and thus every single step from choosing a seed to sowing is being carefully inspected by the drone used in Precision Agriculture. Certainly drones that are equipped with Artificial Intelligence can perform regular checks on the seeds and can monitor their health day by day and can be connected with the farmer by employing Internet of Things without the need that to be checked by the farmer in person. They do just can update every information by the sensors and software embedded on them and according to the task they are being deployed (Lohit et al., 2021).

Spraying Drones

This specifically interesting as well as an important function of this drone is to control the agents that can destroy a crop cluster or even destroy a field completely. Pests and weeds that are a severe threat to crop that are to be taken into account while seeking the methods to be intended in Precision Agriculture. Pesticide and weedicide spraying drones carrying specific concentrated chemicals at higher levels and when connected to a tether that can transport water can be used in combination to dilute the chemical and spray them over the field. The advantage of carrying high concentrated levels of chemicals and a tether to separately carry water is to increase the drones accessibility range as well as reduce the turnaround time of the drone to refill them again. Such spraying drones can much more efficient in saving the time taken in undertaking the same task conventionally by a human. Combined with pesticide and weedicide spraying drones, they can also be employed in irrigation as well as fertilizer spraying tasks. Irrigation based drones can be a great deal to cope up with the shortage and scarcity of water at some parts of the World. Irrigation based drones can be even more efficient than sprinklers at the garden and can be deployed in variety of areas from fields to even gardens. Fertilizer spraying drones can be one of the foremost task that can be replaced by drones since it requires the most man-hour when compared to other tasks mentioned above. The drones must be capable of carrying solid-payloads over medium distances and can be spread over larger area. The height of the drone flying can be varied to change the intensity of fertilizer needed to sprayed over a certain area. Thus the spraying drones can be a good alternative to eliminate the inefficiencies caused during conventional method of irrigation methods (Ay et al., 2015).

Special Cases

One of the most sophisticated aspect of farming is cattle herding. A animal herding drone can be a advanced way to quickly collect a herd of animal to the destined area. GPS-enabled drones can be particularly suited for this task to rear a herd. Deploying automation and Artificial Intelligence is a vital step for drones that engage in cattle herding. Apart from cattle even sheep and lamb can be heard in such a way that this new technological approach is a way new concept that needs to be perfected before commercialization. This animal herding via drones is a new method of livestock farming that can not only supervise a herd in mobility but can also feed and even check their health and hygiene without human intervention. Bulk animal farm which doesn't have enough workers can start bringing this way of technological approach into the livestock management. Terrain mapping is essential to herd animal and with Deep Learning the customized drone can frame its own route to herd the animals. Even the grass level can be monitored to enhance the path taken by the drones to feed the cattle or other animal herd to avoid over-grazing over a particular field for a longer time. Such new key innovations can be done via autonomous drones that can perform the task under Artificial Intelligence Algorithm and the limiting factor is the scaling issues that will be discussed further below in this paper.

Bees, Insects and Hummingbirds that are tiny but effective in pollinating a flowering plant can be challenging to do so if the environment in contained in a Greenhouse enclosure. Such incidents require additional effort by humans to allow a natural process to be done via artificial procedure or approach. One such approach is the intervention of Nano-Engineered drones that gets the crown in as a pollinating agent at a nano scale. Such nano-copters are very efficient and require lesser power and hundred such nano-copters can be connected via a central topology to exchange information with each other and to plan the expected pollination site after a particular area is completely pollinated. These nano-copters can carry pollen between two plant by analysing the quality of the pollen which implies that the quality of pollination can be determined with the help of nano-copters and thus allowing us to decide and control the quality of end product from the plant (Liu et al., 2021).

METHODS TO IMPLEMENT TECHNOLOGICAL AID IN PRECISION AGRICULTURE

IoT Data Analytics and Cloud Computing in Precision Agriculture

IoT also referred to as Internet of Things nearly in reach possibly in all major tech related industries and thus can also be handled in Precision Agriculture. IoT data analytics can provide new benefits to increase the output from the crops as well as the quality of the crops that the industry standards expect. A novel yet technological approach in computing data related to agriculture requires IoT and its power to make sure that there is no need of a farmer to be regularly in the agriculture feed instead the crop health, pest detection and infestation, soil health and moisture management and even fertilizer spraying and irrigation management can all be intimated via IoT. IoT data analytics do analyse the data via dedicated sensors primarily incorporated for Precision Agriculture and can store the data in cloud for later usage and can update the farmers about the specific details mentioned above to keep track of the crop health and condition precisely. Certain uncertain parameters require constant monitoring which can be complex at present scenario but incorporating IoT infused drones and cloud computing embedded systems can reduce the uncertainty levels to an expectable graph. Wireless sensor networks are a basic objective of IoT being employed in Precision Agriculture and scheduled checks via drones can probably decrease any devastating effects such as soil degradation and pest damage that would be catastrophic if aren't accounted or taken care. Agricultural fields that expand over hectares can possibly expand the IoT networks in their precision based farming. Crop management and growth of crop must require day to day information update which requires hefty data being stored and transferred that can be solved solely with the help of cloud computing. Cloud computing servers can allocate specific systems for agricultural purposes which can range from soil health to crop health and additionally local farmer's details and even the e-commerce and retailer info graphs. Such solutions provided by cloud computing is so helpful is marinating a database that is synchronized with both dealer and the farmer and can be allowed to be accessible by public to know the local farming constraints and other details. Such cloud computing based details can be accessible from the information provided by the sensor that can be utilized in future needs (Andrew et al., 2018).

Artificial Intelligence and Deep Learning Embedded Systems

AI and deep learning neural networks embedded in drones can analyse various data that are being interpreted by the sensors embedded in them to obtain specific parameters that needed to be taken into account. Such data can range from crop health monitoring where the Drone's AI must be trained enough to differ a healthy plant from any infested plant. Also pest damaged plants must be spotted by the drones and they needed to calibrated to attack such pests by spraying pesticide in the suitable level neither exceeding nor below the level needed to wipe out such damage. Artificial Intelligence and Deep Learning based approach can be used to train drones to perform terrain mapping and analyse the certain terrain for possible irrigation and as well as other activities to be extended. Assessment of statistical data and performance analysis is a key factor in determining the algorithm to be used in Precision Agriculture. Machine learning can be utilized to further make the Precision Agriculture a compatible one. Here AI and Machine Learning can be demonstrated using IRIS data set which contains values to train the AI and Machine Learning Model. Iris Flower shown in Figure 2.

Figure 2. Iris flower

Table 2. Iris dataset

	Sepal Length	Sepal Width	Petal Length	Petal Width	Class Label
1	5.1 cm	3.5 cm	1.4 cm	0.2 cm	Setosa
2	4.9 cm	3.0 cm	1.4 cm	0.2 cm	Setosa
50	6.4 cm	3.5 cm	4.5cm	1.2 cm	Versicolor
150	5.9 cm	3.0 cm	5.0 cm	1.8 cm	Virginica

From the Table 2, it can be inferred that with the Iris dataset embedded with AI and Machine Learning capabilities, drones can evaluate the parameters for each iris species and form a AI model to regulate the growth of individual iris species more effectively forming a basis for precision agriculture. Image to be trained by the AI model is shown in Figure 3.

Figure 3. Training input: Iris flower

The parameters taken into account are the petal length, petal width, sepal length, sepal width. With these parameters and some values to train the AI model, the AI and Machine Learning Algorithm and can easily do tedious jobs that are considered too complex by humans. Parameters mentioned above can be taken via high clarity cameras that can access the species variant via the trained values. Clear View of iris flower aspect ratio is shown in Figure 4.

Figure 4. Clear view of iris flower aspect ratio

From the above the depicted species of Iris flowers, the Trained drones embedded with AI and machine learning algorithm can detect any abnormalities in each species of flowers and if any abnormalities in the aspect ratio of the flower will trigger the supervising authority to seek any potential threat to the cluster. Once trained the AI and Machine Learning embedded systems can regulate the health of hundreds of species without the necessity of human intervention in the inspection phase. Sepal and Petal view of iris flower is shown in Figure 5.

Figure 5. Clear view of iris flower aspect ratio

RESULTS

Samples of Iris data are preloaded onto the AI and machine learning algorithm in the Drones, which trains itself to find all other Iris flowers related to the same species. New species identified via this method can be useful in training the AI model further more extensively. Species of certain endangered category can be monitored closely which can reduce the chances of the certain endangered species from becoming extinct.

This Iris dataset is an example of how AI and Machine Learning can help in any abnormality detection and crop health monitoring. Further customizing for each separate variant and species and other plants can help farmers benefit from the technological advantage of Drones embedded with AI model (Sharma et al., n.d.).

LIMITATIONS

Challenges of Deploying Drones in Precision Agriculture

Some of the current issues faced by farmers in employing Precision Agriculture over traditional methods constitutes hacking the drones via GPS-Spoofing, low knowledge over such new technologies, scaling issues, not flexible for common people to issue and higher costs associated with employing such technologies at a larger scale. It is to be known that technology adoption is the major hurdle over Precision Agriculture added that the cost of employing is way to higher than labour costs. Farmers aren't advanced enough to seek such technological advantages, which requires specific guidance towards them. The first major hurdle is that without a standardized technology the networking might get challenged. Connectivity between technical team and the farmers might be there 24x7 as a service in need. Data management could arise as a major threat that can only be solved by involving third party influence and the scaling issues constitutes the lion's share of the major hurdle in deploying Precision Agriculture over rural areas (Marios et al., 2017).

RECOMMENDATIONS

Ways to Overcome the Challenges Possessed in Drones Used in Precision Agriculture

The first major challenge discussed is the scalability issue which can overcome by using portable drones. Portable drones can leased among a group of farmers by third party leaser's who can lease the drones for certain period of time frame which can cut costs for overall period of time and can even gain profits rather than buying drones for a hefty price tag. Portable drones can also be deployed in multiple terrain which can be efficient so that certain drones can be used in certain seasons and changing equipment fitted in drones can make it capable to be fitted across all terrain. A single drone could be used as a multi-functional drone which can be working all along a year with demand skyrocketing during sowing times and can be used at all climate dependent crops which makes it even more profitable all along the year. Portable drones if deployed can be advantageous to both the leaser as well as the farmer if done so. The third parties could take over the burden of technological aid so that farmers could be not reluctant to engage in such new Precision Agriculture farming techniques. Large data can be stored in cloud computing servers that keep track of information provided by drones to analyse and monitor the crop growth, health, etc. GPS-Anti-Spoofing techniques can be used to drones that do employ GPS-Coordinates for their location information to avoid any spoofing attacks intended for theft threat. Swarms of drone networks can be beneficial to maintain safety and integrity among Precision Agriculture drones Various Terrain images are shown in Figure 6 and Figure 7 (Hell et al., 2018).

Figure 6. Contour ploughing

Figure 7. Plain land farming

PERFORMANCE ANALYSIS OF AGRICULTURAL DRONES AVAILABLE ON CURRENT MARKET

Table 3 is based upon agricultural drones that are capable of undertaking the agricultural work more precisely and yet economically cheaper while not compromising the payload capacity. The drones produced by two separate Chinese entities Dji and EFT duopoly makes the models comparable with each other.

Table 3. Performance analysis of Dji and EFT's agricultural drones

Drone Model	Dji Agras T-10	EFT G-410
Max Payload	8 kg	8 kg
MTOW (Max Take Off Weight)	24.8 kg	27 kg
Spray Tank Capacity	10 L	10 L
Radar Type	Spherical Radar	Radio Altimeter
Waterproof level	IP67	IP65
Classification Of UAV	VTOL (Vertical Take off and Landing)	VTOL (Vertical Take off and Landing)
Equipped Camera	FPV Cameras	FPV Cameras

Dji uses advanced systems when compared to EFT models but EFT takes the advantage of greater MTOW which could result in greater battery storage capacity and hence longer flight time than the counterpart. They procure the supplies from similar suppliers which could lead to hardware components and sensors to be similar which leads the duopoly to increase the performance capabilities of their models which are given in Table 3.

CONCLUSION AND FUTURE WORK

This chapter concludes with the adoption of Precision Agriculture as a technological aid that farmers need to know and take the required measures to implement in parallel to traditional methods as a way to boost the efficiency, increase crop output rate and decrease pest infestation damages. Precision farming being a successor to smart farming is a way more intelligent approach for the computational machines in the agricultural sector. Cloud computing, Data science, Artificial Intelligence and Machine Learning would make precision agriculture a more effective way of cultivating plants with higher grade of output growth, possibly outperforming traditional ways of cultivating plants. In the future, sustainable power generation alongside precision agriculture would be established to enhance the work.

REFERENCES

Abutalipov, R. N., Bolgov, Y. V., & Senov, H. M. (2016). Flowering plants pollination robotic system for greenhouses by means of nano copter (drone aircraft). *2016 IEEE Conference on Quality Management, Transport and Information Security, Information Technologies (IT&MQ&IS)*, 7-9. 10.1109/ITMQIS.2016.7751907

Andrew, R. C., Malekian, R., & Bogatinoska, D. C. (2018). IoT solutions for Precision Agriculture. *2018 41st International Convention on Information and Communication Technology, Electronics and Microelectronics (MIPRO)*, 345-349. 10.23919/MIPRO.2018.8400066

Ay, F., & İnce, G. (2015). Application of pesticide using unmanned aerial vehicle. *2015 23nd Signal Processing and Communications Applications Conference (SIU)*, 1268-1271. 10.1109/SIU.2015.7130069

Donka, A. K., Seerapu, J. A. R., Verma, S., Sao, A., Shukla, G., & Tripathi, S. (2020). Unmanned Aerial Vehicle for Precision Agriculture in a Modular Approach. *2020 IEEE 7th Uttar Pradesh Section International Conference on Electrical, Electronics and Computer Engineering (UPCON),* 1-5. 10.1109/UPCON50219.2020.9376521

Hell, M., Bolam, R. C., Vagapov, Y., & Anuchin, A. (2018). Design of a portable drone for educational purposes. *2018 25th International Workshop on Electric Drives: Optimization in Control of Electric Drives (IWED),* 1-5. 10.1109/IWED.2018.8321377

Liu, C., Jian, Z., Xie, M., & Cheng, I. (2021). A Real-Time Mobile Application for Cattle Tracking using Video Captured from a Drone. *2021 International Symposium on Networks, Computers and Communications (ISNCC),* 1-6. 10.1109/ISNCC52172.2021.9615648

Lohit, G. V. S., & Bisht, D. (2021). Seed Dispenser using Drones and Deep Learning Techniques for Reforestation. *2021 5th International Conference on Computing Methodologies and Communication (ICCMC),* 1275-1283. 10.1109/ICCMC51019.2021.9418227

Marios, S., & Georgiou, J. (2017). Precision Agriculture: Challenges in sensors and electronics for real-time soil and plant monitoring. *2017 IEEE Biomedical Circuits and Systems Conference (BioCAS),* 1-4. 10.1109/BIOCAS.2017.8325180

Petkovics, I., Simon, J., Petkovics, Á., & Čović, Z. (2017). Selection of unmanned aerial vehicle for Precision Agriculture with multi-criteria decision making algorithm. *2017 IEEE 15th International Symposium on Intelligent Systems and Informatics (SISY),* 151-156. 10.1109/SISY.2017.8080543

Sharma, A., Jain, A., Gupta, P., & Chowdary, V. (2021). Machine Learning Applications for Precision Agriculture: A Comprehensive Review. *IEEE Access : Practical Innovations, Open Solutions, 9,* 4843–4873. doi:10.1109/ACCESS.2020.3048415

Vanitha, N. (2021). *A Study on the Applications of Drones in Smart Farming*. Academic Press.

Vanitha, N., & Padmavathi, G. (2018). A Comparative Study on Communication Architecture of Unmanned Aerial Vehicles and Security Analysis of False Data Dissemination Attacks. *2018 International Conference on Current Trends towards Converging Technologies (ICCTCT),* 1-8. 10.1109/ICCTCT.2018.8550873

Vanitha, N., & Padmavathi, G. (2018). A Study on Various Cyber-Attacks and their Classification in UAV Assisted Vehicular Ad-Hoc Networks. In G. Ganapathi, A. Subramaniam, M. Graña, S. Balusamy, R. Natarajan, & P. Ramanathan (Eds.), *Computational Intelligence, Cyber Security and Computational Models. Models and Techniques for Intelligent Systems and Automation. ICC3 2017. Communications in Computer and Information Science* (Vol. 844). Springer. doi:10.1007/978-981-13-0716-4_11

Vanitha, N., Vinodhini, V., & Rekha, S. (2016). A study on agriculture UAV for identifying the plant damage after plantation. *International Journal of Engineering and Management Research, 6*(6), 310–313.

Conclusion

In this book, titled *Contemporary Developments in Agricultural Cyber-Physical Systems*, a comprehensive exploration of the transformative potential of cyber-physical systems (CPS) in agriculture has been undertaken. Through the integration of advanced technologies such as IoT, robotics, AI, data analytics, and automation, CPS has revolutionized the approach to agriculture, offering innovative solutions to address industry challenges.

Throughout the chapters, fundamental concepts and principles that underpin agricultural CPS have been highlighted, including architectural frameworks, communication protocols, and interoperability standards. The role of IoT devices and sensors in data collection and monitoring has been explored, as well as the utilization of robotics and automation for precision farming. The power of AI and machine learning algorithms in optimizing agricultural operations has been emphasized, alongside the challenges and opportunities associated with big data analytics and cloud computing in agriculture.

Furthermore, the book has examined how agricultural CPS contributes to sustainable farming practices, addressing resource management, water-efficient irrigation systems, smart greenhouse technologies, and resilient crop cultivation techniques. By leveraging CPS, resource consumption can be minimized, productivity enhanced, and environmental impacts reduced, fostering a more sustainable and resilient agricultural sector.

In envisioning the future of agricultural CPS, emerging trends and potential applications have been explored. The integration of blockchain technology for supply chain transparency, the use of drones and satellite imagery for precision agriculture, and the implications of CPS for rural development and social equity are just a few examples of the exciting directions this field is heading towards.

As the book concludes, it is crucial to acknowledge the challenges that lie ahead. The adoption of agricultural CPS requires addressing issues related to data security and privacy, connectivity in rural areas, and the digital divide among farmers. Additionally, the ethical considerations surrounding the use of AI, robotics, and automation in agriculture must be carefully navigated to ensure fair and responsible deployment.

Moving forward, collaboration and knowledge-sharing among researchers, practitioners, policymakers, and stakeholders will be key to unlocking the full potential of agricultural CPS. Fostering interdisciplinary approaches, where experts from various fields such as agriculture, engineering, computer science, and policy work together to design and implement CPS solutions tailored to the specific needs of different regions and farming systems, is essential.

Embracing the advancements in agricultural CPS offers the opportunity to shape a more sustainable, efficient, and resilient future for agriculture. The integration of digital technologies and the physical realm of farming holds immense promise for addressing global challenges such as climate change, food security, and resource scarcity.

As the journey through the contemporary developments in agricultural CPS concludes, readers are invited to continue exploring, researching, and implementing these innovative solutions. By staying at the forefront of technological advancements and fostering a culture of continuous learning and adaptation, the transformative potential of agricultural cyber-physical systems can be unlocked, cultivating a more prosperous and sustainable future for agriculture, ensuring that the needs of both present and future generations are met.

G. S. Karthick
Department of Software Systems, PSG College of Arts and Science, India

Compilation of References

Aasha Nandhini, S., Hemalatha, R., Radha, S., & Indumathi, K. (2018). Web enabled plant disease detection system for agricultural applications using WMSN. *Wireless Personal Communications*, *102*(2), 725–740. doi:10.100711277-017-5092-4

Abbas, A., Jain, S., Gour, M., & Vankudothu, S. (2021). Tomato plant disease detection using transfer learning with C-GAN synthetic images. *Computers and Electronics in Agriculture*, *187*, 106279. doi:10.1016/j.compag.2021.106279

Abbasi, R., Martinez, P., & Ahmad, R. (2022). The digitization of agricultural industry – a systematic literature review on agriculture 4.0 smart agricultural technology Volume-2. *Technology in Society*, *69*, 100042.

Abdallah, E. B., Grati, R., Fredj, M., & Boukadi, K. (2022). A Machine Learning Approach for a Robust Irrigation Prediction via Regression and Feature Selection. In *International Conference on Advanced Information Networking and Applications [AINA 2022: Advanced Information Networking and Applications]*, (*vol. 449*, pp.491-502). Springer. 10.1007/978-3-030-99584-3_43

Abdollahi, A., Pradhan, B., Gite, S., & Alamri, A. (2020). Building footprint extraction from high-resolution aerial images using generative adversarial network (GAN) architecture. *IEEE Access : Practical Innovations, Open Solutions*, *8*, 209517–209527. doi:10.1109/ACCESS.2020.3038225

Abutalipov, R. N., Bolgov, Y. V., & Senov, H. M. (2016). Flowering plants pollination robotic system for greenhouses by means of nano copter (drone aircraft). *2016 IEEE Conference on Quality Management, Transport and Information Security, Information Technologies (IT&MQ&IS)*, 7-9. 10.1109/ITMQIS.2016.7751907

Abuzanouneh, K. I. M., Al-Wesabi, F. N., Albraikan, A. A., Duhayyim, M. A., Al-Shabi, M., Hilal, A. M., Hamza, M. A., Zamani, A. S., & Muthulakshmi, K. (2022). Design of machine learning based smart irrigation system for precision agriculture. [Tech Science Press]. *Computers, Materials & Continua*, *72*(1), 109–124. doi:10.32604/cmc.2022.022648

Adafruit. (n.d.). *Digital temperature sensor DS18B20*. Adafruit. https://www.adafruit.com/product/381

Adafruit. (n.d.). *Temperature Humidity sensor. AM2315C - Encased I2C Temperature/Humidity Sensor*. Adafruit. https://www.adafruit.com/product/5182

Adão, T., Hruška, J., Pádua, L., Bessa, J., Peres, E., Morais, R., & Sousa, J. J. (2017). Hyperspectral imaging: A review on UAV-based sensors, data processing and applications for agriculture and forestry. *Remote Sensing (Basel)*, *9*(11), 1110. doi:10.3390/rs9111110

Agarwal, M., Singh, A., Arjaria, S., Sinha, A., & Gupta, S. (2020). ToLeD: Tomato leaf disease detection using convolution neural network. *Procedia Computer Science*, *167*, 293–301. doi:10.1016/j.procs.2020.03.225

Aggarwal, S., & Kumar, N. (2020). Path planning techniques for unmanned aerial vehicles: A review, solutions, and challenges. *Computer Communications*, *149*, 270–299. doi:10.1016/j.comcom.2019.10.014

Ahmed, G. & Kim, D. (2013),*Cyber Physical System: Architecture, applications, and research challenges*, 11-15. IFIP Wireless Days.

Ahmed, S. H., Kim, G., & Kim, D. (2013). Cyber Physical System: Architecture, Applications and Research Challenges. IEEE IFIP Wire- less Days (WD), (pp. 1-5). IEEE.

Ahmed, N., De, D., & Hussain, I. (2018). Internet of Things (IoT) for Smart Precision Agriculture and Farming in Rural Areas. *IEEE Internet of Things Journal*, *5*(6), 4890–4899. doi:10.1109/JIOT.2018.2879579

Akram, V. K., & Challenger, M. (2021). A Smart Home Agriculture System Based on Internet of Things. In *2021 10th Mediterranean Conference on Embedded Computing (MECO)* (pp. 1-4). IEEE. 10.1109/MECO52532.2021.9460276

Akshya, J., & Priyadarsini, P. L. K. (2020). A Comparison of Machine Learning Approaches for Classifying Flood-Hit Areas in Aerial Images. In *International Conference on Innovative Computing and Communications: Proceedings of ICICC 2019* (vol. 1, pp. 407-415). Springer Singapore. 10.1007/978-981-15-1286-5_34

Akshya, J., & Priyadarsini, P. L. K. (2020). Graph-based path planning for intelligent UAVs in area coverage applications. *Journal of Intelligent & Fuzzy Systems*, *39*(6), 8191–8203. doi:10.3233/JIFS-189140

Akyildiz, I. F., & Kasimoglu, I. H. (2004). Wireless sensor and actor networks: Research challenges. *Ad Hoc Networks*, *2*(4), 351–367. doi:10.1016/j.adhoc.2004.04.003

Aldhyani, T. H. H., & Alkahtani, H. (2023). Cybersecurity for detecting distributed denial of service attacks in Agriculture 4.0: Deep learning model. *Mathematics*, *11*(1), 233. doi:10.3390/math11010233

Alguliyev, R., Imamverdiyev, Y., & Sukhostat, L. (2018). *Cyber-physical systems and their security issues -. Computers in Industry*, *100*, 212–223. doi:10.1016/j.compind.2018.04.017

Alharbi, H. A., & Aldossary, M. (2021, July 30). Energy-efficient edge-fog-cloud architecture for IoT-based smart agriculture environment. *IEEE Access : Practical Innovations, Open Solutions*, *9*, 110480–110492. doi:10.1109/ACCESS.2021.3101397

Alotaibi, K. D., Arcand, M., & Ziadi, N. (2021). Effect of biochar addition on legacy phosphorus availability in long-term cultivated arid soil. *Chemical and Biological Technologies in Agriculture*, *8*(1), 47. Advance online publication. doi:10.118640538-021-00249-0

Al-Sarawi, S., Anbar, M., Alieyan, K., & Alzubaidi, M. (2017). Internet of Things (IoT) communication protocols: Review. In 2017 8th International Conference on Information Technology (ICIT), (pp. 685-690).IEEE.

Amitrano, C., Arena, C., Rouphael, Y., De Pascale, S., & De Micco, V. (2019). Vapour pressure deficit: The hidden driver behind plant morphofunctional traits in controlled environments. *Annals of Applied Biology*, *175*(3), 313–325. doi:10.1111/aab.12544

An, W., Wu, D., Ci, S., Luo, H., Adamchuk, V., & Z. Xu, Z. (2017). Agriculture Cyber-Physical Systems. *Cyber-Physical Systems Foundations, Principles and Applications Intelligent Data-Centric Systems*, 399-417. Elsevier.

An, C., Sun, C., Li, N., Huang, B., Jiang, J., Shen, Y., Wang, C., Zhao, X., Cui, B., Wang, C., Li, X., Zhan, S., Gao, F., Zeng, Z., Cui, H., & Wang, Y. (2022). Nanomaterials and nanotechnology for the delivery of agrochemicals: Strategies towards sustainable agriculture. *Journal of Nanobiotechnology*, *20*(1), 1–19. doi:10.118612951-021-01214-7 PMID:34983545

Andreev, R. A., Andreeva, P. A., Krotov, L. N., & Krotova, E. L. (2018). *Review of blockchain technology: types of blockchain and their application*. Intellekt. Sist. Proizv.

Andrew, R. C., Malekian, R., & Bogatinoska, D. C. (2018). IoT solutions for Precision Agriculture. *2018 41st International Convention on Information and Communication Technology, Electronics and Microelectronics (MIPRO)*, 345-349. 10.23919/MIPRO.2018.8400066

Ang, J. B. (2019). Agricultural legacy and individualistic culture. *Journal of Economic Growth*, *24*(4), 397–425. doi:10.100710887-019-09171-5

Anitha, Muthusamy, & Thiruvenkadam. (2023). An Intelligent Agro-Cloud Framework For Efficient Crop Cultivation. *Journal of Pharmaceutical Negative Results*, 754–763.

Archana, K. S., Srinivasan, S., Bharathi, S. P., Balamurugan, R., Prabakar, T. N., & Britto, A. (2022). A novel method to improve computational and classification performance of rice plant disease identification. *The Journal of Supercomputing*, *78*(6), 8925–8945. doi:10.100711227-021-04245-x

Ashifuddin Mondal, M., & Rehena, Z. (2018). Iot based intelligent agriculture field monitoring system. In *2018 8th International Conference on Cloud Computing, Data Science & Engineering (Confluence)* (pp. 625-629). IEEE.

Ashwinkumar, S., Rajagopal, S., Manimaran, V., & Jegajothi, B. (2022). Automated plant leaf disease detection and classification using optimal MobileNet-based convolutional neural networks. *Materials Today: Proceedings*, *51*, 480–487. doi:10.1016/j.matpr.2021.05.584

Atila, U., Uçar, M., Akyol, K., & Uçar, E. (2021). Plant leaf disease classification using EfficientNet deep learning model. *Ecological Informatics*, *61*, 101182. doi:10.1016/j.ecoinf.2020.101182

Atlas Scientific. (2022). Ezo-CO2. *Embedded Carbon Dioxide Sensor*. Atlas Scientific. https://atlas-scientific.com/probes/co2-sensor/

Atlas Scientific. (2022). Ezo-Pump. *Embedded Dosing Pump*. Atlas Scientific. https://atlas-scientific.com/peristaltic/ezo-pmp/

Attaran, A., & Rashidzadeh, R. (2016, July 11). Chipless radio frequency identification tag for IoT applications. *IEEE Internet of Things Journal*, *3*(6), 1310–1318. doi:10.1109/JIOT.2016.2589928

Avgoustaki, D. D., & Xydis, G. (2020). Indoor vertical farming in the Urban nexus context: Business growth and resource savings. *Sustainability (Basel)*, *12*(5), 1965. doi:10.3390u12051965

Awad, M., & Khanna, R. (2015). Support Vector Regression. In *Efficient Learning Machines*. Apress. doi:10.1007/978-1-4302-5990-9_4

Ay, F., & İnce, G. (2015). Application of pesticide using unmanned aerial vehicle. *2015 23nd Signal Processing and Communications Applications Conference (SIU)*, 1268-1271. 10.1109/SIU.2015.7130069

Ayaz, M., Ammad-Uddin, M., Sharif, Z., Mansour, A., & Aggoune, E.-H. M. (2019). Internet-of-Things (IoT)-based smart agriculture: Toward making the fields talk. *IEEE Access : Practical Innovations, Open Solutions*, *7*, 129551–129583. doi:10.1109/ACCESS.2019.2932609

Azgomi, H., Haredasht, F. R., & Motlagh, M. R. S. (2023). Diagnosis of some apple fruit diseases by using image processing and artificial neural network. *Food Control*, *145*, 109484. doi:10.1016/j.foodcont.2022.109484

Babu, B. S., Kamalakannan, J., Meenatchi, N., Karthik, S., & Boopathi, S. (2022). Economic impacts and reliability evaluation of battery by adopting Electric Vehicle. *IEEE Explore*, (pp. 1–6). IEEE.

Babu, J. S., Mohan, G. K., & Praveena, N. A. (2022). Flow of Cyber-Physical systems—A Comprehensive Survey. Smart Technologies in Data Science and Communication. *Proceedings of SMART-DSC*. Taylor and Francis.

Babu, S., Maravarman, M., & Pitchai, R. (2022). Detection of Rice Plant Disease Using Deep Learning Techniques. *Journal of Mobile Multimedia*, 757-770.

Bagha, H., Yavari, A., & Georgakopoulos, D. (2022). Hybrid Sensing Platform for IoT-Based Precision Agriculture. *Future Internet*, *14*(8), 233. doi:10.3390/fi14080233

Balafoutis, A. T., Van Evert, F. K., & Fountas, S. (2020). Smart farming technology trends: Economic and environmental effects, labor impact, and adoption readiness. *Agronomy (Basel)*, *10*(5), 743. doi:10.3390/agronomy10050743

Balasooriya, T. N., Mantri, P., & Suriyampola, P. (2020). IoT-Based Smart Watering System Towards Improving the Efficiency of Agricultural Irrigation. In *2020 IEEE Global Conference on Artificial Intelligence and Internet of Things (GCAIoT)* (pp. 1-7). IEEE. 10.1109/GCAIoT51063.2020.9345902

Banaseka, F. K., Katsriku, F., Abdulai, J. D., Adu-Manu, K. S., & Engmann, F. N. A. (2021). Signal propagation models in soil medium for the study of wireless underground sensor networks: A review of current trends. *Wireless Communications and Mobile Computing*, *2021*, 1–12. doi:10.1155/2021/8836426

Bannerjee, G., Sarkar, U., Das, S., & Ghosh, I. (2018). Artificial intelligence in agriculture: A literature survey. *International Journal of Scientific Research in Computer Science Applications and Management Studies*, *7*(3), 1–6.

Barbosa, G., Gadelha, F., Kublik, N., Proctor, A., Reichelm, L., Weissinger, E., Wohlleb, G., & Halden, R. (2015). Comparison of land, water, and energy requirements of lettuce grown using hydroponic vs. conventional agricultural methods. *International Journal of Environmental Research and Public Health*, *12*(6), 6879–6891. doi:10.3390/ijerph120606879 PMID:26086708

Barla, S., Salachas, G., & Abeliotis, K. (2020). Assessment of the greenhouse gas emissions from aeroponic lettuce cultivation in Greece. *Euro-Mediterranean Journal for Environmental Integration*, *5*(2), 29. doi:10.100741207-020-00168-w

Bedi, P., & Gole, P. (2021). Plant disease detection using hybrid model based on convolutional autoencoder and convolutional neural network. *Artificial Intelligence in Agriculture*, *5*, 90–101. doi:10.1016/j.aiia.2021.05.002

Bello-Cerezo, R., Bianconi, F., Di Maria, F., Napoletano, P., & Smeraldi, F. (2019). Comparative evaluation of hand-crafted image descriptors vs. off-the-shelf CNN-based features for color texture classification under ideal and realistic conditions. *Applied Sciences (Basel, Switzerland)*, *9*(4), 738. doi:10.3390/app9040738

Bennett, E., Carpenter, S., Gordon, L., Ramankutty, N., Balvanera, P., Campbell, B., Cramer, W., Foley, J., Folke, C., & Karlberg, L. (2014). Toward a more resilient agriculture. *Solutions*, *5*(5), 65–75.

Bharati, S., & Mondal, M. R. H. (2021). 12 Applications and challenges of AI-driven IoHT for combating pandemics: A review. *Computational Intelligence for Managing Pandemics*, *5*, 213–230. doi:10.1515/9783110712254-012

Bhati, H., & Rathore, M. 2022. Diagnosis and Detection of Plant Diseases Using Data Mining Techniques. In Proceedings of Data Analytics and Management: ICDAM 2021, Volume 1 (pp. 451-458). Springer Singapore. doi:10.1007/978-981-16-6289-8_38

Bhat, M. A., Adil, A. W., Sikander, B. M., Lone, Y., & Malik, J. A. (2019). Waste Management Technology for Sustainable Agriculture. In *Waste Management Technology for Sustainable Agriculture* (pp. 156–176). IGI Global. doi:10.4018/978-1-7998-0031-6.ch009

Bhatnagar, V., Poonia, R. C., & Sunda, S. (2019). State of the art and gap analysis of precision agriculture: A case study of Indian farmers. [IJAEIS]. *International Journal of Agricultural and Environmental Information Systems*, *10*(3), 72–92. doi:10.4018/IJAEIS.2019070105

Bhoi, A., Nayak, R. P., Bhoi, S. K., & Sethi, S. (2021). Automated Precision Irrigation System Using Machine Learning and IoT. In Intelligent Systems, Lecture Notes in Networks and Systems, (vol 185, pp. 275-282). Springer. doi:10.1007/978-981-33-6081-5_24

Bibri, S. E. (2018). The IoT for smart sustainable cities of the future: An analytical framework for sensor-based big data applications for environmental sustainability. *Sustainable Cities and Society*, *38*, 230–253. doi:10.1016/j.scs.2017.12.034

Bonakdarpour, B. (2008). Challenges in Transformation of Existing Real-Time Embedded Systems to Cyber-Physical Systems. *ACM SIGBED Review*, *5*(1), 1–2. doi:10.1145/1366283.1366294

Bongiovanni, R., & Lowenberg-DeBoer, J. (2004, August). Precision agriculture and sustainability. *Precision Agriculture*, *5*(4), 359–387. doi:10.1023/B:PRAG.0000040806.39604.aa

Boopathi, S. (2021). Improving of Green Sand-Mould Quality using Taguchi Technique. *Journal of Engineering Research*, 1–12. doi:10.36909/jer.14079

Boopathi, S., & Khare, R. KG, J. C., Muni, T. V., & Khare, S. (2023). Additive Manufacturing Developments in the Medical Engineering Field. In Development, Properties, and Industrial Applications of 3D Printed Polymer Composites (pp. 86–106). IGI Global.

Boopathi, S., & Myilsamy, S. (2021). Material removal rate and surface roughness study on Near-dry wire electrical discharge Machining process. *Materials Today: Proceedings*, *45*(xxxx), 8149–8156. doi:10.1016/j.matpr.2021.02.267

Boopathi, S., Arigela, S. H., Raman, R., Indhumathi, C., Kavitha, V., & Bhatt, B. C. (2022). Prominent Rule Control-based Internet of Things: Poultry Farm Management System. *IEEE Explore*, (pp. 1–6). IEEE.

Boopathi, S., Siva Kumar, P. K., & Meena, R. S. J., S. I., P., S. K., & Sudhakar, M. (2023). Sustainable Developments of Modern Soil-Less Agro-Cultivation Systems. In Human Agro-Energy Optimization for Business and Industry (pp. 69–87). IGI Global. doi:10.4018/978-1-6684-4118-3.ch004

Boopathi, S. (2019). Experimental investigation and parameter analysis of LPG refrigeration system using Taguchi method. *SN Applied Sciences*, *1*(8), 892. doi:10.100742452-019-0925-2

Boopathi, S. (2022a). An experimental investigation of Quench Polish Quench (QPQ) coating on AISI 4150 steel. *Engineering Research Express*, *4*(4), 045009. doi:10.1088/2631-8695/ac9ddd

Boopathi, S. (2022b). An investigation on gas emission concentration and relative emission rate of the near-dry wire-cut electrical discharge machining process. *Environmental Science and Pollution Research International*, *29*(57), 86237–86246. doi:10.100711356-021-17658-1 PMID:34837614

Boopathi, S. (2022b). Experimental investigation and multi-objective optimization of cryogenic Friction-stir-welding of AA2014 and AZ31B alloys using MOORA technique. *Materials Today. Communications*, *33*, 104937. doi:10.1016/j.mtcomm.2022.104937

Boopathi, S. (2022c). (in press). Cryogenically treated and untreated stainless steel grade 317 in sustainable wire electrical discharge machining process: A comparative study. *Environmental Science and Pollution Research International*, 1–10. doi:10.100711356-022-22843-x PMID:36057706

Boopathi, S. (2022d). Performance Improvement of Eco-Friendly Near-Dry Wire-Cut Electrical Discharge Machining Process Using Coconut Oil-Mist Dielectric Fluid. *Journal of Advanced Manufacturing Systems*, 1–20. doi:10.1142/S0219686723500178

Boopathi, S. (2023). An Investigation on Friction Stir Processing of Aluminum Alloy-Boron Carbide Surface Composite. In *Materials Horizons: From Nature to Nanomaterials* (pp. 249–257). Springer. doi:10.1007/978-981-19-7146-4_14

Boopathi, S., Balasubramani, V., & Sanjeev Kumar, R. (2023). Influences of various natural fibers on the mechanical and drilling characteristics of coir-fiber-based hybrid epoxy composites. *Engineering Research Express*, *5*(1), 15002–15020. doi:10.1088/2631-8695/acb132

Boopathi, S., Haribalaji, V., Mageswari, M., & Asif, M. M. (2022). Influences of Boron Carbide Particles on the Wear Rate and Tensile Strength of Aa2014 Surface Composite Fabricated By Friction-Stir Processing. *Materiali in Tehnologije*, *56*(3), 263–270. doi:10.17222/mit.2022.409

Boopathi, S., Jeyakumar, M., Singh, G. R., King, F. L., Pandian, M., Subbiah, R., & Haribalaji, V. (2022). An experimental study on friction stir processing of aluminium alloy (AA-2024) and boron nitride (BNp) surface composite. *Materials Today: Proceedings*, *59*, 1094–1099. doi:10.1016/j.matpr.2022.02.435

Boopathi, S., Kumaresan, A., Manohar, N., & Krishna Moorthi, R. (2017). Review on Effect of Process Parameters-Friction Stir Welding Process. [IRJET]. *International Research Journal of Engineering and Technology*, *4*(7), 272–278.

Boopathi, S., Lewise, K. A. S., Subbiah, R., & Sivaraman, G. (2021). Near-dry wire-cut electrical discharge machining process using water-air-mist dielectric fluid: An experimental study. *Materials Today: Proceedings*, *49*(5), 1885–1890. doi:10.1016/j.matpr.2021.08.077

Boopathi, S., & Sivakumar, K. (2013). Experimental investigation and parameter optimization of near-dry wire-cut electrical discharge machining using multi-objective evolutionary algorithm. *International Journal of Advanced Manufacturing Technology*, *67*(9–12), 2639–2655. doi:10.100700170-012-4680-4

Boopathi, S., & Sivakumar, K. (2014). Study of water assisted dry wire-cut electrical discharge machining. *Indian Journal of Engineering and Materials Sciences*, *21*(1), 75–82. https://doi.org/http://nopr.niscair.res.in/handle/123456789/27453

Boopathi, S., Sivakumar, K., & Kalidas, R. (2012). Parametric Study of Dry WEDM Using Taguchi Method. *International Journal of Engineering Research and Development*, *2*(4), 1–6.

Boopathi, S., Thillaivanan, A., Azeem, M. A., Shanmugam, P., & Pramod, V. R. (2022). Experimental investigation on abrasive water jet machining of neem wood plastic composite. *Functional Composites and Structures*, *4*(2), 025001. doi:10.1088/2631-6331/ac6152

Boopathi, S., Venkatesan, G., & Anton Savio Lewise, K. (2023). Mechanical Properties Analysis of Kenaf–Grewia–Hair Fiber-Reinforced Composite. In *Lecture Notes in Mechanical Engineering* (pp. 101–110). Springer. doi:10.1007/978-981-16-9057-0_11

Bouali, E. T., Abid, M. R., Boufounas, E. M., Hamed, T. A., & Benhaddou, D. (2021). Renewable energy integration into cloud & IoT-based smart agriculture. *IEEE Access : Practical Innovations, Open Solutions*, *10*, 1175–1191. doi:10.1109/ACCESS.2021.3138160

Bouzembrak, Y., Klüche, M., Gavai, A., & Marvin, H. J. P. (2019). Internet of Things in food safety: Literature review and a bibliometric analysis. *Trends in Food Science & Technology*, *94*, 54–64. doi:10.1016/j.tifs.2019.11.002

Brillante, L., Mathieu, O., Bois, B., Van Leeuwen, C., & Lévêque, J. (2015). The use of soil electrical resistivity to monitor plant and soil water relationships in vineyards. *Soil (Göttingen)*, *1*(1), 273–286. doi:10.5194oil-1-273-2015

Cancela, J. J., Fandiño, M., Rey, B. J., & Martínez, E. M. (2015). Automatic irrigation system based on dual crop coefficient, soil and plant water status for Vitis vinifera (cv Godello and cv Mencía). *Agricultural Water Management*, *151*, 52–63. doi:10.1016/j.agwat.2014.10.020

Caro, Miguel Pincheira, Muhammad Salek Ali, Massimo Vecchio, and Raffaele Giaffreda. (2018).Blockchain-based traceability in Agri-Food supply chain management: A practical implementation. In 2018 *IoT Vertical and Topical Summit on Agriculture-Tuscany*, p.1-4.

Casino, F., Kanakaris, V., Dasaklis, T. K., Moschuris, S., & Rachaniotis, N. P. (2019). Modeling food supply chain traceability based on blockchain technology. *IFAC-Papers On Line,* Volume 52, Issue 13, ISSN 2405-8963, https://doi.org/, p.2728-2733. doi:10.1016/j.ifacol.2019.11.620.0

Chakravarthi, P. K., Yuvaraj, D., & Venkataramanan, V. (2022). IoT-based smart energy meter for smart grids. *ICDCS 2022 - 2022 6th International Conference on Devices, Circuits and Systems*, (pp. 360–363). IEEE. 10.1109/ICDCS54290.2022.9780714

Chen, H. (2017). Applications of Cyber Physical Systems: A Literature Review. *Journal of Industrial Integration and Management, 12*(3), 1–28. doi:10.1142/S2424862217500129

Chen, Z., Wu, R., Lin, Y., Li, C., Chen, S., Yuan, Z., Chen, S., & Zou, X. (2022). Plant disease recognition model based on improved YOLOv5. *Agronomy (Basel), 12*(2), 365. doi:10.3390/agronomy12020365

Choubisa, M., & Doshi, R. (2022). Crop Protection Using Cyber-Physical Systems and Machine Learning for Smart Agriculture. In P. M. Rizwan (Ed.), *Handbook of Research on Cyber-Physical Systems for Smart Agriculture* (pp. 211–237). IGI Global.

Chung, S. O., Choi, M. C., Lee, K. H., Kim, Y. J., Hong, S. J., & Li, M. (2016). Sensing technologies for grain crop yield monitoring systems: A review. *Journal of Biosystems Engineering, 41*(4), 408–417. doi:10.5307/JBE.2016.41.4.408

Ciocca, F., Lunati, I., Van de Giesen, N., & Parlange, M. B. (2012). Heated optical fiber for distributed soil-moisture measurements: A lysimeter experiment. *Vadose Zone Journal, 11*(4), vzj2011.0199. doi:10.2136/vzj2011.0199

Comparetti, A. (2011). Precision Agriculture: Past, Present and Future. *Agricultural Engineering and Environment, 16*, 17.

Conesa-Muñoz, J., Bengochea-Guevara, J. M., Andujar, D., & Ribeiro, A. (2016). Route planning for agricultural tasks: A general approach for fleets of autonomous vehicles in site-specific herbicide applications. *Computers and Electronics in Agriculture, 127*, 204–220. doi:10.1016/j.compag.2016.06.012

Corradini, E., De Moura, M. R., & Mattoso, L. H. C. (2010). A preliminary study of the incorparation of NPK fertilizer into chitosan nanoparticles. *Express Polymer Letters, 4*(8), 509–515. doi:10.3144/expresspolymlett.2010.64

Crosson, P. R. (2016). Sustainable agriculture. In *Global development and the environment* (pp. 61–68). Routledge.

Dafflon, B., Moalla, N., & Ouzrout, Y. (2021). The challenges, approaches, and used techniques of CPS for manufacturing in Industry 4.0: A literature review. *International Journal of Advanced Manufacturing Technology, 113*(7-8), 2395–2412. doi:10.100700170-020-06572-4

Dass james, A., & Boopathi, S. (2016). Experimental Study of Eco-friendly Wire-Cut Electrical Discharge Machining Processes. *International Journal of Innovative Research in Science, Engineering and Technology, 5*(1), 666–675.

del Cerro, J., Cruz Ulloa, C., Barrientos, A., & de León Rivas, J. (2021). Unmanned aerial vehicles in agriculture: A survey. *Agronomy (Basel), 11*(2), 203. doi:10.3390/agronomy11020203

DeRosa, M. C., Monreal, C., Schnitzer, M., Walsh, R., & Sultan, Y. (2010). Nanotechnology in fertilizers. *Nature Nanotechnology, 5*(2), 91. doi:10.1038/nnano.2010.2 PMID:20130583

Dev, P., Khandelwal, S., Yadav, S. C., Arya, V., Mali, H. R., Poonam, & Yadav, K. K. (2023). Conservation Agriculture for Sustainable Agriculture. *International Journal of Plant and Soil Science, 35*(5), 1–11. doi:10.9734/ijpss/2023/v35i52828

Dhakate, M., & Ingole, A. B. "Diagnosis of pomegranate plant diseases using neural network." In *Computer Vision, Pattern Recognition, Image Processing and Graphics (NCVPRIPG), 2015 Fifth National Conference on* (pp. 1-4). IEEE. 10.1109/NCVPRIPG.2015.7490056

Dhall, R., & Agrawal, H. (2018). An improved energy efficient duty cycling algorithm for IoT based precision agriculture. *Procedia Computer Science, 141*, 135–142. doi:10.1016/j.procs.2018.10.159

Dhanaraju, M., Chenniappan, P., Ramalingam, K., Pazhanivelan, S., & Kaliaperumal, R. (2022, October 21). Smart Farming: Internet of Things (IoT)-Based Sustainable Agriculture. *Agriculture, 12*(10), 1745. doi:10.3390/agriculture12101745

Diaz, J., Perez, J., Perez, J., & Garbajosa, J. (2016). Conceptualizing a Framework for Cyber-Physical Systems of Systems Development and Deployment. *Proccedings of the 10th European Conference on Software Architecture Workshops (ECSAW),* (pp. 1-7). ACM. 10.1145/2993412.3004852

Dinesh, M., & Saravanan, P. (2011, June). FPGA based real time monitoring system for agricultural field. *International Journal of Electronics and Computer Science Engineering., 1*(3), 1514–1519.

Dineva, K., Parvanov, D., Atanasova, T., Mateeva, G., Petrov, P., & Kostadinov, G. (2021). Towards CPS/IoT System for Livestock Smart Farm Monitoring. *International Conference Automatics and Informatics*. IEEE. 10.1109/ICAI52893.2021.9639460

Divyanth, L. G., Ahmad, A., & Saraswat, D. (2023). A two-stage deep-learning-based segmentation model for crop disease quantification based on corn field imagery. *Smart Agricultural Technology, 3*, 100108. doi:10.1016/j.atech.2022.100108

Domakonda, V. K., Farooq, S., Chinthamreddy, S., Puviarasi, R., Sudhakar, M., & Boopathi, S. (2022). Sustainable Developments of Hybrid Floating Solar Power Plants: Photovoltaic System. In Human Agro-Energy Optimization for Business and Industry (pp. 148–167). IGI Global.

Donka, A. K., Seerapu, J. A. R., Verma, S., Sao, A., Shukla, G., & Tripathi, S. (2020). Unmanned Aerial Vehicle for Precision Agriculture in a Modular Approach. *2020 IEEE 7th Uttar Pradesh Section International Conference on Electrical, Electronics and Computer Engineering (UPCON),* 1-5. 10.1109/UPCON50219.2020.9376521

Douglas, S., White, J., & Gill, C. (2014). Elastic Infrastructure to Sup- port Computing Clouds for Large-Scale Cyber Physical Systems. *7th IEEE International Symposium on Object/Component/Service-Oriented Real-Time Dis- tributed Computing,* (pp. 56-63). IEEE.

EI-Kazzaz, A. A.EI-Kazzaz. (2017). Soilless Agriculture a New and Advanced Method for Agriculture Development: An Introduction. *Agricultural Research & Technology: Open Access Journal, 3*(2). doi:10.19080/ARTOAJ.2017.03.555610

Eldridge, B. M., Manzoni, L. R., Graham, C. A., Rodgers, B., Farmer, J. R., & Dodd, A. N. (2020). Getting to the roots of aeroponic indoor farming. *The New Phytologist, 228*(4), 1183–1192. doi:10.1111/nph.16780 PMID:32578876

Elfatimi, E., Eryigit, R., & Elfatimi, L. (2022). Beans leaf diseases classification using mobilenet models. *IEEE Access : Practical Innovations, Open Solutions, 10*, 9471–9482. doi:10.1109/ACCESS.2022.3142817

ElKadiri, R., Momm, H. G., Bingner, R. L., & Moore, K. (2023). Spatial optimization of conservation practices for sediment load reduction in ungauged agricultural watersheds. *Soil Systems, 7*(1), 4. doi:10.3390oilsystems7010004

Eric, S., Kyoung-sook, K., Ryong L., & de Vaulx , F. (2013). A Vision of Cyber-Physical Cloud Computing for Smart Networked Systems. *Technical Resport NISTR -7951*. NISTR.

Estevez-Ayres, I., Basanta-Val, P., Valls, M. G., Fisteus, J. A., & Almeida, L. (2009). QoS-Aware Real-Time Composition Algorithms for Service-Based Applications. IEEE Transactions on Industrial Informatics, 5(3), 278-288. IEEE. doi:10.1109/TII.2009.2026422

Facchinetti, D., Santoro, S., Galli, L. E., Fontana, G., Fedeli, L., Parisi, S., Bonacchi, L. B., Šušnjar, S., Salvai, F., Coppola, G., Matteucci, M., & Pessina, D. (2021). Reduction of pesticide use in fresh-cut salad production through artificial intelligence. *Applied Sciences (Basel, Switzerland)*, *11*(5), 1992. doi:10.3390/app11051992

Fageria, N. K. (1998, January). Optimizing nutrient use efficiency in crop production. *Revista Brasileira de Engenharia Agrícola e Ambiental*, *2*, 6–16. doi:10.1590/1807-1929/agriambi.v02n01p6-16

Fahad, M., Javid, T., Beenish, H., Siddiqui, A. A., & Ahmed, G. (2021). Extending ONTAgri with Service-Oriented Architecture towards Precision Farming Application. *Sustainability (Basel)*, *13*(17), 9801. doi:10.3390u13179801

FAO. (2018). *Water for sustainable food and agriculture: A report produced for the G20 Presidency of Germany*. Food & Agriculture Org.

FAO. (2022). *Soils for nutrition: state of the art*. FAO EBooks. doi:10.4060/cc0900en

Farooq, M. S., Riaz, S., Abid, A., Abid, K., & Naeem, M. A. (2019, October 25). A Survey on the Role of IoT in Agriculture for the Implementation of Smart Farming. *IEEE Access : Practical Innovations, Open Solutions*, *7*, 156237–156271. doi:10.1109/ACCESS.2019.2949703

Farooq, M. S., Riaz, S., Abid, A., Umer, T., & Zikria, Y. (2020). Role of IoT technology in agriculture: A systematic literature review. *Electronics (Basel)*, *9*(2), 319. doi:10.3390/electronics9020319

Farooq, M. S., Sohail, O. O., Abid, A., & Rasheed, S. (2022, January 13). A survey on the role of iot in agriculture for the implementation of smart livestock environment. *IEEE Access : Practical Innovations, Open Solutions*, *10*, 9483–9505. doi:10.1109/ACCESS.2022.3142848

Ferentinos, K. P. (2018). Deep learning models for plant disease detection and diagnosis. *Computers and Electronics in Agriculture*, *145*, 311–318. doi:10.1016/j.compag.2018.01.009

Frankelius, P., & Rose, M. V. (2023). Towards the fifth innovation era in agricultural innovation. In Agricultural Bioeconomy (pp. 205-229). doi:10.1016/B978-0-323-90569-5.00003-2

Fresco, R., & Ferrari, G. (2018). *Enhancing precision agriculture by internet of things and cyber physical systems*, 53-60. Atti Soc. Tosc. Sci. Nat. Mem. Supplemento.

Gagliardi, G., Lupia, M., Cario, G., Cicchello Gaccio, F., D'Angelo, V., Cosma, A. I. M., & Casavola, A. (2021). An internet of things solution for smart agriculture. *Agronomy (Basel)*, *11*(11), 2140. doi:10.3390/agronomy11112140

Ganesh Kumar, C., Pachayappan, M., & Madanmohan, G. (2017). Agri-food supply chain management: Literature review. *Intelligent Information Management*, *9*(2), 68–96. doi:10.4236/iim.2017.92004

Garg, M., Singhal, S., Sood, K., Rupeika-Apoga, R., & Grima, S. (2023). Price Discovery Mechanism and Volatility Spillover between National Agriculture Market and National Commodity and Derivatives Exchange: The Study of the Indian Agricultural Commodity Market. *Journal of Risk and Financial Management*, *16*(2), 62. doi:10.3390/jrfm16020062

Ge, Y., Dong, Y., & Zhao, H. (2012). A Cyber-Physical Energy Sys- tem Architecture for Electric Vehicles Charging Application. 12th International Conference on Quality Software, (pp. 246-250). ACM.

Geetha, S. K., S. G. R. Preethi, and R. Sreenithi. (2021). Educational certificate verification using blockchain based framework. *Turkish Journal of Physiotherapy and Rehabilitation*, 1833-1843.

Geetha, S. K., & Kalpana, R., M. SK, K. S and P. PS.(2022). Health Record Management Using Blockchain Technology. *8th International Conference on Smart Structures and Systems (ICSSS)*, Chennai, India.

Glória, A., Cardoso, J., & Sebastião, P. (2021). Sustainable Irrigation System for Farming Supported by Machine Learning and Real-Time Sensor Data. *Sensors (Basel), 21*(9), 3079. doi:10.339021093079 PMID:33925142

Goap, A., Sharma, D., Shukla, A. K., & Rama Krishna, C. (2018). An IoT based smart irrigation management system using Machine learning and open source technologies. *Computers and Electronics in Agriculture, 155*(May), 41–49. doi:10.1016/j.compag.2018.09.040

Goldstein, A., Lior Fink, L., Amit Meitin, A., Shiran Bohadana, S., Oscar Lutenberg, O., & Ravid, G. (2018). Applying machine learning on sensor data for irrigation recommendations: Revealing the agronomist's tacit knowledge. *Precision Agriculture, 19*(3), 421–444. doi:10.100711119-017-9527-4

Gondchawar, N., & Kawitkar, R. S. (2016, June). IoT based smart agriculture. *International Journal of Advanced Research in Computer and Communication Engineering, 5*(6), 838–842.

Gour, V. Reddy, V. M., S. N., Vishuvardhan, & Ram, V. T. (2020). IoT based Farming Techniques in Indoor Environment: A Brief Survey. *2020 5th International Conference on Communication and Electronics Systems (ICCES)* (pp. 790-795). IEEE. 10.1109/ICCES48766.2020.9137950

Goyal, Prabha, Rai, & Singh. (2016). Indian Agriculture and Farmers-Problems and Reforms. In Indian Agriculture and Farmers (pp. 79-87). Poddar Publication.

Gunasekaran, K., Boopathi, S., & Sureshkumar, M. (2022). Analysis of a Cryogenically Cooled Near-Dry Wedm Process Using Different Dielectrics. *Materiali in Tehnologije, 56*(2), 179–186. doi:10.17222/mit.2022.397

Gunes, V., Pete, S., Givargi, T., & Vahid, F. (2014). A Survey on Con- cepts, Applications and Challenges in Cyber Physical Systems. *KSII Transactions on Internet and Information Systems, 8*(2), 4242–4268.

Gupta, M., Abdelsalam, M., Khorsandroo, S., & Mittal, S. (2020). Security and privacy in smart farming: Challenges and opportunities. *IEEE Access : Practical Innovations, Open Solutions, 8*, 34564–34584. doi:10.1109/ACCESS.2020.2975142

Gupta, S., Sinha, S., & Bhushan, B. (2020). Emergence of blockchain technology: Fundamentals, working and its various implementations. *In Proceedings of the International Conference on Innovative Computing & Communications.* 10.2139srn.3569577

Gutiérrez, J., Villa-Medina, J. F., Nieto-Garibay, A., & Porta-Gándara, M. Á. (2014). Automated Irrigation System Using a Wireless Sensor Network and GPRS Module. *IEEE Transactions on Instrumentation and Measurement, 63*(1), 166–176. doi:10.1109/TIM.2013.2276487

Haider, R., & Joshy, J. K. (2022). An effect of farm mechanization on Indian agriculture with respect to rice yield. *South Asian Food and Agriculture Research and Policy*, 1-16.

Han, J., Kamber, M., & Pei, J. (2011). Data Transformation and Data Discretization. In *Data Mining: Concepts and Techniques* (pp. 111–118). Elsevier.

Hans, Ronny, Hendrik Zuber, Amr Rizk, and Ralf Steinmetz. (2017). *Blockchain and smart contracts: Disruptive technologies for the insurance market.*

Hao, B., & Yan, Z. (2018). Recovery path planning for an agricultural mobile robot by Dubins-RRT* algorithm. *International Journal of Robotics and Automation, 33*(2). Advance online publication. doi:10.2316/Journal.206.2018.2.206-5443

Haque, M. S. T., Rouf, K. A., Khan, Z. A., Emran, A., & Zishan, M. S. R. (2019). Design and Implementation of an IoT based Automated Agricultural Monitoring and Control System. In *2019 International Conference on Robotics, Electrical and Signal Processing Techniques (ICREST)* (pp. 13-16). IEEE. 10.1109/ICREST.2019.8644212

Haque, S. A., Aziz, S. M., & Rahman, M. (2014). Review of Cyber-Physical System in Healthcare. *International Journal of Distributed Sensor Networks, 10*(4), 217415. doi:10.1155/2014/217415

Harakannanavar, S. S., Rudagi, J. M., Puranikmath, V. I., Siddiqua, A., & Pramodhini, R. (2022). Plant Leaf Disease Detection using Computer Vision and Machine Learning Algorithms. *Global Transitions Proceedings, 3*(1), 305–310. doi:10.1016/j.gltp.2022.03.016

Haribalaji, V., Boopathi, S., Mohammed Asif, M., Yuvaraj, T., Velmurugan, D., Anton Savio Lewise, K., Sudhagar, S., & Suresh, P. (2022). Influences of Mg-Cr filler materials in Friction Stir Process of Aluminium-based dissimilar alloys. *Materials Today: Proceedings, 66*, 948–954. doi:10.1016/j.matpr.2022.04.668

Haribalaji, V., Boopathi, S., & Asif, M. M. (2021). Optimization of friction stir welding process to join dissimilar AA2014 and AA7075 aluminum alloys. *Materials Today: Proceedings, 50*, 2227–2234. doi:10.1016/j.matpr.2021.09.499

Haribalaji, V., Boopathi, S., & Balamurugan, S. (2014). Effect of Welding Processes on Mechanical and Metallurgical Properties of High Strength Low Alloy (HSLA) Steel Joints. *International Journal of Innovation and Scientific Research, 12*(1), 170–179.

Haribalaji, V., Venkatesan, G., Asif, M. M., Pandian, M., Subbiah, R., & Boopathi, S. (2022). Investigation on corrosion and tensile Characteristics: Friction stir welding of AA7075 and AA2014. *Materials Today: Proceedings, 66*, 743–748. doi:10.1016/j.matpr.2022.04.037

Haridasan, A., Thomas, J., & Raj, E. D. (2023). Deep learning system for paddy plant disease detection and classification. *Environmental Monitoring and Assessment, 195*(1), 120. doi:10.100710661-022-10656-x PMID:36399232

Harris, D. R., & Fuller, D. Q. (2014). Agriculture: definition and overview. Encyclopedia of Global Archaeology, 104–113. Springer

Hasan, M., Uddin, K. N., Sayeed, A., & Tasneem, T. (2021). Smart agriculture robotic system based on internet of things to boost crop production. In *2021 2nd International Conference on Robotics, Electrical and Signal Processing Techniques (ICREST)* (pp. 157-162). IEEE. 10.1109/ICREST51555.2021.9331091

Hassan, S. M., & Maji, A. K. (2022). Plant disease identification using a novel convolutional neural network. *IEEE Access : Practical Innovations, Open Solutions, 10*, 5390–5401. doi:10.1109/ACCESS.2022.3141371

Hati, A. J., & Singh, R. R. (2021). Smart Indoor Farms: Leveraging Technological Advancements to Power a Sustainable Agricultural Revolution. In AgriEngineering (Vol. 3, pp. 728–767). MDPI. doi:10.3390/agriengineering3040047

Hatti, V., Raghavendra, M., Singh, Y. V., Sharma, S. K., Halli, H. M., & Goud, B. R. (2020). Application of nano technology in crop production to enhance resource use efficiency with special reference to soil and water conservation: A review. *Journal of Soil and Water Conservation, 19*(4), 375–381. doi:10.5958/2455-7145.2020.00050.8

Haykin, S. (1998). Neural Networks: A Comprehensive Foundation (2 ed.). Prentice Hall.

He, Y. (2023). Knowledge and Technology System of Traditional Agriculture. In The Agricultural Civilization of Ancient China. doi:10.1007/978-981-19-8469-3_4

Heideker, A., Ottolini, D., Zyrianoff, I., Neto, A. T., Salmon Cinotti, T., & Kamienski, C. (2020). IoT-based Measurement for Smart Agriculture. *2020 IEEE International Workshop on Metrology for Agriculture and Forestry, MetroAgriFor 2020 - Proceedings*, (pp. 68–72). IEEE. 10.1109/MetroAgriFor50201.2020.9277546

Hell, M., Bolam, R. C., Vagapov, Y., & Anuchin, A. (2018). Design of a portable drone for educational purposes. *2018 25th International Workshop on Electric Drives: Optimization in Control of Electric Drives (IWED)*, 1-5. 10.1109/IWED.2018.8321377

Herlitzius, T. (2017). *Automation and Robotics - The Trend Towards Cyber Physical Systems in Agriculture Business.* SAE Technical Paper 2017-01-1932.

Hirsch, C., Bartocci, E., & Grosu, R. (2019). Capacitive soil moisture sensor node for IoT in agriculture and home. In *2019 IEEE 23rd International Symposium on Consumer Technologies (ISCT)* (pp. 97-102). IEEE. 10.1109/ISCE.2019.8901012

Horvath, I., & Gerritsen, B. H. M. (2012). Cyber Physical Systems: Concepts, Technologies and Implementation Principles. *Proceedings of TMCE, Voume, 1*, 7–11.

Huang, C.-H., Chen, P.-J., Lin, Y.-J., Chen, B.-W., & Zheng, J. X. (2021). A Robot-Based Intelligent Management Design For Agricultural Cyber-Physical Systems. *Computers and Electronics in Agriculture, 181*, 105967. doi:10.1016/j.compag.2020.105967

Ikram, A., Aslam, W., Aziz, R. H., Noor, F., Mallah, G. A., Ikram, S., Ahmad, M. S., Abdullah, A. M., & Ullah, I. (2022, May 17). Crop Yield Maximization Using an IoT-Based Smart Decision. *Journal of Sensors, 2022*, 2022. doi:10.1155/2022/2022923

Imran, S., Ferdous, T., Adeel, M., & Faheem, M. (2013). Existing land record management system of Pakistan and proposed Arc GIS parcel data model. *In 2013 International Conference on Aerospace Science & Engineering (ICASE)*, pp. 1-3 . IEEE. 10.1109/ICASE.2013.6785567

Iyer, V., Shah, K., Rane, S., & Shankarmani, R. (2021). Decentralized Peer-to-Peer Crop Insurance. *In Proceedings of the 3rd ACM International Symposium on Blockchain and Secure Critical Infrastructure*, p.3-12 10.1145/3457337.3457837

Jackulin, C., & Murugavalli, S. (2022). A comprehensive review of the detection of plant disease using machine learning and deep learning approaches. Measurement. *Sensors (Basel)*, •••, 100441.

Jain, A., & Kumar, A. (2020). Smart agriculture monitoring system using IoT. *International Journal for Research in Applied Science and Engineering Technology, 8*(7), 366–372. doi:10.22214/ijraset.2020.7060

Javidan, S. M., Banakar, A., Vakilian, K. A., & Ampatzidis, Y. (2023). Diagnosis of grape leaf diseases using automatic K-means clustering and machine learning. *Smart Agricultural Technology, 3*, 100081. doi:10.1016/j.atech.2022.100081

Jeevanantham, Y. A., Saravanan, A., Vanitha, V., Boopathi, S., & Kumar, D. P. (2022). Implementation of Internet-of Things (IoT) in Soil Irrigation System. *IEEE Explore*, (pp. 1–5). IEEE.

Jha, K., Doshi, A., Patel, P., & Shah, M. (2019). A comprehensive review on automation in agriculture using artificial intelligence. *Artificial Intelligence in Agriculture, 2*, 1–12. doi:10.1016/j.aiia.2019.05.004

Jha, N., Prashar, D., Khalaf, O. I., Alotaibi, Y., Alsufyani, A., & Alghamdi, S. (2021). Blockchain Based Crop Insurance: A Decentralized Insurance System for Modernization of Indian Farmers. *Sustainability (Basel), 13*(16), 8921. doi:10.3390u13168921

Jiang, P., Chen, Y., Liu, B., He, D., & Liang, C. (2019). Real-time detection of apple leaf diseases using deep learning approach based on improved convolutional neural networks. *IEEE Access : Practical Innovations, Open Solutions, 7*, 59069–59080. doi:10.1109/ACCESS.2019.2914929

Jing Wu, F., Fen Kao, Y., & Tseng, Y. (2011). From Wireless Sensor Network to wards Cyber Physical Systems. *Pervasive and Mobile Computing, 7,* 397413 2011 doi:10.1016/j.pmcj.2011.03.003

Ju, C., & Son, H. I. (2018). Multiple UAV systems for agricultural applications: Control, implementation, and evaluation. *Electronics (Basel), 7*(9), 162. doi:10.3390/electronics7090162

Kalasapur, S., Kumar, M., & Shirazi, B. A. (2007). Dynamic Service Composition in Pervasive Computing. *IEEE Transactions on Parallel and Distributed Systems, 18*(7), 907–918. doi:10.1109/TPDS.2007.1039

Kamilaris, A., Fonts, A., & Prenafeta-Boldύ, F. X. (2019). The rise of blockchain technology in agriculture and food supply chains. *Trends in Food Science & Technology, 91*, 640–652. doi:10.1016/j.tifs.2019.07.034

Kang, M., Wang, X., Wang, H., Hua, J., de Reffye, P., & Wang, F.-Y. (2022). The Development of AgriVerse: Past, Present, and Future. *IEEE Transactions on Systems, Man, and Cybernetics. Systems.* Advance online publication.

Kannan, E., Trabelsi, Y., Boopathi, S., & Alagesan, S. (2022). Influences of cryogenically treated work material on near-dry wire-cut electrical discharge machining process. *Surface Topography : Metrology and Properties, 10*(1), 015027. doi:10.1088/2051-672X/ac53e1

Katarya, R., Raturi, A., Mehndiratta, A., & Thapper, A. (2020). Impact of Machine Learning Techniques in Precision Agriculture. In *2020 3rd International Conference on Emerging Technologies in Computer Engineering: Machine Learning and Internet of Things (ICETCE),* (pp. 1-6). IEEE. 10.1109/ICETCE48199.2020.9091741

Kavitha, C., Malini, P. S. G., Charan, V., Manoj, N., Verma, A., & Boopathi, S. (2022). (in press). An experimental study on the hardness and wear rate of carbonitride coated stainless steel. *Materials Today: Proceedings.* doi:10.1016/j.matpr.2022.09.524

Khaitan, S. K., & McCalley, J. D. (2014). Design Techniques and Ap- plications of Cyber Physical Systems: A Survey. *IEEE Systems Journal*, 1–15.

Khan, A., Aziz, S., Bashir, M., & Khan, M. U. (2020). IoT and wireless sensor network based autonomous farming robot. In *2020 International Conference on Emerging Trends in Smart Technologies (ICETST)* (pp. 1-5). IEEE. 10.1109/ICETST49965.2020.9080736

Khirade, S. D., & Patil, A. B. (2015). Plant Disease Detection Using Image Processing. In *Computing Communication Control and Automation (ICCUBEA), 2015 International Conference on* (pp. 768-771). IEEE.

Kim, M., Stehr, M.-O., Kim, J., & Ha, S. (2010). An Applica- tion Framework for Loosely Coupled Networked Cyber Physical Systems. *Proceedings IEEE/IFIP International Conference on Embedded and Ubiquitous Com- puting (EUC),* (pp. 144-153). IEEE.

Kim, K.-D., & Kumar, P. R. (2013). An Overview and Some Challenges in Cyber- Physical Systems [A Multidisciplinary Reviews Journal]. *Journal of the Indian Institute of Science, 93*(3), 342–352.

Kim, W. S., Lee, W. S., & Kim, Y. J. (2020, December). A review of the applications of the internet of things (IoT) for agricultural automation. *Journal of Biosystems Engineering, 45*(4), 385–400. doi:10.100742853-020-00078-3

King, A. (2017). Technology: The Future of Agriculture. *Nature, 544*(7651), S21–S23. doi:10.1038/544S21a PMID:28445450

Kozai T., & Niu G. (2015). *Challenges for the Next-Generation PFAL, Plant Factory: An Indoor Vertical Farming System for Efficient Quality Food Production*, 387-393. Academic Press. doi:10.1016/B978-0-12-801775-3.00027-5

Kozai, T., & Niu, G. (2019). Water use efficiency, In Plant Factory: An Indoor Vertical Farming System for Efficient Quality Food Production, pp. 97-99. Research Gate.

Kumar, A., Lee, B. G., Lee, H., & Kumari, A. (2012). Secure storage and access of data in cloud computing. *2012 International Conference on ICT Convergence (ICTC)*, 336–339. 10.1109/ICTC.2012.6386854

Kumar, A., Singh, S., Gaurav, A. K., Srivastava, S., & Verma, J. P. (2020). Plant Growth-Promoting Bacteria: Biological Tools for the Mitigation of Salinity Stress in Plants. In Frontiers in Microbiology (Vol. 11). Frontiers Media S.A. doi:10.3389/fmicb.2020.01216

Kumar, J. C. R., & Majid, M. A. (2023). Optimizing Energy Efficiencies of IoT-based Wireless Sensor Network Components for Metaverse Sustainable Development using Carry Resist Adder based Booth Recoder (CRABRA). In *2023 20th Learning and Technology Conference (L&T)* (pp. 91-96). IEEE.

Kumar, J. C. R., Kumar, D. V., Arunsi, M., & Baskar, D. (2021). Energy-Efficient and Secure IoT architecture based on a Wireless Sensor Network Using Machine Learning to Predict Mortality Risk of patients with CoVID-19. In *2021 6th International Conference on Communication and Electronics Systems (ICCES)* (pp. 1853-1861).IEEE.

Kumara, V., Mohanaprakash, T. A., Fairooz, S., Jamal, K., Babu, T., & B., S. (2023). Experimental Study on a Reliable Smart Hydroponics System. In *Human Agro-Energy Optimization for Business and Industry* (pp. 27–45). IGI Global. doi:10.4018/978-1-6684-4118-3.ch002

Kumar, A. (2012). Citrus greening disease detection using aerial hyperspectral and multispectral imaging techniques. *Journal of Applied Remote Sensing*, *6*(1), 63542. doi:10.1117/1.JRS.6.063542

Kumar, A., Joseph, S., Graber, E. R., Taherysoosavi, S., Mitchell, D. R. G., Munroe, P., Tsechansky, L., Lerdahl, O., Aker, W., & Sæbø, M. (2021). Fertilizing behavior of extract of organomineral-activated biochar: Low-dose foliar application for promoting lettuce growth. *Chemical and Biological Technologies in Agriculture*, *8*(1), 21. doi:10.118640538-021-00222-x

Kumar, G., Anwar, A., Dikshit, A., Poddar, A., Soni, U., & Song, W. K. (2022). Obstacle avoidance for a swarm of unmanned aerial vehicles operating on particle swarm optimization: A swarm intelligence approach for search and rescue missions. *Journal of the Brazilian Society of Mechanical Sciences and Engineering*, *44*(2), 56. doi:10.100740430-022-03362-9

Kumar, J. C. R., & Kumar, D. V. (2022). Energy-efficient, high-performance and memory efficient FIR adaptive filter architecture of wireless sensor networks for IoT applications. *Sadhana*, *47*(248), 248. doi:10.100712046-022-02013-y

Kumar, R., Singh, M. P., Kumar, P., & Singh, J. P. (2015). Crop Selection Method to maximize crop yield rate using machine learning technique. *2015 International Conference on Smart Technologies and Management for Computing, Communication, Controls, Energy and Materials (ICSTM)*, (pp. 138–145). IEEE. 10.1109/ICSTM.2015.7225403

Lakhiar, I. Gao J., Syed, T., Chandio, F., Buttar, N., & Qureshi, W. (2018). Monitoring and Control Systems in Agriculture Using Intelligent Sensor Techniques: A Review of the Aeroponic System. *Journal of Sensors*. doi:10.1155/2018/8672769

Lal, R. (2004). Soil carbon sequestration to mitigate climate change. In Geoderma (Vol. 123, pp. 1–22). doi:10.1016/j.geoderma.2004.01.032

Lask, J., Kam, J., Weik, J., Kiesel, A., Wagner, M., & Lewandowski, I. (2021). A parsimonious model for calculating the greenhouse gas emissions of miscanthus cultivation using current commercial practice in the United Kingdom. *Global Change Biology. Bioenergy*, *13*(7), 1087–1098. doi:10.1111/gcbb.12840

Latha, S., Gomathi, T., Pavithra, S., Sudha, P. N., Nalluri, A., & Bhardwaj, P. (2023). Chitosan-Based Porous Carbon Materials for Agriculture and Agro-waste Applications. In A. N. Grace, P. Sonar, P. Bhardwaj, & A. Chakravorty (Eds.), *Handbook of Porous Carbon Materials. Materials Horizons: From Nature to Nanomaterials* (pp. 1–20). doi:10.1007/978-981-19-7188-4_28

Lee, C.-L., Strong, R., & Dooley, K. E. (2021). Analyzing precision agriculture adoption across the globe: A systematic review of scholarship from 1999–2020. *Sustainability (Basel)*, *13*(18), 10295. doi:10.3390u131810295

Lee, E. (2015). The Past, Present and Future of Cyber-Physical Systems: A Focus on Models. *Sensors (Basel), 15*(3), 4837–4869. doi:10.3390150304837 PMID:25730486

Lejdel, B. (2022). *Optimization of energy in smart farms using a genetic algorithm. In Driving Factors for Venture Creation and Success in Agricultural Entrepreneurship*. IGI Global.

Lencsés, E., Takács, I., & Takács-György, K. (2014). Farmers' perception of precision farming technology among Hungarian farmers. *Sustainability (Basel), 6*(12), 8452–8465. doi:10.3390u6128452

Le, V. N. T., Ahderom, S., Apopei, B., & Alameh, K. (2020). A novel method for detecting morphologically similar crops and weeds based on the combination of contour masks and filtered Local Binary Pattern, operators. *GigaScience, 9*(3), giaa017. doi:10.1093/gigascience/giaa017 PMID:32129847

Li, X., Zhao, Y., Zhang, J., & Dong, Y. (2016, November). A hybrid PSO algorithm based flight path optimization for multiple agricultural UAVs. In *2016 IEEE 28th international conference on tools with artificial intelligence (ICTAI)* (pp. 691-697). IEEE. 10.1109/ICTAI.2016.0110

Li, H., Lee, W. S., Wang, K., Ehsani, R., & Yang, C. (2014). 'Extended spectral angle mapping (ESAM)' for citrus greening disease detection using airborne hyperspectral imaging. *Precision Agriculture, 15*(2), 162–183. doi:10.100711119-013-9325-6

Lin, K., & Hao, T. (2020). Link quality analysis of wireless sensor networks for underground infrastructure monitoring: A non-backfilled scenario. *IEEE Sensors Journal, 21*(5), 7006–7014. doi:10.1109/JSEN.2020.3041644

Liu, C., Jian, Z., Xie, M., & Cheng, I. (2021). A Real-Time Mobile Application for Cattle Tracking using Video Captured from a Drone. *2021 International Symposium on Networks, Computers and Communications (ISNCC),* 1-6. 10.1109/ISNCC52172.2021.9615648

Liu, Z., Bashir, R. N., Iqbal, S., Shahid, M. M. A., Tausif, M., & Umer, Q. (2022). Internet of Things (IoT) and Machine Learning Model of Plant Disease Prediction–Blister Blight for Tea Plant. *IEEE Access : Practical Innovations, Open Solutions, 10*, 44934–44944. doi:10.1109/ACCESS.2022.3169147

Lohit, G. V. S., & Bisht, D. (2021). Seed Dispenser using Drones and Deep Learning Techniques for Reforestation. *2021 5th International Conference on Computing Methodologies and Communication (ICCMC),* 1275-1283. 10.1109/ICCMC51019.2021.9418227

Lu, J., Zhou, M., Gao, Y., & Jiang, H. (2018). Using hyperspectral imaging to discriminate yellow leaf curl disease in tomato leaves. *Precision Agriculture, 19*(3), 379–394. doi:10.100711119-017-9524-7

Luo, Z., Yang, W., Gou, R., & Yuan, Y. (2023). TransAttention U-Net for Semantic Segmentation of Poppy. *Electronics (Basel), 12*(3), 487. doi:10.3390/electronics12030487

M, G., B, R., S, S., & Boopathi, S. (2018). Fabrication of Solar Irrigation System. *International Research Journal of Engineering and Technology (IRJET), 5*(2), 2106–2108.

Macana, C. A., Quijano, N., & Nava, E. M. (2011). A Survey on Cyber Physical Energy Systems and their Applications on Smart Grids. IEEE PES Conference on Innovative Smart Grid Technologies (ISGT), (pp. 1-7). IEEE. 10.1109/ISGT-LA.2011.6083194

MacDonald, S. L., Staid, M., Staid, M., & Cooper, M. L. (2016). Remote hyperspectral imaging of grapevine leafroll-associated virus 3 in cabernet sauvignon vineyards. *Computers and Electronics in Agriculture, 130*, 109–117. doi:10.1016/j.compag.2016.10.003

Mahmood, I., Imadi, S. R., Shazadi, K., Gul, A., & Hakeem, K. R. (2016). Effects of pesticides on environment. In *Plant, soil and microbes* (pp. 253–269). Springer. doi:10.1007/978-3-319-27455-3_13

Maksutil, S. (2021). Establishing a Chain of Trust in a Sporadically Connected Cyber-Physical System, *IFIP/IEEE International Symposium on Integrated Network Management*, (pp. 890-895). IEEE.

Malhi, G. S., Kaur, M., & Kaushik, P. (2021). Impact of climate change on agriculture and its mitigation strategies: A review. In Sustainability (Switzerland) (Vol. 13, pp. 1–21). MDPI. doi:10.3390u13031318

Mandal, S. N., & Sinha, P. (2021). Pest and Disease Monitoring and Surveillance for Ensuring Food Security. In *Innovations in Agriculture for a Self-Reliant India* (pp. 293–304). CRC Press. doi:10.1201/9781003245384-19

Maray, M., Albraikan, A. A., Alotaibi, S. S., Alabdan, R., Al Duhayyim, M., & Al-Azzawi, W. K. (2022). Artificial intelligence-enabled coconut tree disease detection and classification model for smart agriculture. *Computers & Electrical Engineering*, *104*, 108399. doi:10.1016/j.compeleceng.2022.108399

Marios, S., & Georgiou, J. (2017). Precision Agriculture: Challenges in sensors and electronics for real-time soil and plant monitoring. *2017 IEEE Biomedical Circuits and Systems Conference (BioCAS),* 1-4. 10.1109/BIOCAS.2017.8325180

Martínez-Carricondo, P., Agüera-Vega, F., & Carvajal-Ramírez, F. (2020). Use of UAV-photogrammetry for quasi-vertical wall surveying. *Remote Sensing (Basel)*, *12*(14), 2221. doi:10.3390/rs12142221

Mat, I., Kassim, M.R., Harun, A.N., & Yusoff, I.M. (2018). Smart agriculture using internet of things. In *2018 IEEE conference on open systems (ICOS)* (pp. 54-59). IEEE.

Mehra, M., Saxena, S., Sankaranarayanan, S., Tom, R. J., & Veeramanikandan, M. (2018). IoT based hydroponics system using Deep Neural Networks. *Computers and Electronics in Agriculture*, *155*, 473–486. doi:10.1016/j.compag.2018.10.015

Mejias, J. H., Salazar, F., Pérez Amaro, L., Hube, S., Rodriguez, M., & Alfaro, M. (2021). Nanofertilizers: A Cutting-Edge Approach to Increase Nitrogen Use Efficiency in Grasslands. In Frontiers in Environmental Science, 9. Frontiers Media S.A. doi:10.3389/fenvs.2021.635114

Mekonnen, Y., Namuduri, S., Burton, L., Sarwat, A., & Bhansali, S. (2020). Review—Machine Learning Techniques in Wireless Sensor Network Based Precision Agriculture. *Journal of the Electrochemical Society*, *167*(3), 037522. doi:10.1149/2.0222003JES

Mendelsohn, R. (2007, March). What causes crop failure? *Climatic Change*, *81*(1), 61–70. doi:10.100710584-005-9009-y

Mendes, J., Pinho, T. M., Neves dos Santos, F., Sousa, J. J., Peres, E., Boaventura-Cunha, J., Cunha, M., & Morais, R. (2020). Smartphone applications targeting precision agriculture practices—A systematic review. *Agronomy (Basel)*, *10*(6), 855. doi:10.3390/agronomy10060855

Mendi, A. F., Sakaklı, K. K., & Çabuk, A. (2020). A blockchain based land registration system proposal for Turkey. *In 2020 4th International Symposium on Multidisciplinary Studies and Innovative Technologies (ISMSIT),* IEEE, p.1-6

Meng, X., Tian, X., Wu, Q., Chen, Y., Pan, J., & Hang, Y. (2023). Meta-baseline based on the deep neuro-fuzzy network for few-shot plant leaf fungal disease recognition. *Imaging Science Journal*, 1–13.

Mikhail,, V., Chester, B., & Allenby, R. (2020). Perspective: The Cyber Frontier and Infrastructure. *IEEE Access : Practical Innovations, Open Solutions.*

Mirkouei, A. (2020). A cyber-physical analyzer system for precision agriculture. *Environmental Science Current Research*, *3*(1), 1–8. doi:10.24966/ESCR-5020/100016

Mitra, A., Singhal, A., Mohanty, S. P., Kougianos, E., & Ray, C. (2022). eCrop: A Novel Framework for Automatic Crop Damage Estimation in Smart Agriculture. *SN Computer Science, 3*(4), 319.

Mohanraj, I., Ashokumar, K., & Naren, J. (2016). Field monitoring and automation using IOT in agriculture domain. *Procedia Computer Science, 93*, 931–939. doi:10.1016/j.procs.2016.07.275

Monica, M., Yeshika, B., Abhishek, G. S., Sanjay, H. A., & Dasiga, S. (2017). IoT based control and automation of smart irrigation system: An automated irrigation system using sensors, GSM, Bluetooth and cloud technology. In *2017 International Conference on Recent Innovations in Signal processing and Embedded Systems (RISE)* (pp. 601-607). IEEE. 10.1109/RISE.2017.8378224

Motroni, A., Rosati, F., Nepa, P., Buffi, A., Pirozzi, M., Donato, L. D., Tomassini, L., & Ferraro, A. (2021). An RFID Tracking System for Agricultural Safety. In *2021 IEEE International Conference on RFID Technology and Applications (RFID-TA)* (pp. 28-31). IEEE. 10.1109/RFID-TA53372.2021.9617356

Mulla, D. J. (n.d.). *Precision Agriculture: Key Advances and Remaining Knowledge Gaps*. ISPA.

Murphy R., Montes-Molina J., Govaerts B., Six J., van Kessel C., Fonte S. (2016). Crop residue retention enhances soil properties and nitrogen cycling in smallholder maize systems of Chiapas, Mexico. Applied Soil Ecology, (2016), 110-116, 103. doi:10.1016/j.apsoil.2016.03.014

Mutschler, P. (2018). *Threats to Precision Agriculture*. Research Gate.

Myilsamy, S., & Sampath, B. (2021). Experimental comparison of near-dry and cryogenically cooled near-dry machining in wire-cut electrical discharge machining processes. *Surface Topography : Metrology and Properties, 9*(3), 035015. doi:10.1088/2051-672X/ac15e0

Nair, J., & Thomas, B. K. (2022). Why is adoption of micro-irrigation slow in India? A review. *Irrigation and Drainage, 71*(3), 413–427.

Namani, S., & Gonen, B. (2020). Smart agriculture based on IoT and cloud computing. In *2020 3rd International Conference on Information and Computer Technologies (ICICT)* (pp. 553-556). IEEE. 10.1109/ICICT50521.2020.00094

Naveenkumar, M., Srithar, S., Rajesh Kumar, B., Alagumuthukrishnan, S., & Baskaran, P. (2021). *InceptionResNetV2 for Plant Leaf Disease Classification*. Fifth International Conference on I-SMAC (IoT in Social, Mobile, Analytics and Cloud) (I-SMAC), Palladam, India. 10.1109/I-SMAC52330.2021.9641025

Nguyen, D.-H., Tuong, N. H., & Pham, H.-A. (2020). Blockchain-based Farming Activities Tracker for Enhancing Trust in the Community Supported Agriculture Model. *International Conference on Information and Communication Technology,* (pp.737-740). IEEE. 10.1109/ICTC49870.2020.9289297

Niranjana Murthy, M., Nithya, B. N., & Jagannatha, S. J. C. C. (2019). Analysis of Blockchain technology: Pros, cons and SWOT. *Cluster Computing, 22*(6), 14743–14757. doi:10.100710586-018-2387-5

Node-RED. (n.d.). *Features*. Node-RED. https://nodered.org/about/

Noguchi, N., Reid, J. F., Zhang, Q., & Tian, L. F. (1998). Vision intelligence for precision farming using fuzzy logic optimized genetic algorithm and artificial neural network. American Society of Agricultural Engineers.

Nolan, P., Paley, D. A., & Kroeger, K. (2017, March). Multi-UAS path planning for non-uniform data collection in precision agriculture. In *2017 IEEE Aerospace Conference* (pp. 1-12). IEEE. 10.1109/AERO.2017.7943794

O'Shaughnessy, S. A., Evett, S. R., Colaizzi, P. D., & Howell, T. A. (2012). A crop water stress index and time threshold for automatic irrigation scheduling of grain sorghum. *Agricultural Water Management*, *107*, 122–132. doi:10.1016/j.agwat.2012.01.018

Obi, F., Ugwuishiwu, B., & Nwakaire, J. (2016). Agricultural Waste Concept, Generation, Utilization and Management. *Nigerian Journal of Technology*, *35*(4), 957. doi:10.4314/njt.v35i4.34

Orchi, H., Sadik, M., & Khaldoun, M. (2022). On using artificial intelligence and the internet of things for crop disease detection: A contemporary survey. *Agriculture*, *12*(1), 9. doi:10.3390/agriculture12010009

Orsini (2020). Sustainable use of resources in indoor farms with artificial lighting. *European Journal of Horticultural Science*, *85*, 5. doi:10.17660/eJHS.2020/85.5.1

Osama Bin, S. (2023). A connective framework to minimize the anxiety of collaborative Cyber-Physical System. *International Journal of Computer Integrated Manufacturing* 1-19.

Osroosh, Y., Peters, R. T., Campbell, C. S., & Zhang, Q. (2016). Comparison of irrigation automation algorithms for drip-irrigated apple trees. *Computers and Electronics in Agriculture*, *128*, 87–99. doi:10.1016/j.compag.2016.08.013

Ozguven, M. M., & Adem, K. (2019). Automatic detection and classification of leaf spot disease in sugar beet using deep learning algorithms. *Physica A*, *535*, 122537. doi:10.1016/j.physa.2019.122537

Padol, P. B., & Yadav, A. A. (2016, June). SVM classifier based grape leaf disease detection. In *2016 Conference on advances in signals processing (CASP)* (pp. 175-179). IEEE. 10.1109/CASP.2016.7746160

Palaniappan, M., Tirlangi, S., Mohamed, M. J. S., Moorthy, R. M. S., Valeti, S. V., & Boopathi, S. (2023). Fused Deposition Modelling of Polylactic Acid (PLA)-Based Polymer Composites: A Case Study. In Development, Properties, and Industrial Applications of 3D Printed Polymer Composites (pp. 66–85). IGI Global.

Pandey, G., Weber, R. J., & Kumar, R. (2018). Agricultural Cyber-Physical System: In-Situ Soil Moisture and Salinity Estimation by Dielectric Mixing. *IEEE Access : Practical Innovations, Open Solutions*, *6*, 43179–43191. doi:10.1109/ACCESS.2018.2862634

Pandian, J. A., Kumar, V. D., Geman, O., Hnatiuc, M., Arif, M., & Kanchanadevi, K. (2022). Plant disease detection using deep convolutional neural network. *Applied Sciences (Basel, Switzerland)*, *12*(14), 6982. doi:10.3390/app12146982

Pantazi, X. E., Moshou, D., & Tamouridou, A. A. (2019). Automated leaf disease detection in different crop species through image features analysis and One Class Classifiers. *Computers and Electronics in Agriculture*, *156*, 96–104. doi:10.1016/j.compag.2018.11.005

Park, K.-J., Zheng, R., & Liu, X. (2012). Cyber-Physical Systems: Mile- stones and Research Challenges. *Computer Communications*, *36*(1), 1–7. doi:10.1016/j.comcom.2012.09.006

Partel, V., Kakarla, S. C., & Ampatzidis, Y. (2019). Development and evaluation of a low-cost and smart technology for precision weed management utilizing artificial intelligence. *Computers and Electronics in Agriculture*, *157*, 339–350. doi:10.1016/j.compag.2018.12.048

Patil, M. A., Adamuthe, A. C., & Umbarkar, A. J. (2019). Smartphone and IoT Based System for Integrated Farm Monitoring. In *Proceedings of the 2nd International Conference on Advanced Technologies for Societal Applications* (vol. 1, pp. 471-478).Springer.

Patrício, D. I., & Rieder, R. (2018). Computer vision and artificial intelligence in precision agriculture for grain crops: A systematic review. *Computers and Electronics in Agriculture*, *153*, 69–81. doi:10.1016/j.compag.2018.08.001

Paul, P., & Saavedra, M. (2020). Agro informatics vis-à-vis Internet of Things (IoT) integration & potentialities—An analysis. *Agro Economist-An International Journal*, *7*(1), 13–20. doi:10.30954/2394-8159.01.2020.2

Petkovics, I., Simon, J., Petkovics, Á., & Čović, Z. (2017). Selection of unmanned aerial vehicle for Precision Agriculture with multi-criteria decision making algorithm. *2017 IEEE 15th International Symposium on Intelligent Systems and Informatics (SISY),* 151-156. 10.1109/SISY.2017.8080543

Pham, T. H., Bestaoui, Y., & Mammar, S. (2017, October). Aerial robot coverage path planning approach with concave obstacles in precision agriculture. In *2017 Workshop on Research, Education and Development of Unmanned Aerial Systems (RED-UAS)* (pp. 43-48). IEEE. 10.1109/RED-UAS.2017.8101641

Podder, P., Mondal, M., Bharati, S., & Paul, P. K. (2021). *Review on the security threats of internet of things.* ArXiv Preprint ArXiv:2101.05614.

Podder, A. K., Al Bukhari, A., Islam, S., Mia, S., Mohammed, M. A., Kumar, N. M., Cengiz, K., & Abdulkareem, K. H. (2021, April 1). IoT based smart agrotech system for verification of Urban farming parameters. *Microprocessors and Microsystems*, *82*, 104025. doi:10.1016/j.micpro.2021.104025

Pooja, V., Das, R., & Kanchana, V. (2017, April). Identification of plant leaf diseases using image processing techniques. In *2017 IEEE Technological Innovations in ICT for Agriculture and Rural Development (TIAR)* (pp. 130-133). IEEE.

Poonia, R. C., Gao, X.-Z., Raja, L., Sharma, S., & Vyas, S. (2018). *Smart Farming Technologies for Sustainable Agricultural Development*. IGI Global.

Popa, A., Hnatiuc, M., Paun, M., Geman, O., Hemanth, D. J., Dorcea, D., Son, L. H., & Ghita, S. (2019). An intelligent IoT-based food quality monitoring approach using low-cost sensors. *Symmetry*, *11*(3), 374. doi:10.3390ym11030374

Prakash, S. (2020). Zigbee based Wireless Sensor Network Architecture for Agriculture Applications. In *2020 Third International Conference on Smart Systems and Inventive Technology (ICSSIT)* (pp. 709-712).IEEE. 10.1109/ICSSIT48917.2020.9214086

Prakash, S., Sharma, A., & Sahu, S. S. (2018). Soil Moisture Prediction Using Machine Learning. In *2018 Second International Conference on Inventive Communication and Computational Technologies (ICICCT)* (pp. 1-6).IEEE.

Prathibha, S.R., Hongal, A., & Jyothi, M.P. (2017). IoT based monitoring system in smart agriculture. In *2017 international conference on recent advances in electronics and communication technology (ICRAECT)* (pp. 81-84). IEEE.

Priyadarsini, P. L. K. (2021, January). Area Partitioning by Intelligent UAVs for effective path planning using Evolutionary algorithms. In *2021 International Conference on Computer Communication and Informatics (ICCCI)* (pp. 1-6). IEEE.

Priyadarsini, P. L. K., Gadupudi, P., & Paladugula, K. (2022, April). Path Planning of Intelligent UAVs using Computational Geometric Techniques. In *2022 International Conference on Electronic Systems and Intelligent Computing (ICESIC)* (pp. 190-193). IEEE.

Putri, D. A., Saputro, R. R., & Budiyono, B. (2012). Biogas production from cow manure. *International Journal of Renewable Energy Development*, *1*(2), 61–64. doi:10.14710/ijred.1.2.61-64

Rad, C., Hancu, O., Takacs, I. A., & Teanu, G. (2015). Smart Monitoring of Potato Crop: A Cyber-Physical System Architecture Model in the Field of Precision Agriculture. *Agriculture and Agricultural Science Procedi*. Research Gate.

Rad, C.-R., Hancu, O., Takacs, I.-A., & Olteanu, G. (2015). Smart monitoring of potato crop: A cyber-physical system architecture model in the field of precision agriculture. *Agriculture and Agricultural Science Procedia*, *6*, 73–79. doi:10.1016/j.aaspro.2015.08.041

Ragaveena, S., Shirly Edward, A., & Surendran, U. (2021). Smart controlled environment agriculture methods: a holistic review. In Reviews in Environmental Science and Biotechnology (Vol. 20, Issue 4, pp. 887–913). Springer Science and Business Media B.V. https://doi.org/ doi:10.100711157-021-09591-z

Rahman, S. U., Xuebin, Q., Riaz, L., Yasin, G., Shah, A. N., Shahzad, U., Jahan, M. S., Ditta, A., Bashir, M. A., Rehim, A., & Du, Z. (2021). The interactive effect of pH variation and cadmium stress on wheat (Triticum aestivum L.) growth, physiological and biochemical parameters. *PLoS ONE, 16*. doi:10.1371/journal.pone.0253798

Rai, V., Acharya, S., & Dey, N. (2012). *Implications of nanobiosensors in agriculture*.

Rajakumar, G., Sankari, M. S., Shunmugapriya, D., & Maheswari, S. U. (2018, April). IoT based smart agricultural monitoring system. *Asian J. Appl. Sci. Technol., 2*, 474–480.

Raj, E. F. I., Appadurai, M., & Athiappan, K. (2021). Precision Farming in Modern Agriculture. In A. Choudhury, A. Biswas, T. P. Singh, & S. K. Ghosh (Eds.), *Smart Agriculture Automation Using Advanced Technologies*. Transactions on Computer Systems and Networks. doi:10.1007/978-981-16-6124-2_4

Rana, R. L., Tricase, C., & De Cesare, L. (2021). Blockchain technology for a sustainable agri-food supply chain. *British Food Journal, 123*(11), 3471–3485. doi:10.1108/BFJ-09-2020-0832

Rani, J. V., Jakka, A., & Kanuru, H. (2022). Disease Detection in Crop Management Using Ensemble Machine Learning. *Sentimental Analysis and Deep Learning Proceedings of ICSADL, 2021*, 907–916.

Raut, P., & Ghare, P. H. (2020). Analysis of wireless channel parameters for the different types of soil in the wusns. *2020 IEEE International Students' Conference on Electrical, Electronics and Computer Science (SCEECS)*, (pp. 1–4). IEEE.

Reijers, N., Wang, Y.-C., Shih, C.-S., Hsu, J. Y., & Lin, K.-J. (2011). Building Intelligent Middleware for Large Scale CPS Systems. *Proceedings of IEEE International Conference on Service-Oriented Computing and Applications (SOCA)*, (pp. 1-4). IEEE. 10.1109/SOCA.2011.6166238

Rencher, A. C., & Christensen, W. F. (2012). Chapter 10, Multivariate regression – Section 10.1, Introduction. Methods of Multivariate Analysis, Wiley Series in Probability and Statistics (3rd ed.), John Wiley & Sons, p. 19

Richa, A., Touil, S., Fizir, M., & Martinez, V. (2020). Recent advances and perspectives in the treatment of hydroponic wastewater: a review. In Reviews in Environmental Science and Biotechnology, 19(4) 945–966. Springer Science and Business Media B.V. doi:10.100711157-020-09555-9

Richa, A., Fizir, M., & Touil, S. (2021). Advanced monitoring of hydroponic solutions using ion-selective electrodes and the internet of things: A review. *Environmental Chemistry Letters, 19*(4), 3445–3463. doi:10.100710311-021-01233-8

Rokach, L. Maimon, O. (2014). Data mining with decision trees: theory and applications, (2nd ed.) World Scientific Pub.

Romeo, D., Vea, E., & Thomsen, M. (2018). Environmental impacts of urban hydroponics in Europe: A case study in Lyon. *Procedia CIRP, 69*, 540–545. doi:10.1016/j.procir.2017.11.048

Rozantsev, A., Sinha, S. N., Dey, D., & Fua, P. (2017). Flight dynamics-based recovery of a UAV trajectory using ground cameras. In *Proceedings of the IEEE Conference on Computer Vision and Pattern Recognition* (pp. 6030-6039). 10.1109/CVPR.2017.266

Rubala, J. I., & Anitha, D. (2017). Agriculture Field Monitoring using Wireless Sensor Networks to Improving Crop Production. *International Journal of Engineering Science*.

S., P. K., Sampath, B., R., S. K., Babu, B. H., & N., A. (2022). Hydroponics, Aeroponics, and Aquaponics Technologies in Modern Agricultural Cultivation. In *Trends, Paradigms, and Advances in Mechatronics Engineering* (pp. 223–241). IGI Global. doi:10.4018/978-1-6684-5887-7.ch012

Sadeghi, S., Soltanmohammadlou, N., & Nasirzadeh, F. (2022). Applications of wireless sensor networks to improve occupational safety and health in underground mines. *Journal of Safety Research*, *83*, 8–25. doi:10.1016/j.jsr.2022.07.016 PMID:36481040

Sadiku, M. N. O., Wang, Y., Cui, S., & Musa, S. M. (2017). Cyber-Physical Systems: A Literature Review. *European Scientific Journal*, *13*(36), 52. doi:10.19044/esj.2017.v13n36p52

Safdar, H., Amin, A., Shafiq, Y., Ali, A., & Yasin, R. (2019). Abbas Shoukat, Maqsood Ul Hussan, Muhammad Ishtiaq Sarwar. A review: Impact of salinity on plant growth. *Nature and Science*, *17*(1), 34–40. doi:10.7537/marsnsj170119.06

Sah, R. R., Khatri, R., Basotra, T., & Chauhan, A. (2021). IoT Based Irrigation System -A Farmer Friendly Version. In *2021 3rd International Conference on Advances in Computing, Communication Control and Networking (ICAC3N)* (pp. 677-681). IEEE.

Sajja, G. S., Rane, K. P., Phasinam, K., Kassanuk, T., Okoronkwo, E., & Prabhu, P. (2021). Towards applicability of blockchain in agriculture sector. *Materials Today: Proceedings.*

Sales, N., Remédios, O., & Arsenio, A. (2015). Wireless sensor and actuator system for smart irrigation on the cloud. In *2015 IEEE 2nd World Forum on Internet of Things (WF-IoT)* (pp. 693-698). IEEE. 10.1109/WF-IoT.2015.7389138

Samikannu, R., Koshariya, A. K., Poornima, E., Ramesh, S., Kumar, A., & Boopathi, S. (2022). Sustainable Development in Modern Aquaponics Cultivation Systems Using IoT Technologies. In *Human Agro-Energy Optimization for Business and Industry* (pp. 105–127). IGI Global.

Sammut, C., & Webb, G. I. (2011). *Encyclopedia of Machine Learning.* Springer US.

Sampath, B. C. S., & Myilsamy, S. (2022). Application of TOPSIS Optimization Technique in the Micro-Machining Process. In Trends, Paradigms, and Advances in Mechatronics Engineering (pp. 162–187). IGI Global. doi:10.4018/978-1-6684-5887-7.ch009

Sampath, B., & Myilsamy, S. (2021). Experimental investigation of a cryogenically cooled oxygen-mist near-dry wire-cut electrical discharge machining process. *Strojniski Vestnik. Jixie Gongcheng Xuebao*, *67*(6), 322–330. doi:10.5545v-jme.2021.7161

Sanislav, T., Mois, G., Folea, S., Miclea, L., Gambardella, G., & Prinetto, P. (2014). *A cloud-based cyber-physical system for environmental monitoring.* IEEE. doi:10.1109/MECO.2014.6862654

Saravanan, A., Venkatasubramanian, R., Khare, R., Surakasi, R., Boopathi, S., Ray, S., & Sudhakar, M. (2022). POLICY TRENDS OF RENEWABLE ENERGY AND NON. *Renewable Energy.*

Saravanan, M., Vasanth, M., Boopathi, S., Sureshkumar, M., & Haribalaji, V. (2022). Optimization of Quench Polish Quench (QPQ) Coating Process Using Taguchi Method. *Key Engineering Materials*, *935*, 83–91. doi:10.4028/p-z569vy

Sardogan, M., Tuncer, A., & Ozen, Y. (2018, September). Plant leaf disease detection and classification based on CNN with LVQ algorithm. In *2018 3rd international conference on computer science and engineering (UBMK)* (pp. 382-385). IEEE. 10.1109/UBMK.2018.8566635

Sathiyamurthi, S., Ramya, M., Saravanan, S., & Subramani, T. (2023). Estimation of soil erosion for a semi-urban watershed in Tamil Nadu, India using RUSLE and geospatial techniques. *Urban Climate*, *37*, 101424. doi:10.1016/j.uclim.2023.101424

Sehgal, V., & Patrick, A. (2014). A Comparative Study of Cyber Physical Cloud, Cloud of Sensors and Internet of Things: Their Ideology, Similarities and Differences. IEEE International Advance Computing Conference (IACC), (pp. 708-716). IEEE. 10.1109/IAdCC.2014.6779411

Selmani., A., Oubehar., H., Outanoute, M., & Ed-Dahhak, A. (2018). Agricultural cyber-physical system enabled for remote management of solar-powered precision irrigation. *Biosystems Engineering.*

Selmani, A., Oubehar, H., Outanoute, M., Ed-Dahhak, A., Guerbaoui, M., Lachhab, A., & Bouchikhi, B. (2019). Agricultural cyber-physical system enabled for remote management of solar-powered precision irrigation. *Biosystems Engineering*, *177*, 18–30. doi:10.1016/j.biosystemseng.2018.06.007

Senthil Kumar, P., & Janet Joshiba, G. (2019). Water footprint of agricultural products. In *Environmental Water Footprints* (pp. 1–19). Springer.

Senthil, T. S., Puviyarasan, M., Babu, S. R., Surakasi, R., Sampath, B., & Associates. (2023). Industrial Robot-Integrated Fused Deposition Modelling for the 3D Printing Process. In Development, Properties, and Industrial Applications of 3D Printed Polymer Composites (pp. 188–210). IGI Global.

Shanbhag, N., Tambosker, S., Bhagat, R., Gaonkar, V., Rodrigues, D., & Leitao, F. (2020). Design of Sensor Nodes for Agricultural field monitoring Using IoT and BLE mesh technology. In *2020 IEEE 7th Uttar Pradesh Section International Conference on Electrical, Electronics and Computer Engineering (UPCON)* (pp. 1-6). IEEE. 10.1109/UPCON50219.2020.9376426

Sharma, R., Parhi, S., & Shishodia, A. (2020). Industry 4.0 applications in agriculture: cyber-physical agricultural systems (CPASs). In *Advances in Mechanical Engineering: Select Proceedings of ICAME 2020* (pp. 807-813). Singapore: Springer Singapore.

Sharma, A., Jain, A., Gupta, P., & Chowdary, V. (2021). Machine Learning Applications for Precision Agriculture: A Comprehensive Review. *IEEE Access : Practical Innovations, Open Solutions*, *9*, 4843–4873. doi:10.1109/ACCESS.2020.3048415

Sharma, M., Kumar, C. J., & Deka, A. (2022). Early diagnosis of rice plant disease using machine learning techniques. *Archiv für Phytopathologie und Pflanzenschutz*, *55*(3), 259–283. doi:10.1080/03235408.2021.2015866

Sharma, V., Tripathi, A. K., & Mittal, H. (2022). Technological revolutions in smart farming: Current trends, challenges and future directions. *Computers and Electronics in Agriculture*, *201*, 107217. doi:10.1016/j.compag.2022.107217

Sharpley, A. N., Duncan, E. W., & Hoag, D. L. K. (2019). Increasing the effectiveness and adoption of agricultural phosphorus management strategies to minimize water quality impairment. *Journal of Environmental Quality*, *48*(5), 1204–1217. doi:10.2134/jeq2019.03.0114 PMID:31589706

Shein, W. W., Cheng, Z., Tan, Y., & Lim, A. O. (2013). Study of Temperature Control Using Cyber Physical System Approach in Home Environment. *IEEE 1st International Conference on Cyber-Physical Systems, Networks and Ap- plications (CPSNA)*, (*vol. 3*, pp. 78-83). IEEE. 10.1109/CPSNA.2013.6614250

Shen, Q., Qiao, X. V., & Athanasios, V. (2014). Web Service Composition- A Decades Overview. *Informatics Sciences*, *250*, 218–238. doi:10.1016/j.ins.2014.04.054

Shetty, A., Shetty, A. D., Pai, R. Y., Rao, R. R., Bhandary, R., Shetty, J., Nayak, S., Dinesh, T. K., & Dsouza, K. J. (2022). Block Chain Application in Insurance Services: A Systematic Review of the Evidence. *SAGE Open*, *12*(1), 21582440221079877. doi:10.1177/21582440221079877

Shinyoung, L., Chung, L., Han, O., & Kim, J. H. (2011). An Inter- active Cyber Physical System for People with Disability and Frail Elderly People. *ACM Proceedings of the 5th International Conference on Ubiquitous Information Management and Communication*, (pp. 113). ACM.

Shoaib, M., Hussain, T., Shah, B., Ullah, I., Shah, S. M., Ali, F., & Park, S. H. (2022). Deep learning-based segmentation and classification of leaf images for detection of tomato plant disease. *Frontiers in Plant Science*, *13*, 1031748. doi:10.3389/fpls.2022.1031748 PMID:36275583

Shuaib, M., Hassan, N. H., Usman, S., Alam, S., Bhatia, S., Koundal, D., Mashat, A., & Belay, A. (2022). Identity model for blockchain-based land registry system: A comparison. *Wireless Communications and Mobile Computing*, *2022*, 1–17. doi:10.1155/2022/5670714

Singh, D. K., Sobti, R., Jain, A., Malik, P. K., & Le, D. (2022). LoRa based intelligent soil and weather condition monitoring with internet of things for precision agriculture in smart cities. *IET Communications*, *16*(5), 604–618. doi:10.1049/cmu2.12352

Singh, P., Singh, R. P., & Singh, Y. (2022). An optimal energy-throughput efficient cross-layer solution using naked mole rat algorithm for wireless underground sensor networks. *Materials Today: Proceedings*, *48*, 1076–1083. doi:10.1016/j.matpr.2021.07.307

Singh, R. K., Berkvens, R., & Weyn, M. (2021). AgriFusion: An Architecture for IoT and Emerging Technologies Based on a Precision Agriculture Survey. *IEEE Access : Practical Innovations, Open Solutions*, *9*, 136253–136283. doi:10.1109/ACCESS.2021.3116814

Singh, V., & Misra, A. K. (2016). Detection of plant leaf diseases using image segmentation and soft computing techniques. *Information Processing in Agriculture*.

Somasundareswari, D., KH, M.S., & Santhiya, A. (2022). Design and Implementation of Smart Farming System. *International Journal of Progressive Research in Science and Engineering*, *3*(5), 184–187.

Souza, A., Zandonadi, D., Santos, M., Canellas, N., de Paula Soares, C., da Silva Irineu, L., de Rezende, C., Spaccini, R., Piccolo, A., Olivares, F., & Canellas, L. (2021). Acclimation with humic acids enhances maize and tomato tolerance to salinity. *Chemical and Biological Technologies in Agriculture*, *8*(1), 40. doi:10.118640538-021-00239-2

Sridhara, S., Manoj, K. N., Gopakkali, P., Kumar, V., & Krishna, K. R. (2023). Evaluation of machine learning approaches for prediction of pigeon pea yield based on weather parameters in India. *International Journal of Biometeorology*, *67*(1), 165–180. doi:10.100700484-022-02396-x PMID:36323951

Suanpang, P., & Jamjuntr, P. (2019). A smart farm prototype with an Internet of Things (IoT) case study: Thailand. *Technology*, *5*(12), 15.

Suhag, S., Singh, N., Jadaun, S., Johri, P., Shukla, A., & Parashar, N. (2021). IoT based soil nutrition and plant disease detection system for smart agriculture. In *2021 10th IEEE International Conference on Communication Systems and Network Technologies (CSNT)* (pp. 478-483). IEEE. 10.1109/CSNT51715.2021.9509719

Sujatha, R., Chatterjee, J. M., Jhanjhi, N. Z., & Brohi, S. N. (2021). Performance of deep learning vs. machine learning in plant leaf disease detection. *Microprocessors and Microsystems*, *80*, 103615. doi:10.1016/j.micpro.2020.103615

Suma, D. N., Samson, S. R., Saranya, S., Shanmugapriya, G., & Subhashri, R. (2017). IOT Based Smart Agriculture Monitoring System. *International Journal on Recent and Innovation Trends in Computing and Communication*, *5*(2), 177–181.

Sundar, C. B. V., & Asokhan, M. (2023). Pivotal Role of Drones in Agriculture – a Review. *Asian Journal of Agricultural Extension. Economia e Sociologia*, *41*(4), 131–140.

Sunny, M. S. H., Hossain, E., Mimma, T. N., & Hossain, S. (2017, September). An autonomous robot: Using ANN to navigate in a static path. In *2017 4th International Conference on Advances in Electrical Engineering (ICAEE)* (pp. 291-296). IEEE.

Suresh Babu, C. V. (2023). *Artificial Intelligence and Expert System* (1st ed.). Anniyappa Publications.

Suresh Babu, C. V. (2023). *Introduction to Data Science* (1st ed.). Anniyappa Publications.

Suryady, Z., Shaharil, M. H. M., Bakar, K. A., Khoshdelniat, R., Sinniah, G. R., & Sarwar, U. (2011). Performance evaluation of 6LoWPAN-based precision agriculture. In *the 2011 International Conference on Information Networking (ICOIN2011)* (pp. 171-176).IEEE.

Talaviya, T., Shah, D., Patel, N., Yagnik, H., & Shah, M. (2020). Implementation of artificial intelligence in agriculture for optimisation of irrigation and application of pesticides and herbicides. *Artificial Intelligence in Agriculture*, *4*, 58–73. doi:10.1016/j.aiia.2020.04.002

Tamhane, S. A., Kumar, M., Passarella, A., & Conti, M. (2012). Service Composition in Opportunistic Networks. IEEE International Conference on Computing and Communications (GreenCom), (pp. 285-292). IEEE.

Tapan, A., Biswas, A. K., & Kundu, S. (2010). Nano-fertiliser-a new dimension in agriculture. *Indian Journal of Fertilisers*, *6*(8), 22–24.

Tarafdar, J. C., Agrawal, A., Raliya, R., Kumar, P., Burman, U., & Kaul, R. K. (2012). ZnO nanoparticles induced synthesis of polysaccharides and phosphatases by Aspergillus fungi. *Advanced Science, Engineering and Medicine*, *4*(4), 324–328. doi:10.1166/asem.2012.1160

Tarafdar, J. C., Raliya, R., & Rathore, I. (2012). Microbial synthesis of phosphorous nanoparticle from tri-calcium phosphate using Aspergillus tubingensis TFR-5. *Journal of Bionanoscience*, *6*(2), 84–89. doi:10.1166/jbns.2012.1077

Tarafdar, J. C., Xiong, Y., Wang, W.-N., Quinl, D., & Biswas, P. (2012). Standardization of size, shape and concentration of nanoparticle for plant application. *Applied Biological Research*, *14*(2), 138–144.

Tauber, M., Gollan., B., Schmittner, C., & Knopf, P. (2023). Passive Precision Farming Reshapes the Agricultural Sector. *Computer*.

Tephila, M. B., Sri, R. A., Abinaya, R., Lakshmi, J. A., & Divya, V. (2022). Automated Smart Irrigation System using IoT with Sensor Parameter. In *2022 International Conference on Electronics and Renewable Systems (ICEARS)* (pp. 543-549).IEEE 10.1109/ICEARS53579.2022.9751993

Thaher, T., & Ishaq, I. (2020). Cloud-based Internet of Things Approach for Smart Irrigation System: Design and Implementation. In *International Conference on Promising Electronic Technologies (ICPET)* (pp. 32-37). IEEE. 10.1109/ICPET51420.2020.00015

Thakur, V., Doja, M. N., Dwivedi, Y. K., Ahmad, T., & Khadanga, G. (2020). Land records on blockchain for implementation of land titling in India. *International Journal of Information Management*, *52*, 101940. doi:10.1016/j.ijinfomgt.2019.04.013

Theile, M., Bayerlein, H., Nai, R., Gesbert, D., & Caccamo, M. (2020, October). UAV coverage path planning under varying power constraints using deep reinforcement learning. In *2020 IEEE/RSJ International Conference on Intelligent Robots and Systems (IROS)* (pp. 1444-1449). IEEE. 10.1109/IROS45743.2020.9340934

Thenmozhi, K., & Reddy, U. S. (2019). Crop pest classification based on deep convolutional neural network and transfer learning. *Computers and Electronics in Agriculture*, *164*, 104906. doi:10.1016/j.compag.2019.104906

Torbick, N., Chowdhury, D., Salas, W., & Qi, J. (2017). Monitoring rice agriculture across myanmar using time series Sentinel-1 assisted by Landsat-8 and PALSAR-2. *Remote Sensing (Basel)*, *9*(2), 119. doi:10.3390/rs9020119

Torres-TelloJ.VenkatachalamS.MorenoL.KoS.-B. (2020). Yield measurements in aeroponics for six different crops. Mendeley Data. doi:10.17632/wmyktpx9hv.1

Trojovský, P., Dhasarathan, V., & Boopathi, S. (2023). Experimental investigations on cryogenic friction-stir welding of similar ZE42 magnesium alloys. *Alexandria Engineering Journal*, *66*(1), 1–14. doi:10.1016/j.aej.2022.12.007

Tudi, M., Ruan, H. D., Wang, L., Lyu, J., Sadler, R., Connell, D., Chu, C., & Phung, D. T. (2021). Agriculture development, pesticide application and its impact on the environment. In International Journal of Environmental Research and Public Health, 18(3), 1–24. MDPI AG. doi:10.3390/ijerph18031112

Tunio, M. H., Gao, J., Shaikh, S. A., Lakhiar, I. A., Qureshi, W. A., Solangi, K. A., & Chandio, F. A. (2020). Potato production in aeroponics: An emerging food growing system in sustainable agriculture for food security. In Chilean Journal of Agricultural Research, 80(1), 118–132. Instituto de Investigaciones Agropecuarias, INIA. doi:10.4067/S0718-58392020000100118

Turkoglu, M., Yanikoğlu, B., & Hanbay, D. (2022). PlantDiseaseNet: Convolutional neural network ensemble for plant disease and pest detection. *Signal, Image and Video Processing*, *16*(2), 301–309. doi:10.100711760-021-01909-2

Tushar, W., Yuen, C., Saha, T. K., Nizami, S., Alam, M. R., Smith, D. B., & Poor, H. V. (2023). Poor. *A Survey of Cyber-Physical Systems From a Game-Theoretic Perspective. IEEE Access : Practical Innovations, Open Solutions*, *11*, 9799–9834. doi:10.1109/ACCESS.2023.3239834

Tyagi, A. & N. Sreenath, N. (2021). *Cyber Physical Systems: Analyses, challenges and possible solutions*, 22-23. Internet of Things and Cyber-Physical Systems.

Udomkun, P., Nagle, M., Argyropoulos, D., Mahayothee, B., & Müller, J. (2016). Multi-sensor approach to improve optical monitoring of papaya shrinkage during drying. *Journal of Food Engineering*, *189*, 82–89. doi:10.1016/j.jfoodeng.2016.05.014

Udutalapally, V., Mohanty, S. P., Pallagani, V., & Khandelwal, V. (2021). sCrop: A Novel Device for Sustainable Automatic Disease Prediction, Crop Selection, and Irrigation in Internet-of-Agro-Things for Smart Agriculture. *IEEE Sensors Journal*, *21*(16), 17525–17538. doi:10.1109/JSEN.2020.3032438

United Nations Department of Economic and Social Affairs, Population Division (2022). *World Population Prospects 2022: Summary of Results.* UN DESA/POP/2022/TR/NO. 3.

United Nations, Department of Economic and Social Affairs, Population Division (2019). *World Urbanization Prospects 2018: Highlights* (ST/ESA/SER.A/421). UN.

Upton, E., & Halfacree, G. (2016). Meet the Raspberry Pi. In E. Upton & G. Halfacree (Eds.), *Raspberry Pi® User Guide.*, doi:10.1002/9781119415572

Vaishali, S., Suraj, S., Vignesh, G., Dhivya, S., & Udhayakumar, S. (2017). Mobile integrated smart irrigation management and monitoring system using IOT. In *2017 International Conference on Communication and Signal Processing (ICCSP)* (pp. 2164-2167). IEEE. 10.1109/ICCSP.2017.8286792

Vallabhajosyula, S., Sistla, V., & Kolli, V. K. K. (2022). Transfer learning-based deep ensemble neural network for plant leaf disease detection. *Journal of Plant Diseases and Protection*, *129*(3), 545–558. doi:10.100741348-021-00465-8

Van Erp-van der Kooij, E., & Rutter, S. M. (2020). Using precision farming to improve animal welfare. *Perspectives in Agriculture, Veterinary Science, Nutrition and Natural Resources, 15*, 1–10. doi:10.1079/PAVSNNR202015051

Vanitha, N. (2021). *A Study on the Applications of Drones in Smart Farming*. Academic Press.

Vanitha, N., & Padmavathi, G. (2018). A Comparative Study on Communication Architecture of Unmanned Aerial Vehicles and Security Analysis of False Data Dissemination Attacks. *2018 International Conference on Current Trends towards Converging Technologies (ICCTCT),* 1-8. 10.1109/ICCTCT.2018.8550873

Vanitha, N., & Padmavathi, G. (2018). A Study on Various Cyber-Attacks and their Classification in UAV Assisted Vehicular Ad-Hoc Networks. In G. Ganapathi, A. Subramaniam, M. Graña, S. Balusamy, R. Natarajan, & P. Ramanathan (Eds.), *Computational Intelligence, Cyber Security and Computational Models. Models and Techniques for Intelligent Systems and Automation. ICC3 2017. Communications in Computer and Information Science* (Vol. 844). Springer. doi:10.1007/978-981-13-0716-4_11

Vanitha, N., Vinodhini, V., & Rekha, S. (2016). A study on agriculture UAV for identifying the plant damage after plantation. *International Journal of Engineering and Management Research, 6*(6), 310–313.

Vanitha, S. K. R., & Boopathi, S. (2023). Artificial Intelligence Techniques in Water Purification and Utilization. In *Human Agro-Energy Optimization for Business and Industry* (pp. 202–218). IGI Global. doi:10.4018/978-1-6684-4118-3.ch010

Vellidis, G., Tucker, M., Perry, C., Kvien, C., & Bednarz, C. (2008). A real-time wireless smart sensor array for scheduling irrigation. *Computers and Electronics in Agriculture, 61*(1), 44–50. doi:10.1016/j.compag.2007.05.009

Vennila, T., Karuna, M. S., Srivastava, B. K., Venugopal, J., Surakasi, R., & Sampath, B. (2022). New Strategies in Treatment and Enzymatic Processes: Ethanol Production From Sugarcane Bagasse. In Human Agro-Energy Optimization for Business and Industry (pp. 219–240). IGI Global.

Verdouw, C., Wolfert, S., & Tekinerdogan, B. (2016). Internet of Things in agriculture. *CABI Reviews, 14*, 1-2.

Vijayaganth, V., & Naveenkumar, M. (2021). *4 Smart Sensor Based Prognostication of Cardiac Disease Prediction Using Machine Learning Techniques. Applications of Machine Learning in Big-Data Analytics and Cloud Computing*. River Publishers.

Vijayakumar, P. S., Abhilash, O. U., Khan, B. M., & Prasad, B. L. V. (2010). Nanogold-loaded sharp-edged carbon bullets as plant-gene carriers. *Advanced Functional Materials, 20*(15), 2416–2423. doi:10.1002/adfm.200901883

Vincent, D. R., Deepa, N., Elavarasan, D., Srinivasan, K., Chauhdary, S. H., & Iwendi, C. (2019). Sensors driven ai-based agriculture recommendation model for assessing land suitability. *Sensors (Basel), 19*(17), 3667. doi:10.339019173667 PMID:31450772

Vinci, A., Brigante, R., Traini, C., & Farinelli, D. (2023). Geometrical characterization of hazelnut trees in an intensive orchard by an unmanned aerial vehicle (UAV) for precision agriculture application. *Remote Sensing (Basel), 15*(2), 541. doi:10.3390/rs15020541

Viriyasitavat, W., & Hoonsopon, D. (2019). Blockchain characteristics and consensus in modern business processes. *Journal of Industrial Information Integration, 13*, 32–39. doi:10.1016/j.jii.2018.07.004

Vishnoi, V. K., Kumar, K., & Kumar, B. (2021). Plant disease detection using computational intelligence and image processing. *Journal of Plant Diseases and Protection, 128*(1), 19–53. doi:10.100741348-020-00368-0

Vootla, P., Remeithi, F. A., Bariaghabr, S. A., & Mansoori, F. A. (2018). Food waste - A global challenge to sustainability. In *2018 Advances in Science and Engineering Technology International Conferences (ASET),* (pp. 1-5). IEEE.

Wang, E. K., Ye, Y., Xu, X., Yiu, H., & Chow, K. P. (2010). Security Issues and Challenges for CPS. IEEE/ACM International Conference on Green Computing and Communications and on Cyber, Physical and Social Computing, (pp. 733-738). IEEE.

Wang, Z., Walsh, K. B., & Verma, B. (2017). On-tree mango fruit size estimation using RGB-D images. *Sensors (Basel), 17*(12), 2738. doi:10.339017122738 PMID:29182534

Wan, J., Chen, M., Xia, F., Di, L., & Zhou, K. (2013). From Machine-to-Machine Communications towards Cyber Physical Systems. *Computer Science and Information Systems, 10*(3), 1105–1128. doi:10.2298/CSIS120326018W

Willmott, C. J., & Matsuura, K. (2005, December 19). Advantages of the mean absolute error (MAE) over the root mean square error (RMSE) in assessing average model performance. *Climate Research, 30*, 79–82. doi:10.3354/cr030079

Wilson, M. A., Tran, N. H., Milev, A. S., Kannangara, G. S. K., Volk, H., & Lu, G. Q. M. (2008). Nanomaterials in soils. *Geoderma, 146*(1–2), 291–302. doi:10.1016/j.geoderma.2008.06.004

Wiyono, S., Suwardi, & Syakir, A. M. (2023). International Conference on Modern and Sustainable Agriculture (ICOMSA): Building Future Agriculture for a Better World. *IOP Conference Series: Earth and Environmental Science, 1133*, 011001.

Wu, T., Dou, W., Hu, C., & Chen, J. (2017). Service Mining for Trusted Service Composition in Cross-Cloud Environment. *IEEE Systems Journal, 11*(1), 283–294. doi:10.1109/JSYST.2014.2361841

Yadav, V. S., Singh, A. R., Raut, R. D., Mangla, S. K., Luthra, S., & Kumar, A. (2022). Exploring the application of Industry 4.0 technologies in the agricultural food supply chain: A systematic literature review. *Computers & Industrial Engineering, 169*, 108304. doi:10.1016/j.cie.2022.108304

Yang, C. (2020). Remote sensing and precision agriculture technologies for crop disease detection and management with a practical application example. *Engineering (Beijing), 6*(5), 528–532. doi:10.1016/j.eng.2019.10.015

Yoganandan, G., Rahman, A. A. A., Vasan, M., Senthilkumar, S., & Sathya, S. (2022). Evaluating agripreneurs' satisfaction: Exploring the effect of demographics and emporographics. *Journal of Innovation and Entrepreneurship, 11*(2), 2. doi:10.118613731-022-00193-9

Yokoyama, R. (2019). Energy Consumption and Heat Sources in Plant Factories. In Plant Factory Using Artificial Light (pp. 177-184). Elsevier. doi:10.1016/B978-0-12-813973-8.00016-6

Young, M. D., Ros, G. H., & de Vries, W. (2021). Impact of agronomic measures on crops, soil and environmental indicators: A review and synthesis of meta-analysis. In Agriculture, Ecosystems and Environment (Vol. 319). Elsevier B.V. doi:10.1016/j.agee.2021.107551

Yu, C., Jing, S., & Li, X. (2012). An Architecture of Cyber Physical Sys- tem Based on Service. *Proceedings International Conference on Computer Science and Service System (CSSS),* (pp. 1409-1412). IEEE.

Yuan, J., Wang, H., Lin, C., Liu, D., & Yu, D. (2019). A novel GRU-RNN network model for dynamic path planning of mobile robot. *IEEE Access : Practical Innovations, Open Solutions, 7*, 15140–15151. doi:10.1109/ACCESS.2019.2894626

Yupapin, P., Trabelsi, Y., Nattappan, A., & Boopathi, S. (2022). Performance Improvement of Wire-Cut Electrical Discharge Machining Process Using Cryogenically Treated Super-Conductive State of Monel-K500 Alloy. *Iranian Journal of Science and Technology - Transactions of Mechanical Engineering*, (pp. 1–17). IEEE. doi:10.1007/s40997-022-00513-0

Zakzouk, S., Ehab, M., Atef, S., Yousri, R., Tawfik, R. M., & Darweesh, M. S. (2021). A Comparative Study of Machine Learning Techniques for Automatic Rice Crop Irrigation. In *2021 17th International Computer Engineering Conference (ICENCO)* (pp. 46-50). IEEE. 10.1109/ICENCO49852.2021.9698935

Zamani, A. S., Anand, L., Rane, K. P., Prabhu, P., Buttar, A. M., Pallathadka, H., Raghuvanshi, A., & Dugbakie, B. N. (2022). Performance of Machine Learning and Image Processing in Plant Leaf Disease Detection. *Journal of Food Quality*, *2022*, 2022. doi:10.1155/2022/1598796

Zambon, I., Cecchini, M., Egidi, G., Saporito, M. G., & Colantoni, A. (2019). Revolution 4.0: Industry vs. agriculture in a future development for SMEs. *Processes (Basel, Switzerland)*, *7*(1), 36. doi:10.3390/pr7010036

Zha, J. (2020). Artificial intelligence in agriculture. *Journal of Physics: Conference Series*, *1693*(1), 12058. doi:10.1088/1742-6596/1693/1/012058

Zhang, X., Andreyev, A., Zumpf, C., Negri, M. C., Guha, S., & Ghosh, M. (2017). Thoreau: A subterranean wireless sensing network for agriculture and the environment. In *2017 IEEE Conference on Computer Communications Workshops (INFOCOM WKSHPS),* (pp. 78-84). IEEE. 10.1109/INFCOMW.2017.8116356

Zhang, B., Han, C., & Yu, X. B. (2015). A non-destructive method to measure the thermal properties of frozen soils during phase transition. *Journal of Rock Mechanics and Geotechnical Engineering*, *7*(2), 155–162. doi:10.1016/j.jrmge.2015.03.005

Zhang, C., & Kovacs, J. M. (2012, December). The application of small unmanned aerial systems for precision agriculture: A review. *Precision Agriculture*, *13*(6), 693–712. doi:10.100711119-012-9274-5

Zhang, L., He, J., & Yu, W. (2013). Challenges, Promising Solutions and Open Problems of Cyber-Physical Systems. *International Journal of Hybrid Information Technology*, *6*(2), 65–73.

Zhang, Y., Chen, D., Wang, S., & Tian, L. (2018). A promising trend for field information collection: An air-ground multi-sensor monitoring system. *Information Processing in Agriculture*, *5*(2), 224–233. doi:10.1016/j.inpa.2018.02.002

Zhang, Z., Yu, X., Wu, P., & Han, W. (2015). Survey on water-saving agricultural internet of things based on wireless sensor network. *International Journal of Control and Automation*, *8*(4), 229–240. doi:10.14257/ijca.2015.8.4.23

Zhao, Y., Qin, X.-M., Tian, X.-P., Yang, T., Deng, R., & Huang, J. (2021). Effects of continuous cropping of Pinellia ternata (Thunb.) Breit. on soil physicochemical properties, enzyme activities, microbial communities and functional genes. *Chemical and Biological Technologies in Agriculture*, *8*(1), 43. *Advance online publication*. doi:10.118640538-021-00243-6

Zotarelli, L., Dukes, M.D., & Morgan, K.T. (2010). Interpretation of Soil Moisture Content to Determine Soil Field Capacity and Avoid Over-Irrigating Sandy Soils Using Soil Moisture Sensors: AE460/AE460. *EDIS*, (2).

Zotarelli, L., Scholberg, J. M., Dukes, M. D., Muñoz-Carpena, R., & Icerman, J. (2009). Tomato yield, biomass accumulation, root distribution and irrigation water use efficiency on a sandy soil, as affected by nitrogen rate and irrigation scheduling. *Agricultural Water Management*, *96*(1), 23–34. doi:10.1016/j.agwat.2008.06.007

About the Contributors

G. S. Karthick is currently working as an Assistant Professor in the Department of Software Systems, PSG College of Arts & Science, Coimbatore, India. Previously, he worked as a Research Fellow under the DST-ICPS Project from April 2019 to March 2022 in the Department of Computer Science, School of Computer Science and Engineering, Bharathiar University, Coimbatore, India. Also, he worked as an Assistant Professor in the Department of Computer Science, Sri Ramakrishna College of Arts and Science, Coimbatore, India. Previously, he worked as an Assistant Professor in the Department of Computer Science, Sri Ramakrishna College of Arts and Science, Coimbatore, India. He received his Ph.D. in Computer Science from Bharathiar University (University Department) in February 2023. He topped the Post-Graduation with Gold Medal in 2017 from Bharathiar University, Coimbatore, India. He topped Under-Graduation with an exemplary performance from SNR Sons College, Coimbatore, India, and also aced three subjects to receive gold medals in 2015. He has qualified UGC-NET Computer Science and Applications in December 2020. Further, he has visited many institutions and universities as a resource person at various events. He has received an Indian Patent Grant for his innovative contribution to the field of Wireless Networks and the Internet of Things. He has published research papers in international journals and presented papers at International and National Conferences. He has published many book chapters and in addition, he has edited one book and authored two books. He is an active Editorial Member of IJHIS, IJERTCS, and IJEHMC. Also, he is an active reviewer in high-standard journals of Elsevier, IEEE, Springer, IGI Journals, and many peer-reviewed journals. He has received many awards and rewards for his achievements in academic and research careers. His areas of specialization are Machine Learning, the Internet of Things, Wireless Sensor Networks, and Analysis of Algorithms. His current research interests focus on the recent developments in the healthcare Internet of Things and Machine Learning Algorithms.

* * *

Devipriya A. is currently working as an Assistant Professor in the Department of Computing Technologies, SRM Institute of Science and Technology, Kattankulathur. She has received Ph.D degree in Information and Communication Engineering from Anna University, Chennai. She is having more than 17 years of experience in Teaching. She has completed AI & ML 6 months Course in Xebia Academy, Delhi and Design Thinking Course in Taylor's University. Her areas of interest is Artificial Intelligence, Machine Learning, Deep Learning, Data Analytics, and Internet of Things.

Neenu Rose Antony is currently an Assistant professor at Amal Jyothi College of Engineering, Kerala, India in the Department of Electrical and Electronics Engineering. She did her post-graduation in Power Electronics and Power Systems. Her areas of research interests are Operation and Control of Microgrids and Sustainable Engineering.

Ahmed Awouda is a Ph.D. candidate and research assistant in the department of Management and production engineering at Politecnico di Torino. He received a M.Sc. Degree in Mechatronics Engineering from Politecnico di Torino. His Research interests include IoT systems, CEA agriculture, Industrial communication, Collaborative robotics, and Simulation of production systems.

C. V. Suresh Babu is a pioneer in Content Development. A true entrepreneur, He floated Anniyappa Publications, a company which is very active in bringing out books related to Computer Science and Management. Dr. C.V. Suresh Babu has also ventured into SB Institute, a centre for knowledge transfer. He holds Ph.D in Engineering Education from National Institute of Technical Teachers Training & Research, Chennai along with seven Master degrees in various disciplines such as Engineering, Computer Applications, Management, Commerce, Economics, Psychology, Law and Education. He also has the UGC-NET/SET qualifications in the disciplines of Computer Science, Management, Commerce and Education to his credit. Personal blog: .

Kunal Biswas is the corresponding author of the chapter and is a currently working as an 'Assistant Professor (Research)', at the Centre for Nanoscience & Nanotechnology, International Research Centre, Sathyabama Institute of Science and Technology (Deemed to be University), Jeppiaar Nagar, Rajiv Gandhi Salai, Chennai-600 119 from December, 2021 till present. He received his PhD from the Department of Biotechnology, MAKAUT, WB in June, 2021. He received his Master's degree in Nanoscience and Technology from the Department of Physics, Tezpur Central University, Assam, India in 2012. He has been working for more than 8 years in the field of 'carbonaceous nano-materials' and its wide spread applications ranging from material sciences to biotechnological domains. He has been honoured with the prestigious West Bengal state award as 'Outstanding paper award in Engineering' section at the 27 TH West Bengal State Science and Technology Congress, 2020, Govt. of West Bengal. His area of research is 'Nano-biotechnology' and his research interests include Material Engineering, Nano-bio sensors, sensor fabrication, Nanotechnology, nano-bio interface etc. He has published more than 30 peer reviewed SCI indexed (Web of Science, Scopus) journals of International and national reputes like Frontiers in Microbiology, Molecules, IET Nano Biotechnology, IEEE Transactions in NanoBiosciences, Springer, Elsevier publications etc. He has also contributed 7 book chapters till date and recently co-edited two book in Taylor & Francis, CRC Press and Elsevier, written many popular science articles in the national science magazines along with organising several international and national seminars (both online and offline).

S. Boopathi () completed his undergraduate in Mechanical Engineering and postgraduate in the field of Computer-Aided Design. He completed his Ph.D. from Anna University and his field of research includes Manufacturing and optimization. He published 60 more research articles in Internationally Peer-reviewed journals, one Patent grant, and three published patents. He has 16 more years of academic and research experiences in the various Engineering Colleges in Tamilnadu, India.

Giulia Bruno received a M.Sc. degree in Computer Engineering and a Ph.D. in Computer and System Engineering from Politecnico di Torino. She is currently Professor of Manufacturing Technologies and Systems in the Department of Management and Production Engineering of Politecnico di Torino. Her research activity is mainly focused on lean production, process modeling, machine learning and simulation of production systems.

Sasikala Chinnasamy is currently working as an Assistant Professor in the Department of Computing Technologies, KPR Institute of Engineering and Technology, Coimbatore. She is pursing Ph.D degree in Information and Communication Engineering from Anna University, Chennai. My research area include Io.T, Cloud, Machine Learning and Deep Transfer learning. She is having more than 5 years of experience in Teaching. She has published 5 papers in various reputed Journals that includes 1 SCI Indexed and 3 Scopus.

Lakshmi D. is Associate Professor of Department of Electrical and Electronics Engineering, Academy of Maritime Education and Training, Deemed to be University, Chennai, Tamil Nādu, India. She has more than 21 years of expertise in the field of power systems. She has guided more than 30 UG students, 10 PG students and 4 research scholars. She has published six book chapters in wiley, Elsevier, IEEE, etc., and one book on Fundamentals of Drives, 32 National, international journals and nearly 25 National and international conferences proceedings. She received many funds from reputed organizations and AICTE for contributing to recent research and product developments. Her areas of interest are Power system operation and Control, Soft Computing renewable energy systems, Microgrid, Electrical Machines.

Rakshi Anuja Dinesh received the Bachelor of Technology degree in Biotechnology from Sathyabama Institute of Science and Technology, Chennai, Tamil Nadu. She has excelled in her degree with gold medal, for being the meritorious student in 2023. She has also received NPTEL DISCIPLINE star award and has successfully completed lots of online courses, which helped her to have a wide knowledge in various fields. She has done her internship in ICMR-VCRC and ICAR on hard ticks bar coding and antimycotic activity respectively. She has published review articles and book chapters.

Benedetta Fasciolo is a PhD student in Management, Production and Design at Politecnico di Torino, Italy. She received her Master's in Industrial Production Engineering. Her current research activity is focused on tools and methods for the optimization of resources in the production of soilless vegetables.

N. Gireesh is working as Professor and Head of the Department, Electronics and Communication Engineering in Mohan Babu University (Erstwhile Sree Vidyanikethan Engineering College), Tirupati. He obtained his Bachelor Degree in Medical Electronics from Bangalore University, Master's Degree in Instrumentation and Control from Sri Venkateswara University, Tirupati and Doctoral Degree from Sri Venkateswara University, Tirupati with Control Systems specialization. He has a total of 23 years' experience in teaching and research. He published 15 papers in reputed Journals and presented 14 papers in reputed conferences. His areas of interest include Biomedical Instrumentation, Signal Processing, Soft Computing, Process Control and Machine learning. He is a life member of ISTE, IETE and ISOI.

Charles Rajesh Kumar James Paulraj received the Bachelor of Engineering degree in Electronics & Instrumentation Engineering from the Madurai Kamaraj University, India, the Master of Engineering degree in Electronics & Communication Engineering from Anna University, India, and Master of Business Administration degree in Financial Management from the Indira Gandhi National University, India. He was bestowed the Gold medal for excellence in Master of Engineering degree by Anna University in 2008. He has been a faculty member of Engineering at various Universities of repute in India and abroad. He serves as a reviewer and published many research articles in peer-reviewed journals in the Web of Science, Science Citation Index, and Scopus databases. His current research involves VLSI design, System on Chip (SoC), Network on Chip (NoC), Embedded Systems, Wireless sensor networks, Artificial intelligence, Machine Learning, Robotics, Video coding, Image processing, IoT and Renewable energy.

Akshya Jothi is serving as Assistant Professor at SRM Institute of Science and Technology. Pursued PhD from SASTRA University with computational geometry as the research area. Designed effective algorithms for path planning of UAVs. Posses research interest in the field of artificial intelligence and machine learning.

S. N. Kumar is Associate Professor, Dept. of Electrical and Electronics Engg., Amal Jyothi College of Engineering, Kanjirappally. He was awarded the B.E. degree in Electrical and Electronics Engineering in 2007 and M.E. degree in Applied Electronics in 2011 from Anna University, Chennai. He was awarded his Ph.D. from Sathyabama Institute of Science and Technology, Chennai in the Dept. of Electronics and Communication Engg. as a part of a DST-sanctioned funded project. With 15 years of teaching experience, his areas of interest include Medical Image Processing and Embedded System Applications in Telemedicine. He is an active life member of BMESI, ISTE, ISRD, Bernoulli Society, and IAENG. He was the Co-Principal Investigator of the DST IDP-funded project and research fellow of the RCA scheme, NTU Singapore. He has authored 22 peer-reviewed journal publications, ten textbooks for engineering disciplines and research community, editor of 2 books (IGI Global, CRC Press). He also has 3 national patents and one international publication, 30 book chapters from prominent publications, and 40 conference publications to his credit.

Dharani L. is a passionate chemical engineer. She specialised in combining the principles of science with modern technology to create innovative solutions that solve complex problems. My work focuses on finding new ways to improve existing processes, develop sustainable methods, and increase efficiency. She also have a passion for teaching, which allows me to share my knowledge and skills with others.

Naveenkumar M. received his B.E degree in Computer Science and Engineering from Info Institute of Engineering from Anna University in the year 2015 and his M.E degree in Computer Science and Engineering from KPR Institute of Engineering and Technology with First class- Distinction from Anna University in the year 2019. He has 2 years of industry experience and since 2019 he is working as an assistant professor in the department of CSE, KPR Institute of Engineering and Technology. He has published 13 research papers in Scopus Indexed Journals. His research area is Machine learning and image processing.

M. A. Majid received the M.Sc. degree in Electrical Engineering from King Fahd University of Petroleum and Minerals (KFUPM.), Dhahran, Saudi Arabia, in 2002. In 2011, he was awarded his Ph.D. degree in Electronic and Electrical Engineering from the University of Sheffield in the U.K. He was an EPSRC prize Postdoctoral Fellow at the University of Sheffield for one year (2011-2012). From 2012 to 2015, he continued as a postdoctoral fellow in the Electrical Engineering program at King Abdullah University of Science and Technology (KAUST), Saudi Arabia. Currently, he is working as an Assistant Professor in the Electrical and Computer Engineering Department, Effat University, where he is engaged in applying III/V semiconductor quantum dot/quantum well devices to Biomedical Imaging, Optical Communications, and Solid-State Lighting.

Vanitha N. received her Master of Computer Applications from the Avinashilingam Institute for Home Science and Higher Education for Women, Coimbatore, in 2008. In 2008 she was hired as a Software programmer in LMW group, Coimbatore. Since 2011, she has been working as an Assistant Professor in the Computer science department at Dr. N. G. P. Arts and Science College, Coimbatore. In 2013, she received her Master of Philosophy in Computer Science at Dr. N. G. P. Arts and Science College, Coimbatore. She received her Doctor of Philosophy in Computer Science (Part Time), from the Avinashilingam Institute for Home Science and Higher Education for Women, Coimbatore. She writes and presents widely on UAV Networks, Cyber Security, Machine, and Deep Learning.

S. Aasha Nandhini is working as Assistant Professor, Networking and Communications Department, School of Computing, SRM Institute of Science and Technology. She has received her B.E. degree in Electronics and Communication Engineering from Anna University and M.E. degree in Communication Systems from Anna University, Chennai. She obtained her Ph.D. degree from Anna University for her research work on Video compressive sensing. Her research interests include Wireless Multimedia Sensor Networks, Compressed Sensing, IoT, and Artificial Intelligence. She has 9 research publications in international journals, 9 research publications in national and international conferences, 2 book chapters, and one book.

Kavishree P. S. is a student at KPR Institute of Engineering and Technology and she is having a keen interest in writing research and academic papers. Her research fields include IoT, Cloud and Machine Learning. Her research projects are make a bridge between the physical and digital worlds, and help create intelligent systems that are capable of autonomously monitoring, controlling, and optimizing their own behavior.

Karthickmanoj Ravichandran received B.E. (Electronics and Communication Engineering) from Syed Ammal Engineering College in 2007, M.E. (VLSI Design) from SRM Eswari Engineering College, Chennai in 2012 and the Pursuing Doctoral degree from Academy of Maritime Education and Training, Deemed to be University, Chennai from 2019. Having 11 Plus years experience in both teaching and Industry. Currently, working as Assistant Professor in Academy of Maritime Education and Training, Deemed to be University, Chennai. He has published 03 national, 04international journals and nearly 10 National and international conferences Proceedings. He has Received certificate of Appreciation for contributing a chapter in the event Asia book of records. He is a life Member in 6 Professional Bodies, and acting as reviewer for reputed journals. His areas of interest are VLSI,Signal Processing, Robotics and Artificial Intelligence.

Valentino Razza received the M.S. degree in Computer Engineering and the Ph.D. in Information and System Engineering from the Politecnico di Torino, Italy, in 2008 and 2012, respectively. From 2009 to 2013 he worked as a Post-Doctoral Researcher in the System Identification and Control (SIC) group of the Control and Computer Department at the Politecnico di Torino. From January 2014 to June 2015, he was with the Multi-Layer Wireless Solutions group at the Istituto Superiore Mario Boella. He worked as a Research Assistant in the SIC group of the Control and Computer Department at the Politecnico di Torino up to 2020 and was a PostDoc Researcher at the Italian Institute of Technology until 2022. Currently, he holds a position as Assistant Professor at the Department of Management and Production Engineering. His main research interests include identification, robust control and model predictive control of linear and nonlinear systems with application to real-world problems in the automotive, robotics, industrial production and green-energy fields.

Biji Rose is an Assistant Professor in the Department of Information Technology at Dr. N. G. P Institute of Technology. She completed M.E Communication Systems at Kumaraguru College of Technology, Coimbatore from Anna University Chennai. She has a teaching experience of over 14 Years from various reputed colleges. Has attended various Webinars, Seminars, Workshops, Conferences, Short Term Programmes and FDPs. She has organized various Workshops, Guest Lectures, Seminars and Conference. Motivated and guided UG projects and mini projects which has won best project award. Has lead student team for L& T TECHgium and IIT PALS to final rounds. She has published several papers in International and National Conferences as well Journals.

Thanuja S. is a student at the KPR Institute of Engineering and Technology. Her research interest includes the domain of Machine Learning, IoT, and Cybersystems. She has made significant contributions to extensive research aimed at studying the challenges faced by Cyber-Physical systems and developing effective solutions to address them. She is an active member in IEEE (SSIT, WIE), IFERP, IAAC, IEI, and Chief volunteer of NSS which engages her in social activities aimed at promoting the well-being of the society. She aspires to address the most critical challenges faced by our society in developing innovative solutions that make a positive impact on the society.

Geetha S. K. is currently working as an Assistant Professor in the Department of Computer science and Engineering, KPR Institute of Engineering and technology, coimbatore. She completed her Master degree in CSE and pursuing PhD degree in Information and Communication Engineering from Anna University, Chennai. She is having more than 13 years of experience in Teaching. Her areas of interest are Blockchain technologies, Computer networks, and Deep Learning.

Rakshun Selvaa S. K. is an aspiring student currently undertaking my B.Sc. Information Technology course in Dr.N.G.P Arts & Science college, Coimbatore. Currently he have a passionate interest in acquiring knowledge regarding the working principles and philosophies of various civilian and military purposed Unmanned Aerial Vehicles.

Mibin P. Sabu is a final year student currently pursuing B.Tech degree in Electrical and Electronics Engineering at Amal Jyothi College of Engineering, Kottayam.

Anitha S. Sastry is a research scholar (Wireless Communication) from BMSCE, Bengaluru. She is an Assistant Professor in the Department of Electronics and Communication Engineering, Global Academy of Technology, Bengaluru. She has obtained B.E (ECE) and M.Tech (Electronics) degree from Visvesvaraya Technological University. She has been awarded "Outstanding Woman in Engineering" award at VIWA2019 for contributing to the field of Electronics and Communication. She has won best project awards for the projects carried out with students. She is authoring a text book on Computer networks. She is a life time member of KRVP and Institute Of Engineers, Bengaluru and an IEEE member. She has published over 20 articles in international journals and has some articles published in regional language. She has been actively contributing in teaching rural children in different summer camps conducted by KRVP. Her areas of interest include Wireless sensor Networks, Optical fiber communication, IOT, Computer Networking and Artificial Intelligence.

Jayashree Shanmugam is working as an Assistant Professor in the Department of Biotechnology in Sathyabama Institute of Science and Technology , Chennai.

Jayapratha T. is presently an Assistant professor in the Department of Information Technology, Sri Eshwar College of Engineering, Coimbatore, Tamilnadu, India. She has more than 12 years of teaching experience. She received a B.E. degree from Anna University, Chennai, and M.E degree from Karpagam University, Coimbatore. She is pursuing her Ph.D. at Anna University, Chennai. Her research interests include Machine Learning, Wireless Sensor Networks.

Sasilatha T., Professor and Dean, Department of Electrical and Electronics Engineering, has 26 years of experience in teaching, research and administration. She received her B.E Degree from Manonmaniam Sundaranar University, ME and Ph.D. Degrees from Anna University, Chennai. She has received CMI Level 5 Certificate in Management and Leadership from UK. She has published more than 100 articles in refereed International Journals and Conferences. Prior to joining AMET in 2017, she had served as a Faculty and elevated as various academic positions in St. Joseph's group of Institutions from 1996 to 2013. She served as Vice Principal in affiliated colleges. She has performed funded research projects and has participated in industry collaborated projects. She is a member of IEEE professional society, Life member of Indian Society for Technical Education (ISTE) and IAENG. She has guided 5 Ph.D. scholars and guiding 08 research scholars in the area of VLSI Design, Artificial Intelligence, Signal Processing and Wireless Sensor Networks. She has received certificate of appreciation from the International Association of Maritime Universities (IAMU) in 2019. Recently she has organized AICTE funded National conference and AICTE funded Short term training Program.

Rajasekaran Thangaraj received B.E. and M.E. Degrees in Computer Science Engineering from Anna University, India, in 2008 and 2011, respectively. He completed his Ph.D. in the area of Deep Learning at Anna University, Chennai in the year 2022. He has been working as an Assistant Professor at the KPR Institute of Engineering and Technology, India, since 2012. He has published 20 articles in reputed International Journals and presented 18 papers in various International Confereces. Also, he published 4 book chapters with reputed publisher and 5 Indian patents. He acts as the reviewer in various International Journals. His research areas of interest include Machine learning, Deep learning, Computer Vision and the Internet of Things. He is currently working computer vision and deep learning algorithms for object recognition and plant disease identification. He is a Life Member of the Indian Society for Technical Education (ISTE).

Sumathi V. is an Assistant Professor in the Department of Computer Applications at Sri Ramakrishna College of Arts and Science, Coimbatore. She has 13 years of teaching and research experience. Her area of specialization is Image Processing and Data mining. She has published more than 10 papers in National and International Journals and one book publication.

Amal Varghese is a final year student currently pursuing B.Tech degree in Electrical and Electronics Engineering at Amal Jyothi College of Engineering, Kottayam.

Index

N

P

S

T

U

V

W

Y

Z